Cooperative Optimal Control of Hybrid Energy Systems

Dong Yue · Huifeng Zhang · Chunxia Dou

Cooperative Optimal Control of Hybrid Energy Systems

Springer

Dong Yue
Nanjing University of Posts
and Telecommunications
Nanjing, Jiangsu, China

Huifeng Zhang
Nanjing University of Posts
and Telecommunications
Nanjing, Jiangsu, China

Chunxia Dou
Nanjing University of Posts
and Telecommunications
Nanjing, Jiangsu, China

ISBN 978-981-33-6721-0 ISBN 978-981-33-6722-7 (eBook)
https://doi.org/10.1007/978-981-33-6722-7

This Springer imprint is published by the registered company Springer Nature Singapore Pte Ltd.
The registered company address is: 152 Beach Road, #21-01/04 Gateway East, Singapore 189721, Singapore

Preface

With the increasing penetration of renewable energy resources, the optimal control of hybrid energy system becomes a great challenge due to its complicated structure and dynamic uncertainty characteristics, the control methodology has received great attention in the past decade. Compared with the traditional centralized or decentralized optimal control approach, cooperative control can be more flexible to confront possible potential risks, as well as has excellent control performance. Hence, the study of hybrid energy system mainly adopts a cooperative control strategy for tackling with the above mentioned problems, this strategy can be utilized for security control, optimal operation, distributed optimization, game theory, and distributed control to ensure the safety, maximum economic profit/minimum economic cost, and stability of hybrid energy system.

The study of hybrid energy system mainly includes several topics: security control caused by uncertainty issue, multi-objective optimization for economic and emission problem, distributed optimization, and distributed control for AC/DC microgrid. First of all, the most important problem in power system is the security issue, some switching control mechanisms are proposed to ensure the safety of hybrid energy system. After that, economic and emission issue can be achieved with maximum effort in the way of multi-objective optimization for full use of different energy resources. The optimization results can further provide the reference for power generators, which can coordinate with each other to perform well in a stable way combined with proposed distributed optimal control algorithms.

Structure and Readership

In this book, Chap. 1 introduces hybrid energy system: the framework, the overview of security control, optimization and cooperative control; Part I is devoted to cooperative optimal security control of hybrid energy system; Part II is devoted to multi-objective optimization for optimal operation of hybrid energy system; Part III is devoted to distributed optimization for energy management of microgrid; Part IV is devoted to the distributed cooperative control of DC microgrid; Part V is devoted to the distributed cooperative control of islanded AC microgrids.

Part I: Cooperative optimal security control is used in hybrid energy system. Therefore, in Chap. 2, the multiagent system-based event-triggered hybrid controls for high-security hybrid energy generation systems is proposed. In Chap. 3, multi-agent based hierarchical hybrid control for smart microgrid is proposed. Two-stage optimal operation strategy of isolated power system with TSK fuzzy identification of supply-security is proposed in Chap. 4.

Part II: Multi-objective optimization for optimal operation is used in hybrid energy system. Therefore, in Chap. 5, MOEA/D based probabilistic PBI approach for risk-based optimal operation of hybrid energy systems with intermittent power uncertainty is proposed. In Chap. 6, gradient decent based multi-objective cultural differential evolution for short-term hydrothermal optimal scheduling of economic emission with integrating wind power and photovoltaic power is proposed. Event-triggered multi-agent optimization for two-layered model of hybrid energy system with price bidding based demand response is proposed in Chap. 7.

Part III: Distributed optimization is used in energy management of microgrid. Therefore, in Chap. 8, consensus-based economic hierarchical control strategy for islanded MG considering communication path reconstruction is proposed. Multi-agent-system-based BI-level bidding strategy of microgrid with game theory in the electricity market is proposed in Chap. 9.

Part IV: Distributed cooperative control is used in DC microgrids. Therefore, in Chap. 10, multiagent system-based distributed coordinated control for radial DC microgrid considering transmission time delays is proposed. In Chap. 11, MAS-based distributed cooperative control for DC microgrids through switching topology communication network with time-varying delays is proposed. Multiagent system-based integrated design of security control and economic dispatch for interconnected microgrid systems is proposed in Chap. 12.

Part V: Distributed cooperative control is used in islanded AC microgrids. Therefore, in Chap. 13, distributed event-triggered cooperative control for frequency and voltage stability and power sharing in isolated inverter-based microgrid is proposed. In Chap. 14, event-triggered mechanism based distributed optimal frequency regulation of power grid is proposed. A virtual complex impedance based P-V droop method for parallel-connected inverters in low-voltage AC microgrids is proposed in Chap. 15.

Nanjing, China
September 2020

Dong Yue
Huifeng Zhang
Chunxia Dou

Acknowledgements We would like to acknowledge the collaborations with Shengxuan Weng, Jianbo Chen, and Zhijun Zhang on the work of cooperative optimal security control of hybrid energy system, multi-objective optimization for optimal operation of hybrid energy system. This work was supported by Basic research project of Leading Technology of Jiangsu Province under Grant (BK20202011), National Key R&D Program of China under Grant (2018YFA0702200), the National Natural Science key fund under Grant (61533010) and (61833008), and National Natural Science Fund under Grant (61973171).

Contents

Chapter 1
Introduction

This chapter introduces hybrid energy systems, the overview of security control, optimization, and cooperative control. Further, the content of this book is briefed.

1.1 Hybrid Energy Systems

With the development and utilization of wind power, photovoltaics and other renewable energy and the increase of people's demand for electricity, a new energy system—hybrid energy system was proposed. Hybrid energy system refers to the integration of several different types of energy production equipment, including electric power generator, electric energy storage system and renewable energy.

The design, operation and resource allocation models of the hybrid energy system are very advanced. Further research of these mathematical models will help to make better use of the characteristics of the hybrid energy system and make it more versatile and perfect. The development and application of optimal control methods and modern control techniques can further improve the operation performance of hybrid energy system and make it better applied to the optimal allocation of multi-objective energy and the control and utilization of renewable energy. Although the design and optimization procedures of hybrid energy systems are complex, the cost of the system will certainly be reduced in the development and research of these technologies. Similarly, optimal energy allocation based on renewable energy forecasting and load side demand can significantly reduce the operating cost of the system. And system safety and reliability maximization and emission minimization are well worth considering, rather than cost minimization as the sole criterion.

However, since hybrid energy systems consist of different energy resources, intermittent energy resources bring uncertainty of hybrid energy systems, which can even lead security problem. Various of requirements turns optimal operation to a multi-

D. Yue et al., *Cooperative Optimal Control of Hybrid Energy Systems*,
https://doi.org/10.1007/978-981-33-6722-7_1

objective and multi-attribute problem, and heterogeneous generators make cooperative control difficult to tackle with. Thus, the security control, optimal operation and cooperative control of hybrid energy systems can be more challenging, relative researches have been taken for several decades, the literature reviews are presented in following sections.

1.2 The Overview of Security Control in Hybrid Energy System

1.2.1 The Voltage Security Control

The use of renewable energy resources is an effective solution to energy shortage and environment pollution problems [1, 2]. However, single renewable energy resource usually cannot continuously meet load demand, since its operation status highly depends on natural condition, such as wind speed and solar radiation, and thus resulting in an intermittent power output. In order to make full use of the renewable energy resources to meet load demand with high security, hybrid energy generation systems (HEGSs) are increasingly paid attention. A HEGS is desired to connect various kinds of small distributed energy resources (DERs) into an electrical network to guarantee energy supply with high security. It is able to run in both grid-connected and grid-disconnected modes. In the grid-connected mode, with the support of distribution electrical grid, the HEGS is relatively easy to meet the load demand with high reliability. However, in the grid-disconnected mode, to realize the objective above, effective control mechanism needs to be introduced with taking full consideration of the characteristics of HEGSs [3, 4].

The HEGSs present typical characteristics of complex hybrid systems mainly because of the following reasons:

(1) The frequent start-stop of renewable energy resources results in that these resources present hybrid characteristics including both dynamic and multimode switching behaviors [5–7].
(2) To maintain the balance between supply and demand, the storage devices also need to frequently switch their operation modes, and accordingly presenting hybrid characteristics [8, 9];
(3) The switching between multiple control modes of controllable DERs results in the change of their dynamic behavior, also displaying interactive hybrid characteristics.

From the whole system point of view, the HEGSs are multi- source interconnected systems with hybrid characteristics, and thus, their control present great challenge.

With respect to control problem of such a complex HEGS, multiagent system (MAS) technique is certainly one of the most effective means. In the past several decades, a lot of MAS-based control approaches regarding HEGSs have been investigated. However, most of them were focused on dynamic regulation without paying enough attention to the treatment of switching behavior of HEGS [10–13]. On the other hand, a few works only focused on MAS-based switching control [11–13]. In [11], MAS-based energy management issue was studied by means of switching control for four operation modes of HEGS. In [12, 13], the MAS-based switching control for storage devices was presented by using logic judgments and fuzzy- logic-rules. As mentioned above, all the relevant research only focused on one aspect of dynamic regulation and switching control without taking full consideration of both interactive hybrid characteristics.

In addition, with respect to a microgrid, there is a previous research of authors that presented MAS-based event-triggered hybrid control [14]. However, when designing switching control strategies, the trigger duration time, switching sequence, and time interval had not been fully considered. With respect to a smart grids or a wide-area power system, references [15, 16] presented MAS-based event-triggered control schemes. In [15], the study focused on economic dispatch of smart grids based on distributed event-triggered communication network. In [16], an event-triggering load frequency control was proposed for multiarea power systems with communication delays. In comparison with this book, the two references above designed the event-triggered controls from a different point of view to solve entirely different problems.

This study defines several control agents, including one central coordinated control agent (CCCA), which is responsible for implementing coordinated switching control (CSC) to ensure high-security energy supply of overall HEGS. In comparison with previous research, the proposed control scheme is very flexible and scalable.

1.2.2 The Supply-Security Control

With the high penetration of renewable energy resources, the great challenge is to adapt the uncertain operating conditions of power generation and system load [17]. Generally, the objective of optimal operation is to minimize economic cost while satisfying different constraints, including the system load balance, the output limit, spinning reserve constraints and the ramp rate limit. In the literature [18] emphasis is put on the energy storage system and spinning reserve in the economic dispatch model of micro-grids. Also, [19] proposes a distributed economic dispatch strategy for microgrids with multiple energy storage systems, which overcomes the challenges of dynamic couplings among all decision variables and stochastic variables in a centralized dispatching formulation. Besides economic factors, environmental issues can also plays a role. Mohammadi-ivatloo et al. [20] optimizes economic cost and emission rate caused by thermal units simultaneously, and produces a set of Pareto optimal schemes. In most instances, the uncertainty or randomness of intermittent energy resources can be considered a tough problem. There are mainly three approaches: (1) Fuzzy optimization, (2) Stochastic optimization and (3) Robust optimization (RO). The first one depends mainly on the membership function of the decision-makers' experience, which can be subjective and not suitable for real-world applications [21, 22]. Stochastic optimization requires probabilistic information, which is difficult to obtain or not accurate enough for optimization [23, 24]. RO can deal with the uncertainty problem objectively, as well as with less probabilistic information, but somehow it may lead to a conservative problem [25, 26]. For avoiding this problem, a robustness condition with flexible control parameters is adopted, which has been deduced in literature [27]. The uncertainty of intermittent energy and system load can also lead to a potential risk in isolated power system. Security is also an important issue in the optimal operation of isolated power system, especially isolated power system. Some researchers focuses on the security assessment and evaluation in distributed systems, including distributed power generators [28–32]. Usually, the security issue is not taken into consideration at the planning stage, or it exists merely in the communication network. A novel efficient security analysis approach is proposed for overcoming the drawbacks of high computational cost in classical N-k-induced cascading contingency analysis [33]. In the literature [34] approaches are presented for clustering active distribution systems into a set of microgrids with optimized reliability and supply-adequacy indexes. Zhang et al. [35] proposes a novel robust security-constrained optimal power flow method to balance the economy, combined with the security requirements under the uncertainties associated with renewable generation and load demand. Supply-security to describe the reliability of isolated power system is adopted, and the relationship between supply-security and fuzzy sets of power generation and system load is modeled with a TSK fuzzy system, which has been a hot issue for system identification [36–41]. In the literature [36], an interactively recurrent self-evolving fuzzy neural network is proposed for prediction and identification of dynamic systems. Deng et al. [37] discusses a knowledge-leverage-based fuzzy system from the perspective of transfer learning, which not only make full use of data from the current scene, but also effectively leverage the existing knowledge from reference scenes. Teh et al. [39] utilizes a system identification-based framework to develop monotone fuzzy If-Then rules for formulating monotone zero-order TSK fuzzy inference systems. Eyoh et al. [40] presents a novel application of a hybrid learning approach to optimize membership functions of a newly developed interval type-2 intuition fuzzy logic system of TSK fuzzy inference with a neural network learning capacity. This book proposes a two stage optimal operation strategy with TSK fuzzy identification to optimize isolated power system, TSK fuzzy identification approach is utilized to identify the relationship between those uncertainty parameters and supply-security, and then a two-stage optimization strategy with switching mechanism is presented to minimize economic cost and emission rate simultaneously.

1.3 The Overview of Optimization in Hybrid Energy System

1.3.1 The Multiobjective Optimization

The increasing penetration of renewable energy resources imposes significant challenges on the optimal operation of the hybrid energy systems, an important issue in modern electric power systems. The main goal of the hybrid energy system management is to schedule the power generation for each generator to minimize the economic cost or to maximize the economic benefit. As environmental problems are drawing increasing global concerns, adequate electricity is not only required at the cheapest possible price but also at the minimum level of pollutions [42].

Thus, the optimal operation of the hybrid energy systems becomes a multiobjective optimization problem (MOP), and many multiobjective evolutionary algorithms (MOEAs) have been proposed to produce a set of nondominant schemes for decision making, such as nondominated sorting genetic algorithm (NSGA-II) [43, 44], niched Pareto genetic algorithm (NPGA) [45], strength Pareto evolutionary algorithm (SPEA) [46], multiobjective particle swarm optimization (MOPSO) [47], multiobjective differential evolution (MOHDE) [48], etc. These MOEAs mainly adopt Paretodominance-based approaches, which determines the priority of evolutionary individuals with the Pareto-dominance order [49]. MOEAs can be broadly classified into three categories [50]: (1) the Pareto-dominance-based approaches [51, 52]; (2) the indicator-based approaches [53, 54]; and (3) the decomposition-based approaches [55, 56]. The MOEA/D mainly optimizes an MOP by decomposing it into several scalar subproblems and optimizing them coordinately in a single run [56]. Each agent is assigned to a different subproblem, and coordinates with other agents to improve the search ability of MOEA/D. Generally, there are three commonly used MOEA/Ds [57, 58]: (1) the weighted sum approach; (2) the weighted Tchebycheff approach; and (3) the PBI approach. However, the weighted sum approach cannot properly optimize a nonconvex Pareto front, the weighted Tchebycheff approach has difficulties to obtain smooth objective when it deals with nonconvex Pareto front, and the efficiency of PBI approach depends on appropriate weight vectors [59]. To overcome the aforementioned problems, [59] imposes constraints on subproblems, and adaptively adjust the constraint during the search process.

Stochastic or uncertainty of intermittent energy resources is the key issue to handle for optimal operation of the hybrid energy system management [60, 61]. Currently, it mainly includes three approaches: (1) fuzzy programming; (2) robust optimization (RO); and (3) stochastic optimization. Haghifam et al. [62] presented an interesting risk-based scheduling strategy using a fuzzy method to model the uncertainty of wind power generation. Liu et al. [63] has established a fuzzy-based energy and reserve co-optimization model considering the high penetration of renewable energy. Zhao et al. [64] proposed an RO approach that considers the uncertainty of wind power output and demand response. Lorca and Sun [65] presented a new framework using adaptive RO for economic dispatch with high level of wind penetration. However, the choice of fuzzy membership values can be subjective, which cannot ensure the accuracy of the obtained value [21, 22]. RO generally does not consider the accuracy of the system model, tends to be conservative when calculating the optimal value at the minimum risk. Stochastic optimization (SO) has some advantages in accounting for uncertainty and risks [66]. For decreasing optimization conservation, flexible parameters are utilized to split the output of intermittent energy resources into several intervals, scenarios are generated for simulating stochastic process caused by intermittent energy resources with probabilistic characteristics of each interval, and further to acquire stochastic information of each scenario under different uncertainty budgets. Thus, this provides the probabilistic domain for the proposed probabilistic PBI optimization approach. Simultaneously, to ensure the robustness or to avoid possible risks caused by intermittent energy resources, two-step decision-making approach establishes proper uncertainty budget for controlling the disturbance of intermittent energy resources, based on which the best optimal scheme can be selected with the aid of a grid-based decision-making method.

1.3.2 The Distributed Optimization

With increasing penetration of renewable energy resources, it can gradually become a great challenge for hybrid energy management due to randomness or uncertainty of power generation, bi-direction energy flow and price-responsive loads, etc. [67]. Effective optimization strategy for hybrid energy management can be necessary to ensure energy utilization to the maximum extent, especially with deterministic model or without considering demand response [68]. However, stochastic nature of intermittent energy and demand requirement brings great challenge on both dynamic hybrid energy management and optimization methodology [17, 69, 70]. With consideration of uncertainty of intermittent energy, stochastic optimization (SO) [23, 24], fuzzy optimization [21, 22] and robust optimization (RO) strategy [25, 27] are employed to get rid of potential risk to hybrid energy system. Stochastic programming approach depends on probability density function by data sampling, which may cause large deviation when data source is limited. Fuzzy approach determines membership mainly by decision-makers' personal experience, optimal scheme can be subjective. RO can achieve optimal scheme without excessive information, but it can be conservative for exchanging economic expense to robustness.

Besides, demand response (DR) can be another important part in hybrid energy system, it can be dynamic and unpredictable, which also motivates further research on modeling and methodology with DR. With considering benefit of DR, literature [71] evaluates the impact market design and DR on reducing wind power forecast error, while literatures [72, 73] investigates positive benefit on short-term trading of wind power producers. Literature [74] employs one DR program with critical peak price, and investigates the optimal value according to load serving entity that sells wind energy to market. For properly managing demand side requirement, multi-agent architecture has been widely used. The intelligent bidding strategy based continues double auction allows consumers to participate DR programs, and agent-based architecture is developed to manage power with considering DR [75]. With multiple micro-grids including DR and distributed storage, an agent-based approach is utilized to reduce system peak demand and minimize electricity cost [76]. Literature [77] has proposed a two-level architecture of multi-agent system for multiple micro-grids, naive auction algorithm is employed to simulate the biding action of market agents that participate real-time bidding. Those literatures can deal with dynamic characteristics of micro-grids system with DR and bidding problem with considering consumers' behavior, but it lacks effective way for potential risk in optimal operation.

Coordinated optimization with switching strategy is involved to avoid potential risk in a positive and coordinated way, event-triggered mechanism is proposed for positive action from power supply and load demand side. Generally, event-triggered coordination approach can be considered as control theory with network communication [15, 78–82]. Though literature [15] has been successfully implemented in power system, event-triggered strategy still depends network communication.

Event-triggered based multi-agent optimization is proposed to optimize hybrid energy system with considering DR, event-triggered mechanism is designed to avoid potential risk caused by uncertainty from intermittent energy and system load.

In comparison to other optimization strategy, the proposed algorithm is dynamic and systemic, which also means that it can be robust while exempting potential risk with coordination between power supply and system load side, and further keep the stability of hybrid energy system.

1.3.3 The Game Optimization

Nowadays, the increasing interests in achieving a quality electric energy supply, which is also reliable, sustainable, and has lower environmental impact, along with advances in communication and control, are driving the development of microgrids (MGs) in the electric energy industry [83].

Microgrid is one of the solutions to aggregating distributed generations in a low-voltage network, which is usually described as confined clusters of loads, storage devices, and small generators [84]. It is expected that MGs improve energy utilization, reduce losses in transport, provide reliability in the whole system, and enable the integration of renewable resources [85, 86]. In recent years, with the increase in the capacity of microgrid, the advance in microgrid technology, and the decrease in the cost of microgrid operation, the focus of the research has been on how to consume the remaining power properly. One of the solutions to this problem is to take part in competing with other power producers in the entire electricity market.

The restructuring of the power industry mainly aims at abolishing monopoly, reducing cost, and optimizing the allocation of social source. Nowadays, China is reforming in the generation side, which is regarded as the start of the electricity revolution. Microgrid could be considered as an independent power producer when its generation capacity can meet the demand of its own load and have some residue. As a self-financing economic entity, every power producer expects to obtain maximum profits in the electricity market, which can be realized according to a suitable bidding strategy. Strategic bidding, including the amount of power to supply and its offered price, refers not only to the optimization of a generation companies (GENCOs) behavior, but also to other GENCOs behaviors, which are unknown or known partially to other GENCOs [87]. When it comes to a power producer, it has to evaluate the possibility of either selling more power at a low per unit price or selling less at a high price per unit of power.

Many significant researches have been conducted on the overall optimization of electricity market transaction. In [88], authors proposed a probability-based method, for submitting optimal bids, in which participants' revenue is defined as an objective function. Bayesian Nash equilibrium was employed by Fang et al. in [89], using Cournot model and assuming complete information in its algorithm. In [90], an optimal bidding strategy problem is solved using a novel algorithm based on Shuffled Frog Leaping Algorithm (SFLA). Most researches referred bidding strategy of power producer focus on separate bidding and how to realize the maximum profit with no-cooperative model. However, compared to traditional power producer, MGs have some disadvantages which have an adverse effect on bidding strategy.

When MGs participate in electricity market competition with traditional power producers, a variety of problems that we must pay attention to will occur, such as auction realtime, the hierarchy for controlling, responding quickly, and so on. These problems have to be comprehensively investigated. Obviously, with the increase in the number of microgrid and the increasing complexity of the electricity market, the disadvantage of traditional control method is becoming more and more prominent. The MAS technology has great potential for distributed power system research because of its characters such as decentralized control, interaction with each other, intelligence, and so on. MAS is an autonomous system, in which many agents are grouped together and depend on each other to form a community that cooperates or competes to achieve the goals of individuals and of the system as a whole [91].

In MAS, every element is an intelligent agent, which has the three typical characteristics: reactive, proactive, and social abilities [92]. All these characteristics meet the features of the power producer as a self-sustaining economic entity. So, all power producers participate in the market bidding as rational agents. The MAS technology has been widely used in the field of dispersion system. In [93], a smart grid control framework based on MAS is proposed. According to the introduction and applications of MAS in [86, 94, 95], it may be the first choice to model autonomous decision-making entities.

In view of the newly established electricity market and simplified calculation at the same time, some assumptions are represented in this chapter. That is, (i) demand side response submits a inelastic demand, i.e. the electricity is reflected only by market supply; (ii) unified clearing price is the only price mechanism in every subordinate market; (iii) the constraints of transmission network and of control are neglected.

For the reasons above, in order to overcome the shortages of single microgrid mentioned above, and to adapt to the characteristics of distributed systems well, this book does the following original works: (1) A MAS in electricity market environment and a bi-level optimization model of bidding with game theory for solving the bidding problem are proposed, in which MGs will take part in the market in the form of alliance. The actual electricity market will be divided into two layers of the market based on MAS, which are considered as the upper-market and the lower-market. In the upper-market, the MGs will ally to increase their overall market influence to obtain more market share, which is in accord with the aim of improving the proportion of renewable clean energy in the electricity market. When the bidding in the upper-market has been finished, the quantity of electricity obtained by the alliance will be allocated by bidding once again according to bidding once more. (2) The iterative search algorithm [96] is introduced to solve the bidding model and ameliorated in this book. Similar to most studies, game theory is used to construct the bidding model in this book. Some literature has to solve the Nash equilibrium of game by the method of iteration. However, most of them don't consider the special constraints of electricity. In fact, electricity producers, especially MGs have many specific constraint conditions in the process of bidding. In this book, those constraints are also considered as the end condition in the current round iterative and then the iterative process will go into the next round directly. The results obtained by this amelioration are more in line with the characteristics of electricity market bidding. (3) Power balance is one of the security constraints that the electricity market must meet. In the process of solving the bidding model, some other constraint conditions may be in contradiction with be power balance constraints, so multi-objective planning is introduced to modify slightly the final result of bidding in the upper-market. (4) The method of undetermined coefficients is introduced to ascertain the coefficient of MSMAG under several specific principles, which are considered as the parameter of MSMAG when it bids in the upper-market.

1.4 The Overview of Cooperative Control in Hybrid Energy System

1.4.1 The Cooperative Control of DC Microgrid

The increasing penetration of DC renewable energy sources and increasing use of DC loads have motivated a growing interest in DC microgrid (MG) [97, 98]. The DC network can integrate various kinds of small DC or non-60 Hz frequency distributed energy resources (DERs) such as photovoltaic arrays, wind turbine and battery etc. to supply DC loads more efficiently. DC MGs have several advantages over their AC counterparts [99]. For instance, the control regarding reactive power and frequency is not an issue. So DC MGs have been widely regarded as one of the best means of using renewable energy resources to meet DC load demand.

The major concern in the stabilization operation problem of DC MGs is how to design a feasible and efficient control scheme to maintain bus voltage stabilization.

In the last a few years, many researches considered the decentralized "peer to peer" control, which means that each DER subsystem is controlled independently based on its local available information, for example local droop control scheme [100]. This control mode is obviously feasible and efficient [101, 102]. However, the bus voltage of each DER subsystem usually changes with the DER operation state. Therefore, the decentralized "peer to peer" control scheme may lead to a large steady voltage deviation between two DER units. In order to solve this problem, many researchers proposed hierarchical coordinated strategies [4, 103, 104]. The secondary coordinated strategies are ordinarily divided into three types: the centralized communication and control scheme [105], centralized communication and decentralized control scheme [106] and distributed coordinated control scheme [107–112]. Due to huge dimensionality of MG and communication pressure, the first scheme is neither applicable nor economic. Although the second scheme implements secondary adjustment in a decentralized manner, but it still depends on global information, which leads to communication pressure and decreased reliability. On the contrary, as an effective means, the third scheme recently has attracted more attentions. In the distributed coordinated control scheme, a local controller of each DER unit at the first level is helped by a distributed coordinated controller at the secondary level. The secondary-level controller uses remote transmission signals that only come from the adjacent DER unit to synthesize decoupling controllaw in order to improve control performance. So this kind of scheme only depends on low bandwidth communication. However, In a MG, DERs are very dispersive and loosely connected to each other, so even two adjacent DERs are far away. When the control signals that come from adjacent DER unit are transmitted through the transmission line, communication delay is unavoidably introduced [113]. In general, the communication delay can vary from tens to several hundred milliseconds or more [113, 114]. The existence of time delays would deteriorate the control performance. Therefore, when designing the distributed coordinated control scheme, the time delay should be taken into consideration. The time delay stabilization problem can be classified into two types: delay-independent stabilization [115] and delay-dependent one [116]. The delay-independent stabilization is considered more conservative in general than the delay-dependent one, especially for the system where time delay is actually small. Due to communication delays in MG are within the range of several hundred milliseconds [114], so the less conservative delay-dependent stabilization approach is more suitable for designing the distributed coordinated control scheme.

With respect to the two-level distributed coordinated scheme, the multi-agent system (MAS) technique is one of the best choices to make the control much intelligent and flexible [10]. MAS is composed of multiple agents, where each agent is not only autonomous and intelligent, but also interacts with each other in a cooperation manner, so that MAS is able to handle complex problems in a more flexible and intelligent way. So far, the researches regarding management and control of MGs had been done in multi-agent approach [12, 14]. In this book, a two-level MAS based distributed coordinated control scheme is proposed, where each DER unit is associated with a unit control agent at the first level to deal with local control for its unit system, also where there is a distributed coordinated agent at the secondary level to implement the distributed coordinated control.

The research is decomposed into the following steps: (i) The MAS based distributed coordinated control scheme is constructed. Also the structures of two levels of agents are designed, respectively. (ii) According to the bus voltage deviation between any two adjacent DER units, the secondary-level agent determines what kind of control mode it needs to execute. (iii) Both local and distributed coordinated controllers in four kinds of control modes are designed. For purpose of robust stability, the local control is designed as local state-feedback based $H\infty$ robust controller. It is worth mentioning that, the distributed coordinated control consists of local state-feedback control and decoupling coordinated control law that only come from adjacent DER units, so that communication pressure is decreased largely. Moreover, for improving control performance, the distributed coordinated control is designed by means of delay-dependent $H\infty$ robust control method taking into account the communication time delays. The sufficient conditions for controller existence in four kinds of control modes are transformed into line matrix inequality (LMI) convex optimization problem. (iv) The validity of the proposed control scheme is demonstrated by means of simulation study.

MOST of distributed energy resources (DERs), such as PV panels, fuel cells, and storage devices, are DC sources. In addition, DC loads also show an increasing trend. So far, 50 the whole buildings are DC loads. Furthermore, the development and implementation of electric vehicles charging stations imply that the electric vehicles would be widely integrated into the electrical grid, resulting in increasing DC demand. The wide use of DC DERs and increasing trend of DC demand have motivated the growing interest in DC microgrids (MGs) [98, 99, 103]. DC MGs have well-known features as follows: (1) power distribution is more efficient than that in AC MGs, since in DC MGs there is no reactive power and (2) power supply for DC loads is more efficient than that in AC MGs. To supply DC loads, an AC MG must use a two-stage conversion topology: first AC-DC and then DC-DC. Supplying DC loads with a DC MG avoids an unnecessary AC-DC conversion stage, and thus enhancing efficiency.

A hierarchical control scheme is conventionally adopted for MG operation [103, 117–121]. The highest tertiary control is responsible for assigning the MG voltage to implement a scheduled power dispatch for economical operation purpose. To meet the voltage requirement of the tertiary control, the secondary control determines and accordingly updates the voltage set points for the primary controls of all DER units based on real-time voltage measurements throughout an MG. According to the voltage set point, the primary control locally regulates the voltage and current of individual DER unit. The tertiary and secondary controls are typically implemented in a centralized way through a communication network with high connectivity [107, 122]. The primary control is typically implemented in a distributed way. With respect to such a hierarchical control structure, any communication failure might lead to fault of the corresponding unit, and accordingly overstressing of other units, system-level instability and cascaded failures [123].

To overcome the problem above, distributed control has been motivated as an attractive alternative. That is, the distributed control paradigm is extended to the secondary and primary controls [124–126], and thus improving system reliability, enabling easily scalability, and reducing communication pressure [107]. Categorically, the DC MG needs to satisfy two main control objectives: global voltage regulation [127] and proportional load sharing [128]. The global voltage regulation means that the all bus voltages across the DC MG (and only not a specific bus voltage) are regulated according to the global voltage set point determined by the tertiary control. The proportional load sharing means that the loads among participating DER units are assigned in proportion to their rated powers, or the per-unit currents of all DER units are equalized, and thus avoiding circulating current [128] and overstressing of any DER unit [129–131].

However, the conventional distributed droop control suffers from poor voltage regulation and load sharing, particularly when line impedances are not negligible [108, 132–134]. The one reason is that line impedances would cause large voltage drop. Another is output voltage mismatches among different DER units, resulting in a poor load sharing. So far, possible solution to the aforementioned issues is the establishment of both centralized control and a fully connected communication network throughout the MG, where any two nodes are directly connected [129, 135, 136]. For example, in [103], a centralized secondary control determines a voltage restoration term, and accordingly sends it to all DER units through a fully connected communication network across the MG. Despite improved accuracy, the solution dependent on a fully connected communication network is susceptible to communication failure, since any link failure would impair the whole control functionality.

Multiagent system (MAS) is one of the most effective tools to perform a distributed cooperative control [137, 138]. In [139] each DER agent in network can be regarded as a first-level agent that exchanges information with neighbor agents according to some specific communication protocols from an upper secondary-level agent. In [110], Bidram et al. propose an MAS-based secondary voltage cooperative control strategy, where consensus theory is used to synchronize global DERs bus voltages via the sparse communication networks. But the time-vary delays are not considered. In [140], Yu et al. propose a novel secondary control based on MAS, but the detailed relationships among each DER agent are not presented.

Taking into account the all aforementioned problems, the MAS-based distributed cooperative controlfor a DC MG is proposed, it can be more reliable, since any communication failure would not impair the whole control functionality.

1.4.2 The Cooperative Control of AC Microgrid

Renewable energy, such as solar and wind power, is seen as an appropriate solution to sustainable development and is often integrated into utility grid through distributed generators (DGs). To incorporate and coordinate different types of DGs, the microgrid (MG) which can operate in both grid-connected and islanded modes has emerged and been widely accepted [103, 141–144].

In the islanded operation of the MG, the DGs are controlled as voltage sources and parallel connected through the MG. To ensure stability of the MG, the active and reactive powers of the DGs should be shared simultaneously [145]. Besides, each DG should be able to operate independently without communications for the practical and economic concerns. Therefore, the droop control method, which mimics the behaviour of synchronous generators, has been extensively adopted in MG for its ability of fulfilling the aforementioned requirements [103, 145–147].

The conventional droop control methods were proposed for purely inductive feeder impedance. In this scenario, the frequency is a global quantity and thus the active power can be shared accurately. On the contrary, the voltage magnitude profile is deeply influenced by mismatched feeder impedance and the reactive power sharing accuracy is often degraded [103]. To improve the reactive power sharing accuracy, various modified droop methods have been proposed and can be roughly divided into two categories [148]: open-loop [134, 149–154] and closed-loop [133, 155–160]. The closed-loop reactive power sharing methods can be further divided into : small signal injection [155–157], detection of voltage of PCC [158, 159], detection of load change [160] and compensation of impedance voltage drop [133]. The small signal injection method [155, 157] requires a small AC voltage signal added into the output voltage reference, which may reduce the quality of the output voltage and current. Besides, the extraction and process of the small signal increase the complexity of the physical realization. Since the DGs may be installed dispersedly, the detection of voltage of PCC [158, 159] is impractical in this situation. In [160], wavelet transform is adopted to detect the load change, which increases the complexity of the controller. In [133], with estimation of impedance voltage drop effect and compensation of the local load, the output power is shared accurately in both grid-connected and islanded mode. However, the islanded sharing accuracy is dependent on the operation of grid-connected mode in advance.

Compared with the closed-loop methods, open-loop schemes are easier to be implemented in actual MG systems [148]. In general, the open-loop strategies can be classified into two types: virtual impedance [134, 149–152] and $Q - \dot{V}$ method [153, 154]. The original idea of virtual impedance strategy is to incorporate a virtual impedance to reshape the output impedance of DGs to be purely inductive or resistive. Therefore, the active and reactive power is decoupled and can be controlled independently. Besides, the virtual impedance doesn't introduce extra power losses, which is also an appealing advantage. In [149], virtual negative resistance and inductance is adopted to reshape the equivalent impedance between virtual power source (VPS) and PCC to be purely inductive. With proper selection of virtual inductance, the accurate reactive power sharing is achieved. In [150], an adaptive virtual output impedance is constructed such that the impedance at fundamental frequency is inductive while resistive at harmonic orders. Hence, not only the active and reactive power but also the harmonic power is shared accurately. However, the mismatch of the feeder impedance, including transformers, cables and the grid-side inductors, is not considered [161].

Being different from the virtual impedance based methods, a novel $Q - \dot{V}$ droop method is proposed in [153, 162] to improve the reactive power sharing accuracy without modifying the output impedance and a $\dot{V}$ restoration mechanism is adopted to reset $\dot{V}$ to zero at steady state. Since the restoration process of different DGs will never be the same [163], the set point of DGs will be different. Thus, the $Q - \dot{V}$ droop method is not able to tackle the power sharing performance thoroughly. To address the set point deviation problem based on only local information, a modified $\dot{V}$ restoration technique was proposed in [154] to reduce the deviation of Q_{0x} before entering the steady state. Despite the effectiveness of the $Q - \dot{V}$ method under purely inductive cases, its performance is degraded when the equivalent impedance is complex, since the active and reactive power are affected by both voltage magnitude and phase difference.

As for low-voltage microgrids (LVMG) where feeder impedance is mainly resistive, extensive works have been done to reshape the output impedance to be purely inductive [133, 134, 149–157, 159–161] based on the assumption that the output impedance of DG is inductive and larger value of grid-side inductor is adopted. However, this assumption is not always true since the output impedance mainly depends on the control strategy [164] and the grid-side inductor may be small when LCL is designed for active power filter (APF). Compared with purely inductive cases, few attention has been paid to purely resistive case [164, 165]. In [164], the output impedance of parallel DGs is shaped to be resistive, where the Q-V droop equation is replaced by the P-V droop control equation. With the resistive output impedance, the overall system is more damped and the harmonic current is automatically shared [165]. However, the mismatch of feeder impedance is still ignored the same as [150] which will have great limitations in real applications.

In view of extensive purely inductive cases and lack of more practical solutions for resistive scenarios, an open-loop method which can not only improve the active power sharing accuracy, but also easier to be implemented in low-cost micro-controllers is required. Comparing the virtual impedance strategy with $Q - \dot{V}$ droop method, it is concluded that the former is a powerful tool to decouple the active and reactive power with limited ability to compensate the mismatch of feeder impedance while the latter is better at reducing the effect of mismatched line impedance without consideration of power coupling.

In this chapter, a virtual complex impedance based $P - \dot{V}$ droop method which combines the advantages of both virtual impedance based method [164] and $Q - \dot{V}$ droop method [153], is proposed to solve the power sharing problem of LVMG. Besides, in most existing papers, P and Q adopted in droop equations are calculated according to the capacitor voltage and grid-side current, which are not actually decoupled since the physical impedance between PCC and filter capacitor is still complex. The actually decoupled P/Q injected by VPS is adopted in this book, since the impedance between PCC and VPS is reshaped to be purely resistive. In addition to the original $P - \dot{V}$ droop method, a modified $P - \dot{V}$ method is also proposed, where an extra parameter S_p is utilized to accelerate the dynamic regulation performance and active power sharing accuracy at the same time.

1.5 Structure of This Brief

The book is organized as follows: This chapter introduces hybrid energy system: the framework, the overview of security control, optimization and cooperative control; Part I consists of Chaps. 2–4, which is devoted to cooperative optimal security control of hybrid energy system; Part II consists of Chaps. 5–7, which refer to multi-objective optimization for optimal operation of hybrid energy system; Part III includes Chaps. 8 and 9, devoting to distributed optimization for energy management of microgrid; Part IV includes Chaps. 10–12, devoting to the distributed cooperative control of DC microgrid; Part V includes Chaps. 13–15, devoting to the distributed cooperative control of islanded AC microgrids.

In Chap. 2, the multiagent system-based event-triggered hybrid controls for high-security hybrid energy generation systems is proposed.

In Chap. 3, multi-agent based hierarchical hybrid control for smart microgrid is proposed.

In Chap. 4, two-stage optimal operation strategy of isolated power system with TSK fuzzy identification of supply-security is proposed.

In Chap. 5, MOEA/D based probabilistic PBI approach for risk-based optimal operation of hybrid energy systems with intermittent power uncertainty is proposed.

In Chap. 6, gradient decent based multi-objective cultural differential evolution for short-term hydrothermal optimal scheduling of economic emission with integrating wind power and photovoltaic power is proposed.

In Chap. 7, event-triggered multi-agent optimization for two-layered model of hybrid energy system with price bidding based demand response is proposed.

In Chap. 8, consensus-based economic hierarchical control strategy for islanded MG considering communication path reconstruction is proposed.

In Chap. 9, multi-agent-system-based BI-level bidding strategy of microgrid with game theory in the electricity market is proposed.

In Chap. 10, multiagent system-based distributed coordinated control for radial DC microgrid considering transmission time delays is proposed.

In Chap. 11, MAS-based distributed cooperative control for DC microgrid through switching topology communication network with time-varying delays is proposed.

In Chap. 12, multiagent system-based integrated design of security control and economic dispatch for interconnected microgrid systems is proposed.

In Chap. 13, distributed event-triggered cooperative control for frequency and voltage stability and power sharing in isolated inverter-based microgrid is proposed.

In Chap. 14, event-triggered mechanism based distributed optimal frequency regulation of power grid is proposed.

In Chap. 15, a virtual complex impedance based P-V droop method for parallel-connected inverters in low-voltage AC microgrids is proposed.

References

1. S.M. Ashabani, Y.A.I. Mohamed, New family of microgrid control and management strategies in smart distribution grids-analysis, comparison and testing. IEEE Trans. Power Syst. **29**(5), 2257–2269 (2014)
2. A. Khodaei, Provisional microgrids. IEEE Trans. Smart Grid **6**(3), 1107–1115 (2015)
3. J.M. Guerrero, M. Chandorkar, T. Lee, P.C. Loh, Advanced control architectures for intelligent microgrids - part I: decentralized and hierarchical control. IEEE Trans. Ind. Electron. **60**(4), 1254–1262 (2013)
4. A. Bidram, A. Davoudi, Hierarchical structure of microgrids control system. IEEE Trans. Smart Grid **3**(4), 1963–1976 (2012)
5. E. Natsheh, A. Natsheh, A. Albarbar, Intelligent controller for managing power flow within standalone hybrid power systems. Iet Sci. Meas. & Technol. **7**(4), 191–200 (2013)
6. Q. Li, W. Chen, Y. Li, S. Liu, J. Huang, Energy management strategy for fuel cell/battery/ultracapacitor hybrid vehicle based on fuzzy logic. Int. J. Electr. Power & Energy Syst. **43**(1), 514–525 (2012)
7. A. Kaabeche, R. Ibtiouen, Techno-economic optimization of hybrid photovoltaic/wind/diesel/battery generation in a stand-alone power system. Solar Energy **103**, 171–182 (2014)
8. A. Ghazanfari, M. Hamzeh, H. Mokhtari, H. Karimi, Active power management of multihybrid fuel cell/supercapacitor power conversion system in a medium voltage microgrid. IEEE Trans. Smart Grid **3**(4), 1903–1910 (2012)
9. Z. Jiang, L. Gao, R.A. Dougal, Adaptive control strategy for active power sharing in hybrid fuel cell/battery power sources. IEEE Trans. Energy Convers. **22**(2), 183–194 (2007)
10. T. Logenthiran, D. Srinivasan, A.M. Khambadkone, Multi-agent system for energy resource scheduling of integrated microgrids in a distributed system. Electric Power Syst. Res. **81**(1), 138–148 (2011)
11. Z. Jun, L. Junfeng, W. Jie, H.W. Ngan, A multi-agent solution to energy management in hybrid renewable energy generation system. Renew. Energy **36**(5), 1352–1363 (2011)
12. J. Lagorse, D. Paire, A. Miraoui, A multi-agent system for energy management of distributed power sources. Renew. Energy **35**(1), 174–182 (2010)
13. J. Lagorse, M.G. Simoes, A. Miraoui, A multiagent fuzzy-logic-based energy management of hybrid systems. IEEE Trans. Ind. Appl. **45**(6), 2123–2129 (2009)
14. C. Dou, B. Liu, J.M. Guerrero, Event-triggered hybrid control based on multi-agent system for microgrids. Iet Gener. Trans. & Distrib. **8**(12), 1987–1997 (2014)
15. C. Li, X. Yu, W. Yu, T. Huang, Z.-W. Liu, Distributed event-triggered scheme for economic dispatch in smart grids. IEEE Trans. Ind. Inf. **12**(5), 1775–1785 (2016)
16. S. Wen, X. Yu, Z. Zeng, J. Wang, Event-triggering load frequency control for multiarea power systems with communication delays. IEEE Trans. Ind. Electron. **63**(2), 1308–1317 (2016)
17. D.E. Olivares, A. Mehrizi-Sani, A.H. Etemadi, C.A. ñizares, R. Iravani, M. Kazerani, A.H. Hajimiragha, O. Gomis-Bellmunt, M. Saeedifard, R. Palma-Behnke, G.A. Jiménez-Estévez, N.D. Hatziargyriou, Trends in microgrid control. IEEE Trans. Smart Grid **5**(4), 1905–1919 (2014)
18. S.X. Chen, H.B. Gooi, M.Q. Wang, Sizing of energy storage for microgrids. IEEE Trans. Smart Grid **3**(1), 142–151 (2012)
19. M. Mahmoodi, P. Shamsi, B. Fahimi, Economic dispatch of a hybrid microgrid with distributed energy storage. IEEE Trans. Smart Grid **6**(6), 2607–2614 (2015)
20. B. Mohammadi-ivatloo, A. Rabiee, M. Ehsan, Time-varying acceleration coefficients ipso for solving dynamic economic dispatch with non-smooth cost function. Energy Convers. Manag. **56**, 175–183 (2012)
21. R. Liang, J. Liao, A fuzzy-optimization approach for generation scheduling with wind and solar energy systems. IEEE Trans. Power Syst. **22**(4), 1665–1674 (2007)
22. Y.-K. Liu, Convergent results about the use of fuzzy simulation in fuzzy optimization problems. IEEE Trans. Fuzzy Syst. **14**(2), 295–304 (2006)

23. F. Xiao, J.D. McCalley, Risk-based security and economy tradeoff analysis for real-time operation. IEEE Trans. Power Syst. **22**(4), 2287–2288 (2007)
24. J.-C. Lee, W.-M. Lin, G.-C. Liao, T.-P. Tsao, Quantum genetic algorithm for dynamic economic dispatch with valve-point effects and including wind power system. Int. J. Electr. Power & Energy Syst. **33**(2), 189–197 (2011)
25. X. Li, C. Jiang, Short-term operation model and risk management for wind power penetrated system in electricity market. IEEE Trans. Power Syst. **26**(2), 932–939 (2011)
26. C. Chen, Y. Li, G. Huang, Y. Li, A robust optimization method for planning regional-scale electric power systems and managing carbon dioxide. Int. J. Electr. Power & Energy Syst. **40**(1), 70–84 (2012)
27. C. Peng, P. Xie, L. Pan, R. Yu, Flexible robust optimization dispatch for hybrid wind/photovoltaic/hydro/thermal power system. IEEE Trans. Smart Grid **7**(2), 751–762 (2016)
28. S. Conti, R. Nicolosi, S.A. Rizzo, Generalized systematic approach to assess distribution system reliability with renewable distributed generators and microgrids. IEEE Trans. Power Deliv. **27**(1), 261–270 (2012)
29. I. Bae, J. Kim, Reliability evaluation of customers in a microgrid. IEEE Trans. Power Syst. **23**(3), 1416–1422 (2008)
30. R. Palma-Behnke, C. Benavides, F. Lanas, B. Severino, L. Reyes, J. Llanos, D. Sáez, A microgrid energy management system based on the rolling horizon strategy. IEEE Trans. Smart Grid **4**(2), 996–1006 (2013)
31. F. Luo, Z. Dong, G. Chen, Y. Xu, K. Meng, Y. Chen, K. Wong, Advanced pattern discovery-based fuzzy classification method for power system dynamic security assessment. IEEE Trans. Ind. Inf. **11**(2), 416–426 (2015)
32. C. Dou, D. Yue, J.M. Guerrero, Multiagent system-based event-triggered hybrid controls for high-security hybrid energy generation systems. IEEE Trans. Ind. Inf. **13**(2), 584–594 (2017)
33. Y. Jia, Z. Xu, L.L. Lai, K.P. Wong, Risk-based power system security analysis considering cascading outages. IEEE Trans. Ind. Inf. **12**(2), 872–882 (2016)
34. S.A. Arefifar, Y.A.I. Mohamed, Dg mix, reactive sources and energy storage units for optimizing microgrid reliability and supply security. IEEE Trans. Smart Grid **5**(4), 1835–1844 (2014)
35. W. Zhang, Y. Xu, Z. Dong, K.P. Wong, Robust security constrained-optimal power flow using multiple microgrids for corrective control of power systems under uncertainty. IEEE Trans. Ind. Inf. **13**(4), 1704–1713 (2017)
36. Y. Lin, J. Chang, C. Lin, Identification and prediction of dynamic systems using an interactively recurrent self-evolving fuzzy neural network. IEEE Trans. Neural Netw. Learn. Syst. **24**(2), 310–321 (2013)
37. Z. Deng, Y. Jiang, K. Choi, F. Chung, S. Wang, Knowledge-leverage-based tsk fuzzy system modeling. IEEE Trans. Neural Netw. Learn. Syst. **24**(8), 1200–1212 (2013)
38. D.T. Ho, J.M. Garibaldi, Context-dependent fuzzy systems with application to time-series prediction. IEEE Trans. Fuzzy Syst. **22**(4), 778–790 (2014)
39. C.Y. Teh, Y.W. Kerk, K.M. Tay, C.P. Lim, On modeling of data-driven monotone zero-order tsk fuzzy inference systems using a system identification framework. IEEE Trans. Fuzzy Syst. **26**(6), 3860–3874 (2018)
40. I. Eyoh, R. John, G.D. Maere, E. Kayacan, Hybrid learning for interval type-2 intuitionistic fuzzy logic systems as applied to identification and prediction problems. IEEE Trans. Fuzzy Syst. **26**(5), 2672–2685 (2018)
41. X. Gu, F. Chung, S. Wang, Bayesian takagi-sugeno-kang fuzzy classifier. IEEE Trans. Fuzzy Syst. **25**(6), 1655–1671 (2017)
42. J. Zahavi, Economic-environmental power dispatch. IEEE Trans. Syst. **SMC-5**(5), 485–489 (1975)
43. R.T.F.A.K.C.S.R. Deb, Evolutionary multi-objective environmental/economic dispatch: Stochastic versus deterministic approaches. *Lecture Notes in Computer Science*, vol. 3410 (2005)

44. K. Deb, A. Pratap, S. Agarwal, T. Meyarivan, A fast and elitist multiobjective genetic algorithm: Nsga-ii. IEEE Trans. Evol. Comput. **6**(2), 182–197 (2002)
45. M. Abido, A niched pareto genetic algorithm for multiobjective environmental/economic dispatch. Int. J. Electr. Power & Energy Syst. **25**(2), 97–105 (2003). Fuel & Energy Abs. **44**(4), 261 (2003)
46. A.M. Abido, Environmental/economic power dispatch using multiobjective evolutionary algorithms IEEE Trans. Power Syst. (2003)
47. S. Agrawal, B.K. Panigrahi, M. Tiwari, Multiobjective particle swarm algorithm with fuzzy clustering for electrical power dispatch. IEEE Trans. Evol. Comput. **12**(5), 529–541 (2008)
48. H. Zhang, D. Yue, X. Xie, S. Hu, S. Weng, Multi-elite guide hybrid differential evolution with simulated annealing technique for dynamic economic emission dispatch. Appl. Soft Comput. **34**, 312–323 (2015)
49. D.B. Das, C. Patvardhan, New multi-objective stochastic search technique for economic load dispatch. IEE Proc. - Gen. Trans. Distrib. **145**(6), 747–752 (1998)
50. A. Zhou, B.Y. Qu, H. Li, S.Z. Zhao, Q. Zhang, Multiobjective evolutionary algorithms: a survey of the state of the art. Swarm & Evol. Comput. **1**(1), 32–49 (2011)
51. A. Asrari, S. Lotfifard, M.S. Payam, Pareto dominance-based multiobjective optimization method for distribution network reconfiguration. IEEE Trans. Smart Grid **7**(3), 1401–1410 (2016)
52. E.Z.L. Thiele, Spea2: Improving the strength pareto evolutionary algorithm. Technical report 103, computer engineering and networks laboratory (TIK) (2001)
53. E. Zitzler, S. Künzli, Indicator-based selection in multiobjective search. *Parallel Problem Solving From Nature-PPSN VIII* (Springer, Berlin, 2004), pp. 832–842
54. N. Beume, B. Naujoks, M. Emmerich, Sms-emoa: multiobjective selection based on dominated hypervolume. Europ. J. Oper. Res. **181**(3), 1653–1669 (2007)
55. T. Murata, Specification of genetic search directions in cellular multi-objective genetic algorithms, in *Evolutionary Multi-Criterion Optimization-EMO 2001*, vol. 1993 (2001), pp. 82–95
56. Q. Zhang, H. Li, Moea/d: a multiobjective evolutionary algorithm based on decomposition. IEEE Trans. Evol. Comput. **11**(6), 712–731 (2007)
57. I.C.P. Fleming, Economic-environmental power dispatch, in *Evolutionary Multi-Criterion Optimization* (2013), pp. 428–442
58. Q. Kang, X. Song, M.C. Zhou, L. Li, A collaborative resource allocation strategy for decomposition-based multiobjective evolutionary algorithms. IEEE Trans. Syst. Man & Cybern. Syst. **99**, 1–8 (2018)
59. Q.L. Wang, M.A. Zhou, L. Jiao, Constrained subproblems in a decomposition-based multi-objective evolutionary algorithm. IET Control Theory Appl. **20**(3), 475–480 (2016)
60. X. Kong, X. Liu, L. Ma, K.Y. Lee, Hierarchical distributed model predictive control of standalone wind/solar/battery power system. IEEE Trans. Syst. Man Cybern.: Syst. 1570–1581 (2019)
61. C. Dou, D. Yue, Q.L. Han, J.M. Guerrero, Multi-agent system-based event-triggered hybrid control scheme for energy internet. IEEE Access **5**(99), 3263–3272 (2017)
62. M.-R. Haghifam, H. Falaghi, O. Malik, Risk-based distributed generation placement. Iet Gener. Trans. & Distrib. **2**(2), 1 (2008)
63. C. Liu, A. Botterud, Z. Zhou, P. Du, Fuzzy energy and reserve co-optimization with high penetration of renewable energy. IEEE Trans. Sustain. Energy **8**(2), 782–791 (2017)
64. C. Zhao, J. Wang, J. Watson, Y. Guan, Multi-stage robust unit commitment considering wind and demand response uncertainties, in *2014 IEEE PES General Meeting | Conference Exposition* (2014), pp. xx
65. A. Lorca, X.A. Sun, Adaptive robust optimization with dynamic uncertainty sets for multi-period economic dispatch under significant wind. IEEE Trans. Power Syst. **30**(4), 1702–1713 (2015)
66. Q.P. Zheng, J. Wang, A.L. Liu, Stochastic optimization for unit commitment-a review. IEEE Trans. Power Syst. **30**(4), 1913–1924 (2015)

67. Y. Zhang, N. Gatsis, G.B. Giannakis, Robust energy management for microgrids with high-penetration renewables. IEEE Trans. Sustain. Energy **4**(4), 944–953 (2013)
68. Y. Xiang, J. Liu, Y. Liu, Robust energy management of microgrid with uncertain renewable generation and load. IEEE Trans. Smart Grid **7**(2), 1034–1043 (2016)
69. F. Katiraei, R. Iravani, N. Hatziargyriou, A. Dimeas, Microgrids management. IEEE Power Energy Mag. **6**(3), 54–65 (2008)
70. Z. Xu, H. Qu, W. Shao, W. Xu, Virtual power plant-based pricing control for wind/thermal cooperated generation in china. IEEE Trans. Syst. Man Cybern.: Syst. **46**(5), 706–712 (2016)
71. M. Amelin, An evaluation of intraday trading and demand response for a predominantly hydro-wind system under nordic market rules. IEEE Trans. Power Syst. **30**(1), 3–12 (2015)
72. E. Heydarian-Forushani, M.P. Moghaddam, M.K. Sheikh-El-Eslami, M. Shafie-khah, J.P.S. Catalão, Risk-constrained offering strategy of wind power producers considering intraday demand response exchange. IEEE Trans. Sustain. Energy **5**(4), 1036–1047 (2014)
73. J. Mohammadi, A. Rahimi-Kian, M. Ghazizadeh, Aggregated wind power and flexible load offering strategy. IET Renew. Power Gen. **5**(6), 439–447 (2011)
74. X. Zhang, Optimal scheduling of critical peak pricing considering wind commitment. IEEE Trans. Sustain. Energy **5**(2), 637–645 (2014)
75. H.S.V.S.K. Nunna, S. Doolla, Demand response in smart distribution system with multiple microgrids. IEEE Trans. Smart Grid **3**(4), 1641–1649 (2012)
76. H. Nunna, S. Doolla, Energy management in microgrids using demand response and distributed storage-a multiagent approach. IEEE Trans. Power Deliv. **28**(2), 939–947 (2013)
77. H. Nunna, S. Doolla, Multiagent-based distributed-energy-resource management for intelligent microgrids. IEEE Trans. Ind. Electron. **60**(4), 1678–1687 (2013)
78. L. Ding, Q. Han, X. Ge, X. Zhang, An overview of recent advances in event-triggered consensus of multiagent systems. IEEE Trans. Cybern. **48**(4), 1110–1123 (2018)
79. D. Yue, E. Tian, Q.L. Han, A delay system method for designing event-triggered controllers of networked control systems. IEEE Trans. Autom. Control **58**(2), 475–481 (2013)
80. C. Peng, S. Ma, X. Xie, Observer-based non-pdc control for networked t-s fuzzy systems with an event-triggered communication. IEEE Trans. Cybern. **47**(8), 2279–2287 (2017)
81. X. Yi, K. Liu, D.V. Dimarogonas, K.H. Johansson, Dynamic event-triggered and self-triggered control for multi-agent systems. IEEE Trans. Autom. Control 1–1 (2018)
82. W. Hu, L. Liu, G. Feng, Consensus of linear multi-agent systems by distributed event-triggered strategy. IEEE Trans. Cybern. **46**(1), 148–157 (2016)
83. M.A. Lopez, S. Martin, J. Aguado, S.D. La Torre, Market-oriented operation in microgrids using multi-agent systems, in *Proceedings of the 2011 International Conference Power Engineering Energy Electrical* (2011), pp. 1–6
84. E. Barklund, N. Pogaku, M. Prodanovic, C.A. Hernandezaramburo, T.C. Green, Energy management in autonomous microgrid using stability-constrained droop control of inverters. IEEE Trans. Power Electron. **23**(5), 2346–2352 (2008)
85. R. Bhuvaneswari, S.K. Srivastava, C.S. Edrington, D.A. Cartes, S. Subramanian, Intelligent agent based auction by economic generation scheduling for microgrid operation, in *Innovative Smart GridTechnol (ISGT), Gaithersburg, Maryland* (2010), pp. 1–5
86. S.D.J. Mcarthur, E.M. Davidson, V.M. Catterson, A.L. Dimeas, N.D. Hatziargyriou, F. Ponci, T. Ferdinanda, Multi-agent systems for power engineering applications-part ii: technologies, standards, and tools for building multi-agent systems. IEEE Trans. Power Syst. **22**(4), 1753–1759 (2007)
87. A. Mozdawar, B. Khaki, Strategic bidding in pay as bid power market by combined probabilistic and game theory procedures, in *IEEE Bucharest Power Tech* vol. 2009, pp. 1–6 (2009)
88. J.S.H. Rajabi, M. Latifi, Optimal supply bidding with risk management in an electric pay as bid auction, in *15thPower System Computation Conference (PSCC), Liege* (2005), pp. 22–26
89. D. Fang, X. Wang, F. Ouyang, C. Ye, Bayesian nash equilibrium bidding strategies for generation companies, in *2004 IEEE International Conference on Electric Utility Deregulation, Restructuring and Power Technologies. Proceedings*, vol. 2 (2004), pp. 692–697

90. J. Kumar, D. Kumar, Generation bidding strategy in a pool based electricity market using shuffled frog leaping algorithm. Appl. Soft Comput. **21**, 407–414 (2014)
91. Y. Zheng, Y. Zhu, L. Wang, Consensus of heterogeneous multi-agent systems. IET Control Theory Appl. **5**(16), 1881–1888 (2011)
92. C. Dou, D. Hao, B. Jin, W. Wang, N. An, Multi-agent-system-based decentralized coordinated control for large power systems. Int. J. Electr. Power & Energy Syst. **58**, 130–139 (2014)
93. M.K. Kouluri, R.K. Pandey, Intelligent agent based micro grid control, in *2011 2nd International Conference on Intelligent Agent and Multi-Agent Systems (IAMA)* (2011), pp. 62–66
94. T. Logenthiran, D. Srinivasan, A.M. Khambadkone, H.N. Aung, Multi-agent system (mas) for short-term generation scheduling of a microgrid, in *IEEE International Conference on Sustainable Energy Technologies (ICSET)* (2010), pp. 1–6
95. S.D.J. Mcarthur, E.M. Davidson, V.M. Catterson, A.L. Dimeas, N.D. Hatziargyriou, F. Ponci, T. Funabashi, Multi-agent systems for power engineering applications-part i: concepts, approaches, and technical challenges. IEEE Trans. Power Syst. **22**(4), 1743–1752 (2007)
96. A.S. Chuang, F. Wu, P. Varaiya, A game-theoretic model for generation expansion planning: problem formulation and numerical comparisons. IEEE Trans. Power Syst. **16**(4), 885–891 (2001)
97. A. Kwasinski, Quantitative evaluation of dc microgrids availability: Effects of system architecture and converter topology design choices. IEEE Trans. Power Electron. **26**(3), 835–851 (2011)
98. Q. Shafiee, T. Dragicevic, J.C. Vasquez, J.M. Guerrero, Hierarchical control for multiple dc-microgrids clusters. IEEE Trans. Energy Convers. **29**(4), 922–933 (2014)
99. A.T. Elsayed, A. Mohamed, O.A. Mohammed, Dc microgrids and distribution systems: an overview. Electric Power Syst. Res. **119**(119), 407–417 (2015)
100. W. Yao, M. Chen, J. Matas, J.M. Guerrero, Z. Qian, Design and analysis of the droop control method for parallel inverters considering the impact of the complex impedance on the power sharing. IEEE Trans. Ind. Electron. **58**(2), 576–588 (2011)
101. A. Khorsandi, M. Ashourloo, H. Mokhtari, A decentralized control method for a low-voltage dc microgrid. IEEE Trans. Energy Convers. **29**(4), 793–801 (2014)
102. Y. Gu, X. Xiang, W. Li, X. He, Mode-adaptive decentralized control for renewable dc microgrid with enhanced reliability and flexibility. IEEE Trans. Power Electron. **29**(9), 5072–5080 (2014)
103. J.M. Guerrero, J.C. Vasquez, J. Matas, L.G. de Vicuna, M. Castilla, Hierarchical control of droop-controlled ac and dc microgrids-a general approach toward standardization. IEEE Trans. Ind. Electron. **58**(1), 158–172 (2011)
104. X. Lu, J.M. Guerrero, K. Sun, J.C. Vasquez, R. Teodorescu, L. Huang, Hierarchical control of parallel ac-dc converter interfaces for hybrid microgrids. IEEE Trans. Smart Grid **5**(2), 683–692 (2014)
105. J.M. Guerrero, P.C. Loh, T. Lee, M. Chandorkar, Advanced control architectures for intelligent microgrids-part ii: Power quality, energy storage, and ac/dc microgrids. IEEE Trans. Ind. Electron. **60**(4), 1263–1270 (2013)
106. Q. Shafiee, J.M. Guerrero, J.C. Vasquez, Distributed secondary control for islanded microgrids-a novel approach. IEEE Trans. Power Electron. **29**(2), 1018–1031 (2014)
107. S. Anand, B.G. Fernandes, J.M. Guerrero, Distributed control to ensure proportional load sharing and improve voltage regulation in low-voltage dc microgrids. IEEE Trans. Power Electron. **28**(4), 1900–1913 (2013)
108. X. Lu, J.M. Guerrero, K. Sun, J.C. Vasquez, An improved droop control method for dc microgrids based on low bandwidth communication with dc bus voltage restoration and enhanced current sharing accuracy. IEEE Trans. Power Electron. **29**(4), 1800–1812 (2014)
109. Q. Shafiee, C. Stefanovići, T. Dragići, P. Popovski, J.C. Vasquez, J.M. Guerrero, Robust networked control scheme for distributed secondary control of islanded microgrids. IEEE Trans. Ind. Electron. **61**(10), 5363–5374 (2014)
110. A. Bidram, A. Davoudi, F.L. Lewis, J.M. Guerrero, Distributed cooperative secondary control of microgrids using feedback linearization. IEEE Trans. Power Syst. **28**(3), 3462–3470 (2013)

111. V. Nasirian, S. Moayedi, A. Davoudi, F.L. Lewis, Distributed cooperative control of dc microgrids. IEEE Trans. Power Electron. **30**(4), 2288–2303 (2015)
112. V. Nasirian, A. Davoudi, F.L. Lewis, J.M. Guerrero, Distributed adaptive droop control for dc distribution systems. IEEE Trans. Energy Convers. **29**(4), 944–956 (2014)
113. B. Chaudhuri, R. Majumder, B.C. Pal, Wide-area measurement-based stabilizing control of power system considering signal transmission delay. IEEE Trans. Power Syst. **19**(4), 1971–1979 (2004)
114. H. Wu, K. Tsakalis, G.T. Heydt, Evaluation of time delay effects to wide-area power system stabilizer design. IEEE Trans. Power Syst. **19**(4), 1935–1941 (2004)
115. C. Dou, Q. Jia, S. Jin, Z. Bo, Delay-independent decentralized stabilizer design for large interconnected power systems based on wams. Int. J. Electr. Power & Energy Syst. **29**(10), 775–782 (2007)
116. C. Dou, Z. Duan, X. Jia, Delay-dependent $h\infty$ robust control for large power systems based on two-level hierarchical decentralised coordinated control structure. Int. J. Syst. Sci. **44**(2), 329–345 (2013)
117. Z. Zhang, M. Chow, Convergence analysis of the incremental cost consensus algorithm under different communication network topologies in a smart grid. IEEE Trans. Power Syst. **27**(4), 1761–1768 (2012)
118. L. Xu, D. Chen, Control and operation of a dc microgrid with variable generation and energy storage. IEEE Trans. Power Deliv. **26**(4), 2513–2522 (2011)
119. C. Yuen, A. Oudalov, A. Timbus, The provision of frequency control reserves from multiple microgrids. IEEE Trans. Ind. Electron. **58**(1), 173–183 (2011)
120. H. Kanchev, D. Lu, F. Colas, V. Lazarov, B. Francois, Energy management and operational planning of a microgrid with a pv-based active generator for smart grid applications. IEEE Trans. Ind. Electron. **58**(10), 4583–4592 (2011)
121. C. Chen, S. Duan, T. Cai, B. Liu, G. Hu, Optimal allocation and economic analysis of energy storage system in microgrids. IEEE Trans. Power Electron. **26**(10), 2762–2773 (2011)
122. T. Zhou, B. Francois, Energy management and power control of a hybrid active wind generator for distributed power generation and grid integration. IEEE Trans. Ind. Electron. **58**(1), 95–104 (2011)
123. P.C. Loh, D. Li, Y.K. Chai, F. Blaabjerg, Autonomous control of interlinking converter with energy storage in hybrid ac-dc microgrid. IEEE Trans. Ind. Appl. **49**(3), 1374–1382 (2013)
124. J. Schonberger, R. Duke, S. Round, Dc-bus signaling: a distributed control strategy for a hybrid renewable nanogrid. IEEE Trans. Ind. Electron. **53**(5), 1453–1460 (2006)
125. K. Sun, L. Zhang, Y. Xing, J.M. Guerrero, A distributed control strategy based on dc bus signaling for modular photovoltaic generation systems with battery energy storage. IEEE Trans. Power Electron. **26**(10), 3032–3045 (2011)
126. D. Chen, L. Xu, L. Yao, Dc voltage variation based autonomous control of dc microgrids. IEEE Trans. Power Deliv. **28**(2), 637–648 (2013)
127. P. Karlsson, J. Svensson, Dc bus voltage control for a distributed power system. IEEE Trans. Power Electron. **18**(6), 1405–1412 (2003)
128. Y. Ito, Y. Zhongqing, H. Akagi, Dc microgrid based distribution power generation system, in *Power Electronics and Motion Control Conference, 2004. IPEMC 2004. The 4th International* (2004), pp. 1740–1745
129. W. Qiu, Z. Liang, Practical design considerations of current sharing control for parallel vrm applications, in *IEEE Applied Power Electronics Conference & Exposition* (2005), pp. 281–286
130. J.M. Guerrero, L. Hang, J. Uceda, Control of distributed uninterruptible power supply systems. IEEE Trans. Ind. Electron. **55**(8), 2845–2859 (2008)
131. X. Lu, K. Sun, J.M. Guerrero, J.C. Vasquez, L. Huang, State-of-charge balance using adaptive droop control for distributed energy storage systems in dc microgrid applications. IEEE Trans. Ind. Electron. **61**(6), 2804–2815 (2014)
132. S. Anand, B.G. Fernandes, Steady state performance analysis for load sharing in dc distributed generation system, in *Proceedings of the 10th International Conference on Environmental and Electrical Engineering* (2011), pp. 1–4

133. Y.W. Li, C. Kao, An accurate power control strategy for power-electronics-interfaced distributed generation units operating in a low-voltage multibus microgrid. IEEE Trans. Power Electron. **24**(12), 2977–2988 (2009)
134. J. He, Y.W. Li, Analysis, design, and implementation of virtual impedance for power electronics interfaced distributed generation. IEEE Trans. Ind. Appl. **47**(6), 2525–2538 (2011)
135. H. Laaksonen, P. Saari, R. Komulainen, Voltage and frequency control of inverter based weak lv network microgrid, in *International Conference on Future Power Systems* (2006)
136. T.L. Vandoorn, B. Meersman, L. Degroote, B. Renders, L. Vandevelde, A control strategy for islanded microgrids with dc-link voltage control. IEEE Trans. Power Deliv. **26**(2), 703–713 (2011)
137. C. Dou, M. Lv, T. Zhao, Y. Ji, H. Li, Decentralised coordinated control of microgrid based on multi-agent system. IET Gen. Trans. Distrib. **9**(16), 2474–2484 (2015)
138. H. Liang, B.J. Choi, W. Zhuang, X. Shen, A.S.A. Awad, A. Abdr, Multiagent coordination in microgrids via wireless networks. IEEE Wireless Commun. **19**(3), 14–22 (2012)
139. C. Dou, N. Li, D. Yue, T. Liu, Hierarchical hybrid control strategy for micro-grid switching stabilisation during operating mode conversion. IET Gene. Trans. Distrib. **10**(12), 2880–2890 (2016)
140. Z. Yu, Q. Ai, J. Gong, L. Piao, A novel secondary control for microgrid based on synergetic control of multi-agent system. Energies **9**(4), 243 (2016)
141. M.H. Rehmani, M. Reisslein, A. Rachedi, M. Erol-Kantarci, M. Radenkovic, Integrating renewable energy resources into the smart grid: Recent developments in information and communication technologies. IEEE Trans. Ind. Inf. **14**(7), 2814–2825 (2018)
142. R. Bonetto, M. Rossi, S. Tomasin, M. Zorzi, On the interplay of distributed power loss reduction and communication in low voltage microgrids. IEEE Trans. Ind. Inf. **12**(1), 322–337 (2016)
143. Z. Li, C. Zang, P. Zeng, H. Yu, S. Li, Fully distributed hierarchical control of parallel grid-supporting inverters in islanded ac microgrids. IEEE Trans. Ind. Inf. **14**(2), 679–690 (2018)
144. J. Rocabert, A. Luna, F. Blaabjerg, P. Rodríguez, Control of power converters in AC microgrids. IEEE Trans. Power Electron. **27**(11), 4734–4749 (2012)
145. H.Y. Han, E.C.P. Shen, J. Guerrero, Review of active and reactive power sharing strategies in hierarchical controlled microgrids. IEEE Trans. Power Electron. **32**(3), 2427–2451 (2017)
146. M.N. Kabir, Y. Mishra, G. Ledwich, Z.Y. Dong, K.P. Wong, Coordinated control of grid-connected photovoltaic reactive power and battery energy storage systems to improve the voltage profile of a residential distribution feeder. IEEE Trans. Ind. Inf. **10**(2), 967–977 (2014)
147. L. Ding, Q. Han, X. Zhang, Distributed secondary control for active power sharing and frequency regulation in islanded microgrids using an event-triggered communication mechanism. IEEE Trans. Ind. Inf. **15**(7), 3910–3922 (2019)
148. H. Xu, C. Yu, C. Liu, Q. Wang, F. Liu, F. Li, An improved virtual capacitor algorithm for reactive power sharing in multi-paralleled distributed generators. IEEE Trans. Power Electron. **34**(11), 10 786–10 795 (2019)
149. C. Dou, Z. Zhang, D. Yue, M. Song, Improved droop control based on virtual impedance and virtual power source in low-voltage microgrid. IET Gen. Trans. Distrib. **11**(4), 1046–1054 (2017)
150. J.M. Guerrero, J. Matas, L. Garcia De Vicunagarcia De Vicuna, M. Castilla, J. Miret, Wireless-control strategy for parallel operation of distributed-generation inverters. IEEE Trans. Ind. Electron. **53**(5), 1461–1470 (2006)
151. J. He, Y.W. Li, F. Blaabjerg, An enhanced islanding microgrid reactive power, imbalance power, and harmonic power sharing scheme. IEEE Trans. Power Electron. **30**(6), 3389–3401 (2015)
152. R. An, Z. Liu, J. Liu, Modified adaptive virtual impedance method to compensate mismatched line impedances in microgrids, in *2019 IEEE Applied Power Electronics Conference and Exposition (APEC)* (2019), pp. 1109–1114
153. C. Lee, C. Chu, P. Cheng, A new droop control method for the autonomous operation of distributed energy resource interface converters, in *2010 IEEE Energy Conversion Congress and Exposition* (2010), pp. 702–709

154. J. Zhou, P. Cheng, A modified $q - \dot{V}$ droop control for accurate reactive power sharing in distributed generation microgrid. IEEE Trans. Ind. Appl. **55**(4), 4100–4109 (2019)
155. A. Tuladhar, H. Jin, T. Unger, K. Mauch, Control of parallel inverters in distributed AC power systems with consideration of line impedance effect. IEEE Trans. Ind. Appl. **36**(1), 131–138 (2000)
156. J. He, Y.W. Li, An enhanced microgrid load demand sharing strategy. IEEE Trans. Power Electron. **27**(9), 3984–3995 (2012)
157. B. Liu, Z. Liu, J. Liu, R. An, H. Zheng, Y. Shi, An adaptive virtual impedance control scheme based on small-AC-signal injection for unbalanced and harmonic power sharing in islanded microgrids. IEEE Trans. Power Electron. **34**(12), 12 333–12 355 (2019)
158. Q. Zhong, Robust droop controller for accurate proportional load sharing among inverters operated in parallel. IEEE Trans. Ind. Electron. **60**(4), 1281–1290 (2013)
159. J. He, Y.W. Li, J.M. Guerrero, F. Blaabjerg, J.C. Vasquez, An islanding microgrid power sharing approach using enhanced virtual impedance control scheme. IEEE Trans. Power Electron. **28**(11), 5272–5282 (2013)
160. M. Kosari, S.H. Hosseinian, Decentralized reactive power sharing and frequency restoration in islanded microgrid. IEEE Trans. Power Syst. **32**(4), 2901–2912 (2017)
161. H. Mahmood, D. Michaelson, J. Jiang, Accurate reactive power sharing in an islanded microgrid using adaptive virtual impedances. IEEE Trans. Power Electron. **30**(3), 1605–1617 (2015)
162. C. Lee, C. Chu, P. Cheng, A new droop control method for the autonomous operation of distributed energy resource interface converters. IEEE Trans. Power Electron. **28**(4), 1980–1993 (2013)
163. H. Han, X. Hou, J. Yang, J. Wu, M. Su, J.M. Guerrero, Review of power sharing control strategies for islanding operation of AC microgrids. IEEE Trans. Smart Grid **7**(1), 200–215 (2016)
164. J.M. Guerrero, J. Matas, L. Garcia de Vicuna, M. Castilla, J. Miret, Decentralized control for parallel operation of distributed generation inverters using resistive output impedance. IEEE Trans. Ind. Electron. **54**(2), 994–1004 (2007)
165. T.L. Vandoorn, J.D.M. De Kooning, B. Meersman, J.M. Guerrero, L. Vandevelde, Automatic power-sharing modification of $p/\ v$ droop controllers in low-voltage resistive microgrids. IEEE Trans. Power Deliv. **27**(4), 2318–2325 (2012)

Part I
Cooperative Optimal Security Control of Hybrid Energy System

Chapter 2
Multiagent System-Based Event-Triggered Hybrid Controls for High-Security Hybrid Energy Generation Systems

This chapter proposes multiagent system-based event-triggered hybrid controls for guaranteeing energy supply of a hybrid energy generation system with high security. First, a multiagent system is constituted by an upper level central coordinated control agent combined with several lower level unit agents. Each lower level unit agent is responsible for dealing with internal switching control and distributed dynamic regulation for its unit system. The upper level agent implements coordinated switching control to guarantee the power supply of overall system with high security. The internal switching control, distributed dynamic regulation, and coordinated switching control are designed fully dependent on the hybrid behaviors of all distributed energy resources and the logical relationships between them, and interact with each other by means of the multiagent system to form hierarchical hybrid controls. Finally, the validity of the proposed hybrid controls is demonstrated by means of simulation results in different scenarios.

2.1 MAS-Based Control Scheme of HEGS

A HEGS example is shown in Fig. 2.1, which is divided into following two regions:

(1) region 1 that contains a photovoltaic (PV) unit, a battery unit, and a fuel cell (FC) unit combined with a microturbine (MT) unit and
(2) region 2 that contains a wind turbine (WT) unit, an ultracapacitor (WT/UC) unit, and a group of critical and noncritical loads.

With respect to the grid-disconnected HEGS, the MAS is built with the following features:

D. Yue et al., *Cooperative Optimal Control of Hybrid Energy Systems*,
https://doi.org/10.1007/978-981-33-6722-7_2

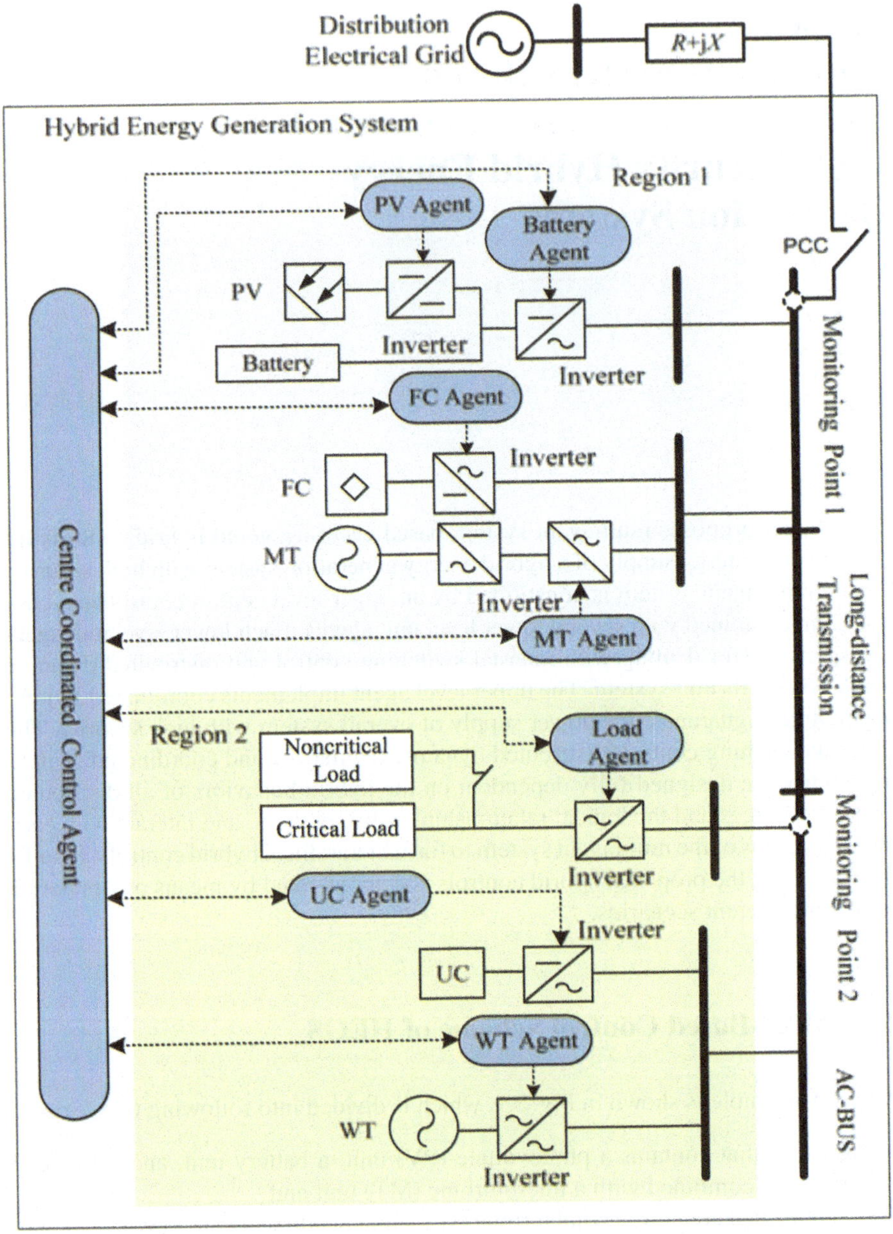

Fig. 2.1 MAS-based control scheme of HEGS

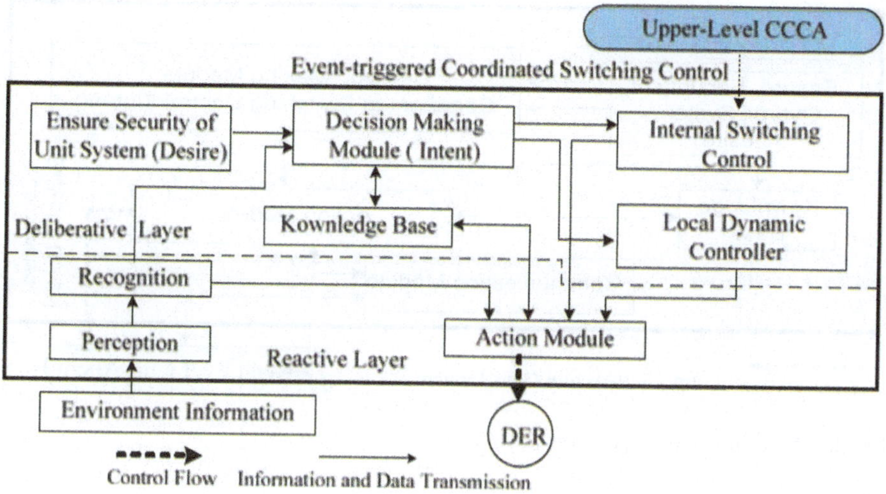

Fig. 2.2 Structure of the lower level unit agent

(1) in the lower level, seven unit agents are mainly responsible for ISCs and distributed dynamic regulation for its unit system;
(2) in the upper level, one CCCA implements CSCs; and
(3) all agents interact with each other according to the designed interactive modes.

Each lower level unit agent is designed as hybrid agent, which is composed of *reactive* and *deliberative* layers as shown in Fig. 2.2. The reactive layer defined as "*recognition, perception and action*," has priority to respond quickly to the emergencies of operation status. The deliberative layer defined as "*belief, desire and intent*", has higher intelligence to control the hybrid behaviors of its unit system. The *belief* comes from environment recognition and knowledge base. The *desire* is to ensure security and dynamic performance of the unit system. The *intent* is to determent local hybrid controls including the ISC and the distributed dynamic control based on the current *belief* and *desire*. Finally, the local hybrid controls is executed by means of the action module. Notice that the time scale of the local hybrid controls is in the level of millisecond.

The upper level CCCA is designed as a *belief-desire-intent* agent, as shown in Fig. 2.3. The *belief* of the agent comes from the information exchanged between agents and knowledge base. The *desire* is to ensure security of the overall system. The *intent* is to determine the CSC based on the current *belief* and *desire* in decision-making module. Finally, the CSC is sent to corresponding unit agent by means of the action module. In this case, the time scale of the CSC is executed in the level of millisecond or second.

In the MAS, the interactive modes between agents are designed as follows:

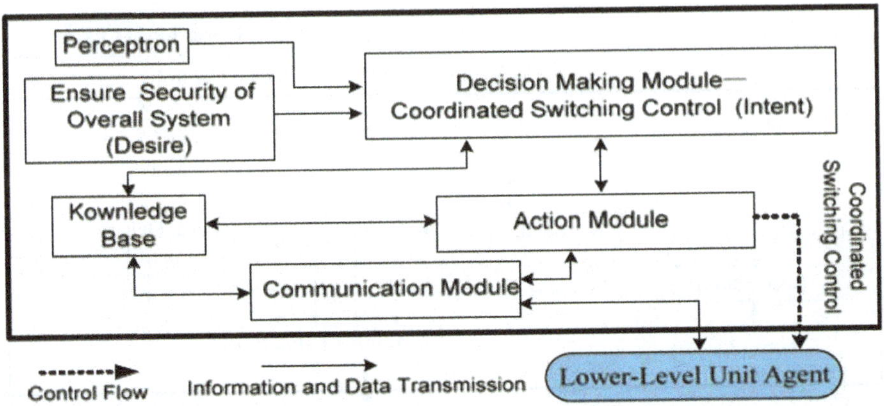

Fig. 2.3 Structure of the upper level CCCA

(1) master-slave mode between the CCCA and the unit agents, which means that request of the upper level CCCA must be responded by the asked lower level unit agent and
(2) nonmaster-slave modes between the unit agents, which implies that the lower level unit agents interact in an equal way.

2.2 Event-Triggered Hybrid Controls

It is well known that the DHPN model is one of the most useful modeling methods for complex hybrid systems [1, 2]. In order to fully describe the hybrid behaviors of DERs and the logical relationships between them, the DHPN modeling of the HEGS is discussed first.

2.2.1 DHPN Model of the HEGS

Corresponding to Fig. 2.1, DHPN model of the HEGS is built as shown in Fig. 2.4. The DHPN model is defined by a nine-tuple (P_D, T_D, P_{DF}, T_{DF}, A_N, P_{re}, P_{os}, Γ, M_0) set, where

$P_D \in \{P_1, P_2, \ldots, P_{21}\}$ is a set of discrete places, which represents operation modes of all DER and load units;

$T_D \in \{T_1, T_2, \ldots, T_{34}\}$ is a set of discrete transitions, which represents event-triggered switching behaviors;

$P_{DF} \in \{P_{1f}, P_{2f}, \ldots, P_{7f}\}$ is a set of continuous places, which describes continuous dynamics of all unit systems;

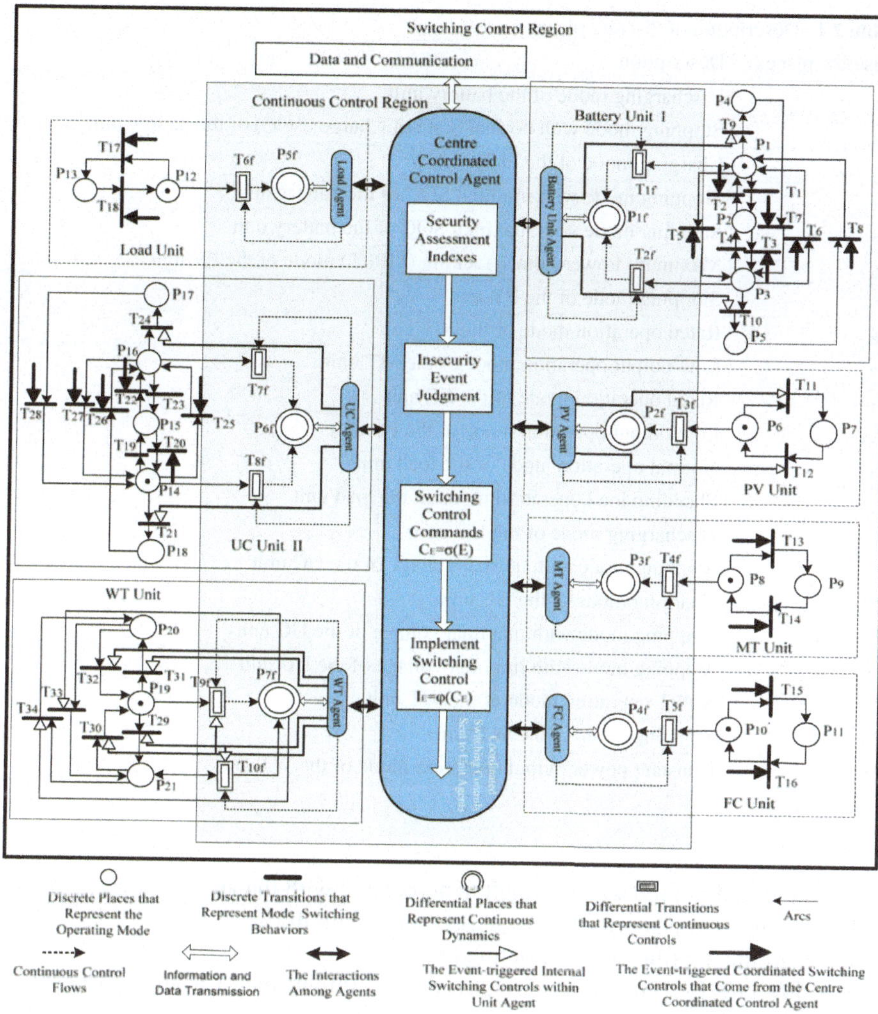

Fig. 2.4 Event-triggered hybrid controls based on the DHPN model

$T_{DF} \in \{T_{1f}, T_{2f}, \ldots, T_{10f}\}$ is a set of continuous transitions, which represents distributed continuous controls of all unit systems;

$$P = P_D \cup P_{DF}, T = T_D \cup T_{DF}, P \cap T = \varphi, P \cup T \neq \varphi.$$

The detailed descriptions regarding places and transitions above are given in Tables 2.1, 2.2, 2.3 and 2.4.

$A_N \subseteq ((P_D \times T_D) \cup (T_D \times P_D))$
$\cup ((P_D \times T_{DF}) \cup (T_{DF} \times P_D))$ is a set of arcs;

Table 2.1 Description of discrete places

Discrete places	Description
P_1	Discharging mode of the battery unit
P_2	Stopping mode with normal State of Charge (SOC) of the battery unit
P_3	Charging mode of the battery unit
P_4	Stopping mode with minimal SOC of the battery unit
P_5	Stopping mode with maximal SOC of the battery unit
P_6	Maximum Power Point Tracking (MPPT) mode of the PV unit
P_7	Stopping mode of the PV unit
P_8	Rated operation mode of the MT unit
P_9	Low-output operation mode of the MT unit
P_{10}	Rated operation mode of the FC unit
P_{11}	Low-output operation mode of the FC unit
P_{12}	Normal operation mode of the load unit
P_{13}	Shedding load operation mode of the load unit
P_{14}	Discharging mode of the UC unit
P_{15}	Stopping mode with normal voltage of the UC unit
P_{16}	Charging mode of the UC unit
P_{17}	Stopping mode with maximal voltage of the UC unit
P_{18}	Stopping mode with minimal voltage of the UC unit
P_{19}	MPPT operation mode of the WT unit
P_{20}	Stopping mode of the WT unit
P_{21}	Constant power output operation mode of the WT unit

P_{re} is defined as predecessor function associated with the arc;
P_{os} is defines as posterior function associated with the arc;
Here, all the arc functions are defined as "1".

$\Gamma \in \{d_{T1}, d_{T2}, \ldots, d_{T34}, d_{T1f}, \ldots, d_{T10f}\}$ is a timing map for all the transients, which defines the triggered time of all the transients.

$M_0 \in \{M_{10}, M_{20}, \ldots, M_{70}\}$ is the initial marking of seven unit systems.

The initial operation mode of each unit system is marked with a token, which is described as discrete place with a black dot in Fig. 2.4. If the operation mode of a unit system is in P_i, then P_i is defined as logic "1", and accordingly other operation modes are certainly logic "0", since in each unit system, only one operation mode is "1". Therefore, corresponding to the DHPN model in Fig. 2.4, the initial markings of all unit systems are as follows:

Battery unit: $M_{10}\,(P_1, P_2, P_3, P_4, P_5) = [1, 0, 0, 0, 0]$;
PV unit: $M_{20}\,(P_6, P_7) = [1, 0]$;
MT unit: $M_{30}\,(P_8, P_9) = [1, 0]$;
FC unit: $M_{40}\,(P_{10}, P_{11}) = [1, 0]$;
Load unit: $M_{50}\,(P_{12}, P_{13}) = [1, 0]$;

Table 2.2 Description of discrete transitions

Discrete transition	Description
T_1	Switch the battery unit from discharging mode to stopping mode with the normal SOC
T_2	Switch the battery unit from stopping mode with the normal SOC to discharging mode
T_3	Switch the battery unit from stopping mode with the normal SOC to charging mode
T_4	Switch the battery unit from charging mode to stopping mode with the normal SOC
T_5	Switch the battery unit from discharging mode to charge mode
T_6	Switch the battery unit from charging mode to discharge mode
T_7	Switch the battery unit from stopping mode with minimal SOC to charge mode
T_8	Switch the battery unit from stopping mode with maximal SOC to discharge mode
T_9	Switch the battery unit to stopping mode because SOC reaches to the minimum
T_{10}	Switch the battery unit to stopping mode because SOC reaches to the maximum
T_{11}	Switch the PV unit from MPPT mode to stopping mode
T_{12}	Switch the PV unit from stopping mode to MPPT mode
T_{13}	Switch the MT unit from rated mode to low-output mode
T_{14}	Switch the MT unit from low-output mode to rated mode
T_{15}	Switch the FC unit from rated mode to low-output mode
T_{16}	Switch the FC unit from low-output mode to rated mode
T_{17}	Switch the load unit from normal mode to shedding load mode
T_{18}	Switch the load unit from shedding load mode to normal mode
T_{19}	Switch the UC unit from discharging mode to stopping mode with the normal voltage
T_{20}	Switch the UC unit from stopping mode with the normal voltage to discharging mode
T_{21}	Switch the UC unit to stopping mode because the voltage reaches to the minimum
T_{22}	Switch the UC unit from stopping mode with the normal voltage to charging mode
T_{23}	Switch the UC unit from charging mode to stopping mode with the normal voltage
T_{24}	Switch the UC unit to stopping mode because the voltage reaches to the maximum
T_{25}	Switch the UC unit from discharging mode to charge mode
T_{26}	Switch the UC unit from stopping mode with minimal voltage to charge mode
T_{27}	Switch the UC unit from charging mode to discharge mode
T_{28}	Switch the UC unit from stopping mode with maximal voltage to discharge mode
T_{29}	Switch the WT unit from MPPT mode to constant output mode
T_{30}	Switch the WT unit from constant output mode to MPPT mode
T_{31}	Switch the WT unit from MPPT mode to stopping mode
T_{32}	Switch the WT unit from stopping mode to MPPT mode
T_{33}	Switch the WT unit from stopping mode to constant output mode
T_{34}	Switch the WT unit from constant output mode to stopping mode

Table 2.3 Description of differential places

Differential places	Description
P_{1f}–P_{10f}	Continuous dynamics of the corresponding units

Table 2.4 Description of differential transitions

Differential transitions	Description
T_{1f} and T_{2f}	Dynamic controls of the battery unit in discharge and charge mode respectively
T_{3f}	Dynamic control of the PV unit in MPPT mode
T_{4f} and T_{5f}	Dynamic controls of the MT and FC units respectively in rated mode
T_{6f}	Dynamic control of the load unit in normal mode
T_{7f} and T_{8f}	Dynamic controls of the UC unit in charge and discharge mode respectively
T_{9f} and T_{10f}	Dynamic controls of the WT unit in MPPT and constant power output mode respectively

UC unit: M_{60} $(P_{14}, P_{15}, P_{16}, P_{17}, P_{18}) = [1, 0, 0, 0, 0]$;
WT unit: M_{70} $(P_{19}, P_{20}, P_{21}) = [1, 0, 0]$.

The switching principle of the DHPN model is as follows: when one transition is triggered, if the predecessor place connected with the transition has token, then the token is transmitted into the posterior place connected with the transition, and accordingly resulting in switching of operation mode. In the light of the principle above, to switch the operation modes of all unit systems in a coordinated way, all transitions of the DHPN model should be triggered by means of designed logical control functions. Therefore, in this chapter, the switching controls for operation mode are innovatively designed as the triggered functions of transitions.

In Fig. 2.4, the hierarchical event-triggered hybrid controls are described as follows:

(1) The ISC in the lower level unit agent, which is responsible for locally switching the operation mode of unit system, and described as

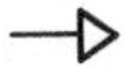

(2) The distributed dynamic control in the lower level unit agent, which is responsible for the dynamic regulation of unit systems, and described as

(3) The CSC in the upper level CCCA, which is responsible for switching operation modes of all unit systems in a coordinated way, and described as

The rest of the chapter is mainly focus on designing the event-triggered hybrid controls.

2.2.2 Internal Switching Control

The ISC is designed as a set of triggered functions of transitions, where CVFs are triggered conditions of them. The design concept is explained as follows:

(1) once a constraint condition is violated, and correspondingly the designed CVF is activated (becomes logic "1");
(2) it triggers the connected transition, according to the switching principle of the DHPN, corresponding operation mode can be switched.

In the PV unit, the ISCs are designed as follows:

$$\begin{aligned}&\text{If } t = t_0,\ \text{CVF}: G_{\text{ing}}(t) \text{ drops to } G_{\text{ing}}(t) \le C,\\&\text{then ISC}\left(\text{T}_{11}\right) = 1\left(t - t_0\right) - 1\left(t - t_0 - d_{\text{T}11}\right)\end{aligned} \tag{2.1}$$

where $G_{ing}(t)$ is the incident irradiance; C is the threshold value; $1\,(t - t_0)$ is unit step function as

$$1\,(t - t_0) = \begin{cases} 1, & t \ge t_0 \\ 0, & t < t_0 \end{cases}$$

The design of ISC(T11) is explained as follows: at $t = t_0$, if $G_{ing}(t)$ drops to $G_{ing}(t) \le C$, and accordingly the CVF is activated, and then triggers the transition T_{11} for the duration time of d_{T11}, resulting in that operation mode of the PV unit is switched from P_6 to P_7. The following designs are in a similar way:

$$\begin{aligned}&\text{If } t = t_0,\ \text{CVF}: G_{\text{ing}}(t) \text{ rises to } G_{\text{ing}}(t) > C,\\&\text{then ISC}\left(\text{T}_{12}\right) = 1\left(t - t_0\right) - 1\left(t - t_0 - d_{\text{T}12}\right).\end{aligned} \tag{2.2}$$

In the WT unit, the ISCs are designed as follows:

$$\begin{aligned}&\text{If } t = t_0,\ \text{CVF}: \nu(t) \text{ drops to } \nu(t) \le \nu_{ci},\\&\text{then ISC}\left(\text{T}_{31}\right) = 1\left(t - t_0\right) - 1\left(t - t_0 - d_{\text{T}31}\right)\end{aligned} \tag{2.3}$$

$$\begin{aligned}&\text{If } t = t_0,\ \text{CVF}: \nu(t) \text{ rises to } \nu_{ci} < \nu(t) \le \nu_R,\\&\text{then ISC}\left(\text{T}_{32}\right) = 1\left(t - t_0\right) - 1\left(t - t_0 - d_{\text{T}32}\right)\end{aligned} \tag{2.4}$$

$$\begin{aligned}&\text{If } t = t_0,\ \text{CVF}: \nu(t) \text{ rises to } \nu_R < \nu(t) \le \nu_{co},\\&\text{then ISC}\left(\text{T}_{29}\right) = 1\left(t - t_0\right) - 1\left(t - t_0 - d_{\text{T}29}\right),\\&\text{and ISC}\left(\text{T}_{10f}\right) = 1\left(t - t_0\right) - 1\left(t - t_0 - d_{\text{T}10f}\right)\end{aligned} \tag{2.5}$$

$$\text{If } t = t_0,\ \text{CVF}: \nu(t) \text{ rises to } \nu(t) > \nu_{co},\\ \text{then ISC}\,(\text{T}_{34}) = 1\,(t - t_0) - 1\,(t - t_0 - d_{\text{T}34}) \tag{2.6}$$

$$\text{If } t = t_0,\ \text{CVF}: \nu(t) \text{ drops to } \nu_R < \nu_t \le \nu_{co},\\ \text{then ISC}\,(\text{T}_{33}) = 1\,(t - t_0) - 1\,(t - t_0 - d_{\text{T}33}) \tag{2.7}$$

$$\text{If } t = t_0,\ \text{CVF}: \nu(t) \text{ drops to } \nu_{ci} < \nu(t) \le \nu_R,\\ \text{then ISC}\,(\text{T}_{30}) = 1\,(t - t_0) - 1\,(t - t_0 - d_{\text{T}30}),\\ \text{and ISC}\,(\text{T}_{9f}) = 1\,(t - t_0) - 1\,(t - t_0 - d_{\text{T}9f}) \tag{2.8}$$

where $\nu(t)$ is the wind speed; ν_{ci} is the cut-in wind speed; ν_{co} is the cutoff wind speed; and ν_R is the rated wind speed.

In the battery unit, the ISCs are designed as follows:

$$\text{If } t = t_0,\ \text{CVF}: \text{SOC(t) drops to SOC(t)} \le \text{SOC}_{\text{min}},\\ \text{then ISC}\,(\text{T}_9) = 1\,(t - t_0) - 1\,(t - t_0 - d_{\text{T}9}) \tag{2.9}$$

$$\text{If } t = t_0,\ \text{CVF}: \text{SOC(t) rises to SOC(t)} \ge \text{SOC}_{\text{max}},\\ \text{then ISC}\,(\text{T}_{10}) = 1\,(t - t_0) - 1\,(t - t_0 - d_{\text{T}10}) \tag{2.10}$$

where SOC(t) is the state of charge; SOC_{max} and SOC_{min} are the maximum and minimum values, respectively.

In the UC unit, the ISCs are designed as follows:

$$\text{If } t = t_0,\ \text{CVF}: U(t) \text{ drops to } U(t) \le U_{min},\\ \text{then ISC}\,(\text{T}_{21}) = 1\,(t - t_0) - 1\,(t - t_0 - d_{\text{T}21}) \tag{2.11}$$

$$\text{If } t = t_0,\ \text{CVF}: U(t) \text{ rises to } U(t) \ge U_{max},\\ \text{then ISC}\,(\text{T}_{24}) = 1\,(t - t_0) - 1\,(t - t_0 - d_{\text{T}24}) \tag{2.12}$$

where $U(t)$ is the voltage of the UC; U_{max} and U_{min} are the maximum and minimum values of $U(t)$, respectively.

Table 2.5 Operation mode switching by means of the CSCS

Orders	Times	Insecure events	Operation mode switching
(1)	9.40	E_5	The transition T_{25} is triggered so that the UC unit is switched from discharging to charging mode. After one switching interval time, T_5 is triggered so that battery unit is switched from the discharging to the charging mode
(2)	11.40	E_2	The transition T_6 is triggered so that the battery unit is switched from the charging to the discharging mode
(3)	12.45	E_1	The transition T_5 is triggered so that the battery unit is switched from the discharging to the charging mode. After one switching interval time, T_{13} and T_{15} are triggered simultaneously so that the FC&MT units are switched from the rated operation to the low-output operation mode
(4)	15.30	E_2	The transition T_6 is triggered so that the battery unit is switched from the charging to the discharging mode. After one switching interval time, T_{14} and T_{16} are triggered simultaneously so that the FC&MT units are switched from the low-output operation to the rated operation mode
(5)	18.15	E_4	The transition T_{27} is triggered so that the UC unit is switched from the charging mode to the discharging mode

2.2.3 *Distributed Dynamic Control*

In different operation modes (except stopping mode), each unit system is regulated in a distributed way by using different dynamic control strategies [3–5]. In the HEGS, the storage unit is regarded as a grid-forming control unit, which is mainly responsible for the frequency/voltage (f-V) regulations by means of f-V droop control strategies as shown in Fig. 2.5. Other DER units, like FC/MT, PV, and WT, are defined as grid-following control units, which are mainly responsible for a desired amount of active/reactive output powers by means of P-Q control strategies, as shown in Fig. 2.6.

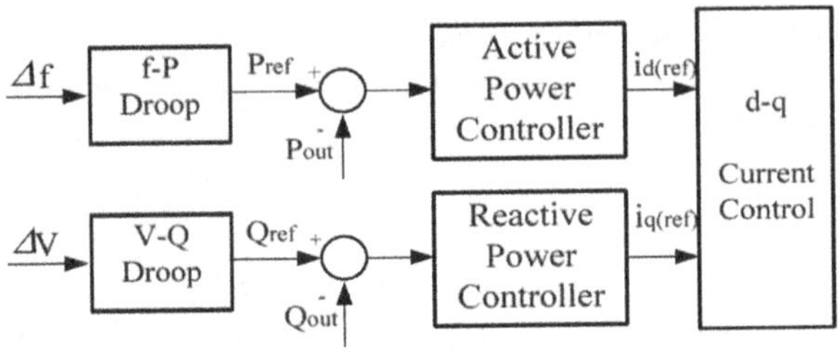

Fig. 2.5 Grid-forming f-V droop control strategy

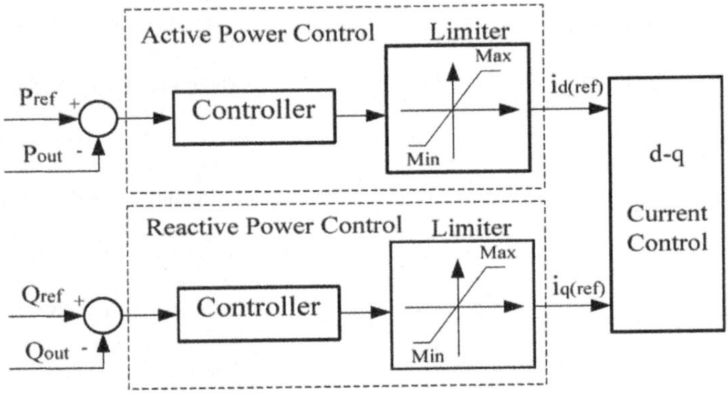

Fig. 2.6 Grid-following P-Q control strategy

With respect to two kinds of control strategies above, outer loop active/reactive power controllers are designed based on different droop characteristics to achieve different control objectives, and inner loop voltage/current controllers are designed as H∞ robust controllers instead of traditional PI controllers to enhance capacity of resisting disturbance. The detailed design regarding the two loops of controllers was presented in author's previous research [6]. Considering the limitation of chapter pages, the design process is here not presented again.

2.2.4 Coordinated Switching Control

The CSC is also designed as a set of triggered functions of transitions, where voltage SAIs are triggered conditions of them. The design concept is explained as follows: (1) once a voltage SAI exceeds the desired secure range, correspondingly an insecure event is estimated to occur, and then its event-triggered function is activated (becomes logic "1") and (2) it triggers the connected transition, according to the switching principle of the DHPN, corresponding operation mode can be switched. The CSC is design according to the following steps:

Step 1: SAIs

The insecure voltage generally is manifested initially as a slow voltage decay following a sharp decline at the point of collapse. Therefore, the voltage magnitude alone cannot be regarded as a reliable SAI. In this chapter, dynamic voltage SAIs are constructed by using the voltage sequences at two monitoring points, which are set at PCC and load nodes in two regions [7].

The voltage sequences at the two monitoring points are represented as $V_i = [V_i^1, V_i^2, \ldots, V_i^N], i \in 1, 2$.

(1) The moving average voltage values at the ith node at the tth instant are obtained by using N available voltage measurements

$$V_i^{(t)} = \sum_{k=1}^{t} V_i^k / t, \quad t \in 1, 2, \ldots, N; \quad \text{if } t \leq N;$$
$$V_i^{(t)} = \sum_{k=t-N+1}^{t} V_i^k / N, \quad t \in N+1, N+2, \ldots, m; \quad \text{if } t > N. \tag{2.13}$$

(2) The percentages of diversity between the real voltage V_i^t and moving average value $V_i^{(t)}$ i at the tth instant are defined as

$$C_i^t = \frac{V_i^t - V_i^{(t)}}{V_i^{(t)}} \times 100, \quad t \in 1, 2, \ldots, m. \tag{2.14}$$

(3) The voltage SAIs at the tth instant are estimated by dividing the area under the percentage of diversity curve

$$U_i^{(t)} = \sum_{k=1}^{t} \left(C_i^k + C_i^{k-1}\right) / 2t, \quad t \in 1, 2, \ldots, N; \text{ if } t \leq N;$$
$$U_i^{(t)} = \sum_{k=t-N+1}^{t} \left(C_i^k + C_i^{k-1}\right) / 2N, \quad t \in N+1, N+2, \ldots, m; \text{ if } t > N. \tag{2.15}$$

Step 2: Insecure event judgment

By means of the voltage SAIs above, insecure events are judged as follows:

(1) Define the voltage SAI at the tth instant at the PCC node as $U_1^{(t)}$, and at the load node as $U_2^{(t)}$.
(2) Set the maximum and maximum thresholds of the voltage SAI at the PCC node as $U_{1\max}$ and $U_{1\min}$, and at the load node as $U_{2\max}$ and $U_{2\min}$.
(3) At the tth instant,
when $U_1^{(t)} > U_{1max}$, define that the event $E_{11}(t)$ occurs;
when $U_1^{(t)} < U_{1min}$, define that the event $E_{12}(t)$ occurs;
when $U_2^{(t)} > U_{2max}$, define that the event $E_{21}(t)$ occurs;
when $U_2^{(t)} < U_{2min}$, define that the event $E_{22}(t)$ occurs.
(4) In this chapter, we set two voltage monitoring points in two regions, according to the above definitions, at the tth instant, there are potentially six kinds of insecure events at least as follows:

$$E_1(t) = \{E_{11}(t)\};\ E_2(t) = \{E_{12}(t)\};\ E_3(t) = \{E_{21}(t)\},$$
$$E_4(t) = \{E_{22}(t)\};\ E_5(t) = \{E_{11}(t), E_{21}(t)\};\ E_6(t) = \{E_{21}(t), E_{22}(t)\}.$$

Step 3: CSCs

Before design the CSCs, we define the following logic relation and switching principle:

(1) If at the tth instant, $\mathrm{E_i}$ insecure event is judged to occur, then $\mathrm{E_i}$(t) is activated and becomes logic "1"; otherwise, $\mathrm{E_i}$(t) is logic "0";
(2) Each CSC is activated not more than three times.

At the $t = t_0$ instant, corresponding to the above six kinds of insecure events, the CSCs are designed as follows:

$$\begin{aligned}
\mathrm{C}(\mathrm{E}_1) = {} & \mathrm{E}_1(t)\mathrm{P}_1\left(1(t-t_0) - 1(t-t_0-d_{\mathrm{T}5})\right) \\
& + \mathrm{P}_2\left(1(t-t_0) - 1(t-t_0-d_{\mathrm{T}3})\right) \\
& + \mathrm{P}_4\left(1(t-t_0) - 1(t-t_0-d_{\mathrm{T}7})\right) \\
& + \mathrm{E}_1(t+\Delta t)\mathrm{P}_8\left(1(t-t_0-\Delta t)\right. \\
& \left. -1(t-t_0-\Delta t-d_{\mathrm{T}13})\right) \\
& + \mathrm{P}_{10}\left(1(t-t_0-\Delta t)\right. \\
& \left. -1(t-t_0-\Delta t-d_{\mathrm{T}15})\right) \\
& + \mathrm{E}_1(t+2\Delta t)\mathrm{P}_{14}\left(1(t-t_0-2\Delta t)\right. \\
& \left. -1(t-t_0-2\Delta t-d_{\mathrm{T}25})\right) \\
& + \mathrm{P}_{15}\left(1(t-t_0-2\Delta t) - 1(t-t_0-2\Delta t-d_{\mathrm{T}22})\right)
\end{aligned} \tag{2.16}$$

$$
\begin{aligned}
\mathrm{C}\left(\mathrm{E}_2\right) = {} & \mathrm{E}_2(t)\mathrm{P}_3\left(1\left(t-t_0\right)-1\left(t-t_0-d_{\mathrm{T}6}\right)\right) \\
& +\mathrm{P}_2\left(1\left(t-t_0\right)\right. \\
& \left.-1\left(t-t_0-d_{\mathrm{T}2}\right)\right) \\
& +\mathrm{P}_5\left(1\left(t-t_0\right)-1\left(t-t_0-d_{\mathrm{T}8}\right)\right) \\
& +\mathrm{E}_2(t+\Delta t)\mathrm{P}_9\left(1\left(t-t_0-\Delta t\right)\right. \\
& \left.-1\left(t-t_0-\Delta t-d_{\mathrm{T}14}\right)\right) \\
& +\mathrm{P}_{11}\left(1\left(t-t_0-\Delta t\right)-1\left(t-t_0-\Delta t-d_{\mathrm{T}16}\right)\right) \\
& +\mathrm{E}_2(t+2\Delta t)\mathrm{P}_{16}\left(1\left(t-t_0-2\Delta t\right)\right. \\
& \left.-1\left(t-t_0-2\Delta t-d_{\mathrm{T}27}\right)\right) \\
& +\mathrm{P}_{15}\left(1\left(t-t_0-2\Delta t\right)-1\left(t-t_0-2\Delta t-d_{\mathrm{T}20}\right)\right) \\
& +\mathrm{P}_{17}\left(1\left(t-t_0-2\Delta t\right)-1\left(t-t_0-2\Delta t-d_{\mathrm{T}28}\right)\right)
\end{aligned} \tag{2.17}
$$

$$
\begin{aligned}
\mathrm{C}\left(\mathrm{E}_3\right) = {} & \mathrm{E}_3(t)\mathrm{P}_{14}\left(1\left(t-t_0\right)-1\left(t-t_0-d_{\mathrm{T}25}\right)\right) \\
& +\mathrm{P}_{15}\left(1\left(t-t_0\right)-1\left(t-t_0-d_{\mathrm{T}22}\right)\right) \\
& +\mathrm{P}_{18}\left(1\left(t-t_0\right)-1\left(t-t_0-d_{\mathrm{T}26}\right)\right) \\
& +\mathrm{E}_3(t+\Delta t)\mathrm{P}_1\left(1\left(t-t_0-\Delta t\right)\right. \\
& \left.-1\left(t-t_0-\Delta t-d_{\mathrm{T}5}\right)\right) \\
& +\mathrm{P}_2\left(1\left(t-t_0-\Delta t\right)-1\left(t-t_0-\Delta t-d_{\mathrm{T}3}\right)\right) \\
& +\mathrm{P}_4\left(1\left(t-t_0-\Delta t\right)-1\left(t-t_0-\Delta t-d_{\mathrm{T}7}\right)\right) \\
& +\mathrm{E}_3(t+2\Delta t)\mathrm{P}_8\left(1\left(t-t_0-2\Delta t\right)\right. \\
& \left.-1\left(t-t_0-2\Delta t-d_{\mathrm{T}13}\right)\right) \\
& +\mathrm{P}_{10}\left(1\left(t-t_0-2\Delta t\right)-1\left(t-t_0-2\Delta t-d_{\mathrm{T}15}\right)\right)
\end{aligned} \tag{2.18}
$$

$$
\begin{aligned}
\mathrm{C}\left(\mathrm{E}_4\right) = {} & \mathrm{E}_4(t)\mathrm{P}_{15}\left(1\left(t-t_0\right)-1\left(t-t_0-d_{\mathrm{T}20}\right)\right) \\
& +\mathrm{P}_{16}\left(1\left(t-t_0\right)-1\left(t-t_0-d_{\mathrm{T}27}\right)\right) \\
& +\mathrm{P}_{17}\left(1\left(t-t_0\right)-1\left(t-t_0-d_{\mathrm{T}28}\right)\right) \\
& +\mathrm{E}_4(t+\Delta t)\mathrm{P}_2\left(1\left(t-t_0-\Delta t\right)\right. \\
& \left.-1\left(t-t_0-\Delta t-d_{\mathrm{T}2}\right)\right) \\
& +\mathrm{P}_3\left(1\left(t-t_0-\Delta t\right)-1\left(t-t_0-\Delta t-d_{\mathrm{T}6}\right)\right) \\
& +\mathrm{P}_5\left(1\left(t-t_0-\Delta t\right)-1\left(t-t_0-\Delta t-d_{\mathrm{T}8}\right)\right) \\
& +\mathrm{E}_4(t+2\Delta t)\mathrm{P}_9\left(1\left(t-t_0-2\Delta t\right)\right. \\
& \left.-1\left(t-t_0-2\Delta t-d_{\mathrm{T}14}\right)\right) \\
& +\mathrm{P}_{11}\left(1\left(t-t_0-2\Delta t\right)-1\left(t-t_0-2\Delta t-d_{\mathrm{T}16}\right)\right)
\end{aligned} \tag{2.19}
$$

$$\begin{aligned}
\mathrm{C}\left(\mathrm{E}_5\right) = {} & \mathrm{E}_5(t)\mathrm{P}_{13}\left(1\left(t-t_0\right)-1\left(t-t_0-d_{\mathrm{T}18}\right)\right) \\
& +\mathrm{E}_5(t+\Delta t)\mathrm{P}_1\left(1\left(t-t_0-\Delta t\right)\right. \\
& \left.-1\left(t-t_0-\Delta t-d_{\mathrm{T}5}\right)\right) \\
& +\mathrm{P}_2\left(1\left(t-t_0-\Delta t\right)-1\left(t-t_0-\Delta t-d_{\mathrm{T}3}\right)\right) \\
& +\mathrm{P}_4\left(1\left(t-t_0-\Delta t\right)-1\left(t-t_0-\Delta t-d_{\mathrm{T}7}\right)\right) \\
& +\mathrm{E}_5(t+\Delta t)\mathrm{P}_{14}\left(1\left(t-t_0-\Delta t\right)\right. \\
& \left.-1\left(t-t_0-\Delta t-d_{\mathrm{T}25}\right)\right) \\
& +\mathrm{P}_{15}\left(1\left(t-t_0-\Delta t\right)-1\left(t-t_0-\Delta t-d_{\mathrm{T}22}\right)\right) \\
& +\mathrm{P}_{18}\left(1\left(t-t_0-\Delta t\right)-1\left(t-t_0-\Delta t-d_{\mathrm{T}26}\right)\right) \\
& +\mathrm{E}_5(t+2\Delta t)\mathrm{P}_8\left(1\left(t-t_0-2\Delta t\right)\right. \\
& \left.-1\left(t-t_0-2\Delta t-d_{\mathrm{T}13}\right)\right) \\
& +\mathrm{P}_{10}\left(1\left(t-t_0-2\Delta t\right)-1\left(t-t_0-2\Delta t-d_{\mathrm{T}15}\right)\right)
\end{aligned} \tag{2.20}$$

$$\begin{aligned}
\mathrm{C}\left(\mathrm{E}_6\right) = {} & \mathrm{E}_6(t)\mathrm{P}_3\left(1\left(t-t_0\right)-1\left(t-t_0-d_{\mathrm{T}6}\right)\right) \\
& +\mathrm{P}_2\left(1\left(t-t_0\right)-1\left(t-t_0-d_{\mathrm{T}2}\right)\right) \\
& +\mathrm{P}_5\left(1\left(t-t_0\right)-1\left(t-t_0-d_{\mathrm{T}8}\right)\right) \\
& +\mathrm{E}_6(t)\mathrm{P}_{16}\left(1\left(t-t_0\right)-1\left(t-t_0-d_{\mathrm{T}27}\right)\right) \\
& +\mathrm{P}_{15}\left(1\left(t-t_0\right)-1\left(t-t_0-d_{\mathrm{T}20}\right)\right) \\
& +\mathrm{P}_{17}\left(1\left(t-t_0\right)-1\left(t-t_0-d_{\mathrm{T}28}\right)\right) \\
& +\mathrm{E}_6(t+\Delta t)\mathrm{P}_9\left(1\left(t-t_0-\Delta t\right)\right. \\
& \left.-1\left(t-t_0-\Delta t-d_{\mathrm{T}14}\right)\right) \\
& +\mathrm{P}_{11}\left(1\left(t-t_0-\Delta t\right)-1\left(t-t_0-\Delta t-d_{\mathrm{T}16}\right)\right) \\
& +\mathrm{E}_6(t+2\Delta t)\mathrm{P}_{12}\left(1\left(t-t_0-2\Delta t\right)\right. \\
& \left.-1\left(t-t_0-2\Delta t-d_{\mathrm{T}17}\right)\right)\}
\end{aligned} \tag{2.21}$$

where $\mathrm{C}(\mathrm{E_i})$ represents the CSC of $\mathrm{E_i}(\mathrm{t})$, $\mathrm{i} \in 1, 2, \ldots, 6$, and Δt is switching interval.

Take $\mathrm{C}(\mathrm{E}_1)$ in (2.16) as an example, the design is explained as follows:

(1) when $t = t_0$, if the insecure event E_1 is judged to occur, and at that moment the battery unit runs in any one of P_1, P_2, and P_4 modes, then $\mathrm{C}(\mathrm{E}_1)$ is activated to become "logic 1".
(2) After the time of Δt, if the event E_1 still exists, and at the moment the MT and FC units, respectively, runs in P_8 and P_{10} modes, then the $\mathrm{C}(\mathrm{E}_1)$ is activated again
(3) After the time of $2\Delta t$, if the event E_1 still exists, and at the moment the UC unit runs in any one of P_{15} and P_{18} modes, then the $\mathrm{C}(\mathrm{E}_1)$ is activated again.

The process above implies that,

(1) the times of activation of CSC is not more than three times;
(2) the time interval of activation is Δt;

(3) the switching sequence is from the battery to FC/MT to UC unit; and
(4) the duration time of activation is set according to the requirement of switching time of different operation mode.

For example, when the battery is in P_1, the duration time of activation of $C(E_1)$ is d_{T5}; while when the battery is in P_2, the duration time of activation of $C(E_1)$ is d_{T3}, The rest of CSCs are explained in a similar way.

Step 4: Implementation of CSC

The above CSCs need to be sent to corresponding unit agents by the CCCA, and are applied to corresponding transitions, and accordingly resulting in switching of operation modes. Implementation strategies of CSCs are designed as follows:

$$\begin{aligned} & I_{E1}(T_5) = I_{E1}(T_3) = I_{E1}(T_7) = I_{E1}(T_{13}) = I_{E1}(T_{15}) \\ & = I_{E1}(T_{25}) = I_{E1}(T_{22}) = I_{E1}(T_{26}) = C(E_1) \end{aligned} \tag{2.22}$$

$$\begin{aligned} & I_{E2}(T_6) = I_{E2}(T_2) = I_{E2}(T_8) = I_{E2}(T_{14}) = I_{E2}(T_{16}) \\ & = I_{E2}(T_{27}) = I_{E2}(T_{20}) = I_{E2}(T_{28}) = C(E_2) \end{aligned} \tag{2.23}$$

$$\begin{aligned} & I_{E3}(T_{25}) = I_{E3}(T_{22}) = I_{E3}(T_{26}) = I_{E3}(T_5) = I_{E3}(T_3) \\ & = I_{E3}(T_7) = I_{E3}(T_{13}) = I_{E3}(T_{15}) = C(E_3) \end{aligned} \tag{2.24}$$

$$\begin{aligned} & I_{E4}(T_{20}) = I_{E4}(T_{27}) = I_{E4}(T_{28}) = I_{E4}(T_2) = I_{E4}(T_6) \\ & = I_{E4}(T_8) = I_{E4}(T_{14}) = I_{E4}(T_{16}) = C(E_4) \end{aligned} \tag{2.25}$$

$$\begin{aligned} & I_{E5}(T_{18}) = I_{E5}(T_5) = I_{E5}(T_3) = I_{E5}(T_7) = I_{E5}(T_{25}) \\ & = I_{E5}(T_{22}) = I_{E5}(T_{26}) = I_{E5}(T_{13})\, I_{E5}(T_{15}) = C(E_5) \end{aligned} \tag{2.26}$$

$$\begin{aligned} & I_{E6}(T_6) = I_{E6}(T_2) = I_{E6}(T_8) = I_{E6}(T_{27}) = I_{E6}(T_{20}) \\ & = I_{E6}(T_{28}) = I_{E6}(T_{14}) = I_{E6}(T_{16}) = I_{E6}(T_{17}) = C(E_6) \end{aligned} \tag{2.27}$$

where $I_{E1}(T_5)$ represents event-triggered function of transition T_5 when the event E_1 is judged to occur. The rest is defined in a similar way.

Take the design of (2.22) as an example, which is explained as follows: when $t = t_0$, the insecure event E_1 is judged to occur, and the $C(E_1)$ is sent to the battery, MT/FC and UC unit agents by the CCCA. Then, it triggers transitions T_5, T_3, T_7, T_{13}, T_{15}, T_{25}, T_{22}, and T_{26} according to different activation sequence of $C(E_1)$, so that corresponding operation modes can be switched in proper order to guarantee power supply in a high-security level. The rest of event-triggered functions are done in a similar way.

2.3 Simulation Results

In order to verify the validity of the proposed approach, the HEGS as shown in Fig. 2.1 was simulated on a typical day from 8 to 24 o'clock.

Case 1: Load following

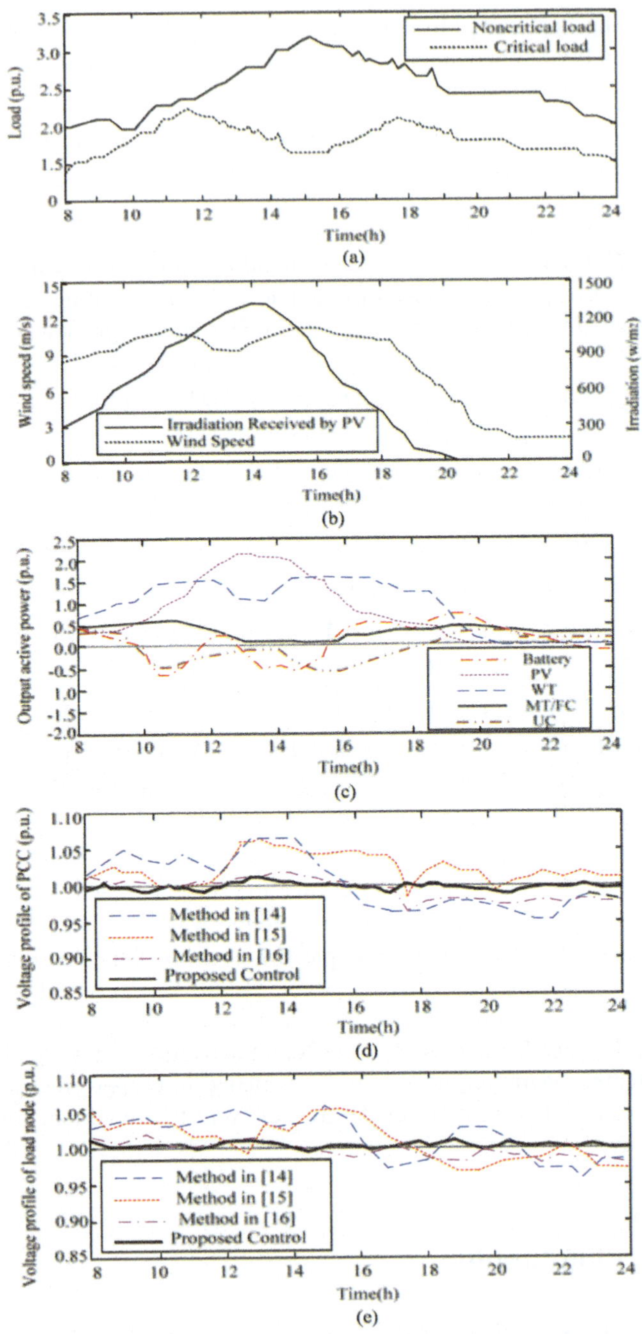

Fig. 2.7 Control performance in the case I. **a** Curves of loads. **b** Wind speed and sun irradiance. **c** Active powers all DER units. **d** Voltage profile on PCC node. **e** Voltage profile on load node

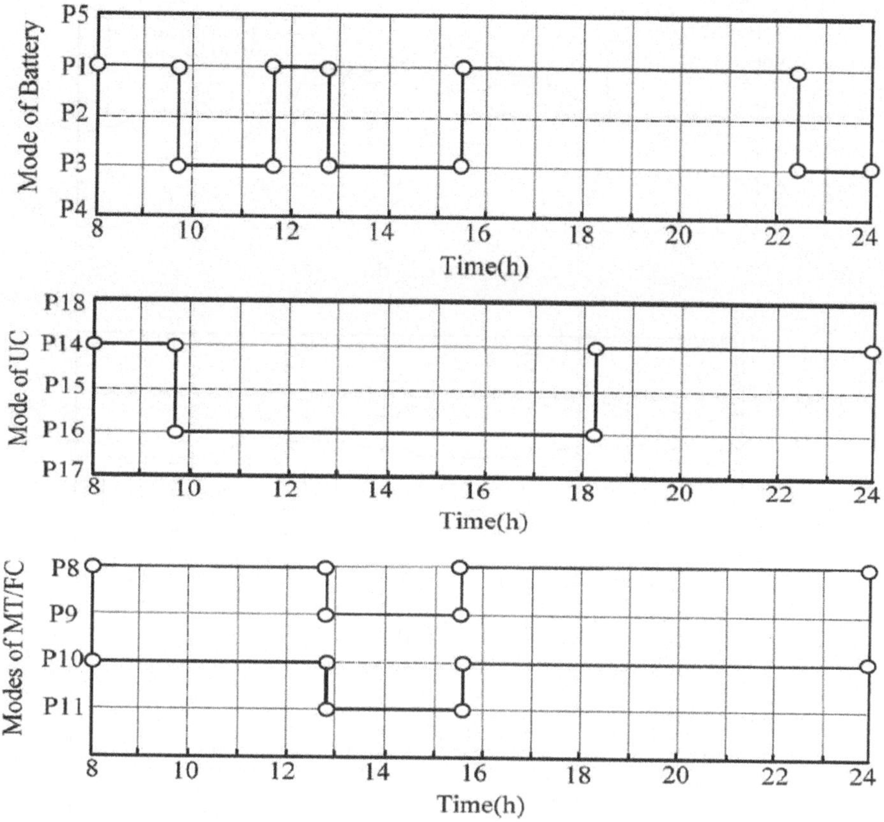

Fig. 2.8 Operation modes of four controllable DER units

The critical and noncritical load demands are shown in Fig. 2.7a, and wind speed and sun irradiance are given in Fig. 2.7b. By means of the proposed MAS-based hybrid controls, the active powers of the six DER units are given in Fig. 2.7c. Moreover, the proposed approach is compared with three switching control schemes:

(1) the logical judgment that is similar to [8];
(2) the fuzzy-logical-rule that is similar to [9];
(3) the authors' previous research [10].

The comparative results are presented in terms of the voltage control performance at the PCC and load nodes, as shown in Fig. 2.7d and e.

By using the proposed MAS-based hybrid controls, the controllable DER units operation modes are switched, as shown in Fig. 2.8.

From Figs. 2.7c and 2.8, it can be seen that the operation of controllable DER units are switched by means of the CSCs, as shown in Table 2.5.

From Fig. 2.7d and e, it can be shown that by means of the first two schemes, the PCC and load voltages are not controlled within the secure range of 0.95–1.05 p.u.

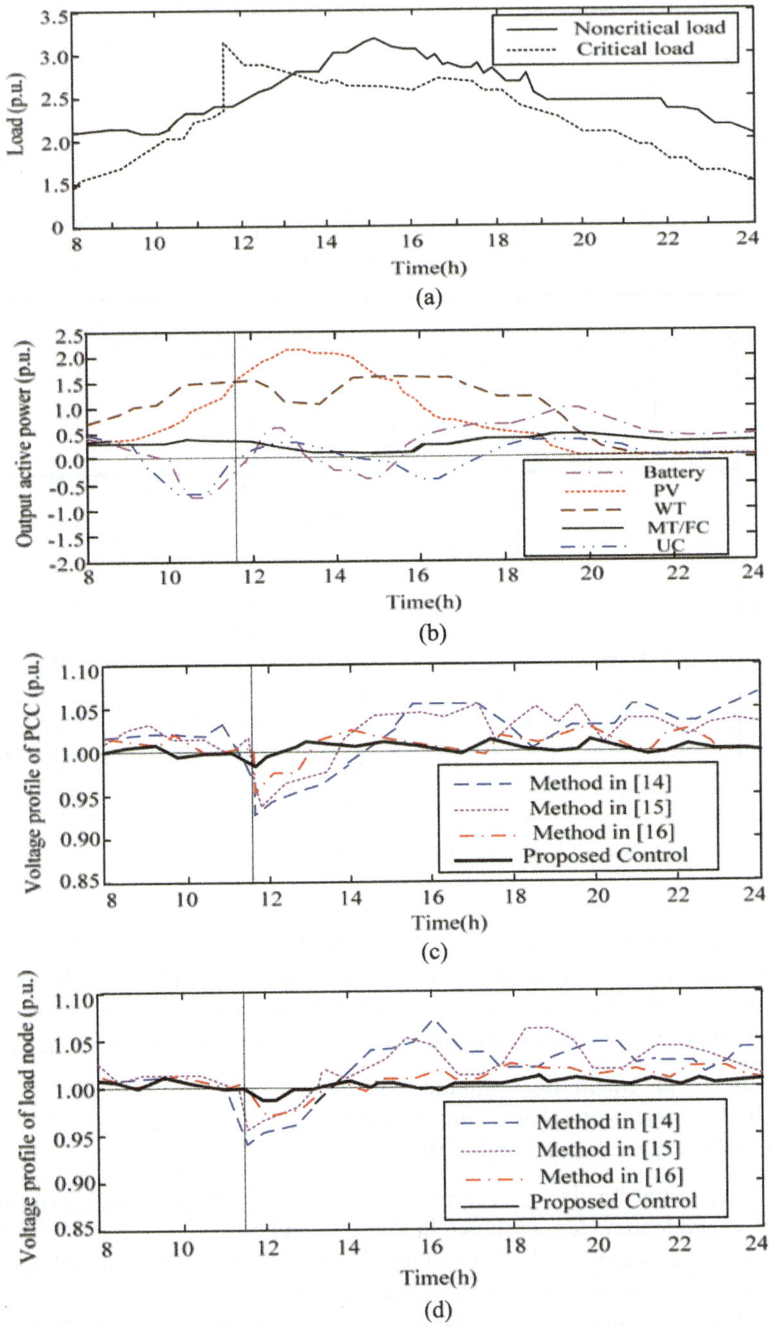

Fig. 2.9 Control performance in the case II. **a** Curves of loads. **b** Active powers all DER units. **c** Voltage profile at PCC node. **d** Voltage profile at load node

The reason is that only by switching the operation modes of the storage units, the unbalanced power cannot be regulated quickly, and thus, resulting in larger voltage fluctuations. By using the authors' previous control scheme in [10], the PCC and load voltages can be maintained within the secure range. However, in comparison with the voltage performance obtained by using the proposed hybrid controls in this chapter, they present somewhat larger fluctuations. The above simulation results show that the proposed MAS-based hybrid controls can meet changed loads with the best voltage performance.

Case 2: Load disturbance

At 11.40 am the critical load increases 50%, as shown in Fig. 2.9a. Figure 2.9b shows the active powers of all DER units by using the proposed hybrid controls. Specially at 11.40 am, the insecurity event E_6 is estimated to occur, and accordingly the triggered function $C(E_6)$ is activated, and then triggers the transitions T_{27} and T_6 simultaneously, so that the UC and battery units are switched from the charging to discharging mode. By means of the CSCs above, the PCC and load voltages can return to the normal range rapidly. Therefore, from Fig. 2.9c and d, it can be seen that, by using the proposed control even if at 11:40 am, the sudden load increase does not lead to severe decline of the voltage. The PCC and load voltages can be regulated rapidly within the range of 0.98–1.02 p.u. From Fig. 2.9c and d, it can be also seen that by means of control schemes in [8, 9], at 11:40 am, the sudden load increase results in sharp voltage drops, afterwards these voltages present larger fluctuations. The reason is that in [8, 9] the switching control is only applied to the storage units, and thus, the imbalance between supply and demand cannot be regulated quickly enough. In addition, in comparison with the proposed control, the voltage performance by using the control scheme in [11] is a little bit worse. The simulation results show that the proposed MAS-based hybrid controls can ensure the best voltage performance when faced with a large load disturbance.

2.4 Conclusion

This chapter proposes MAS-based event-triggered hybrid controls, by which the HEGS can ensure energy supply with high security. As original works, the hierarchical hybrid controls are designed taking full consideration of the hybrid behaviors of all DERs, and then implemented by means of the MASs in a distributed coordinated way. In comparison with the simula- tion results of previous relevant research, the proposed hybrid controls present the best voltage performance.

The proposed MAS-based hybrid controls only rely on sparse communication network. Despite the upper level CCCA needs to know the operation modes of all unit systems, the information can be obtained through both direct and indirect interactions between agents. Therefore, the proposed scheme does not require a fully connected communication network throughout the HEGS.

In addition, the proposed scheme can be easily extended for controlling other energy networks, such as active distribution grids and smart grids, by changing the switching controls according to different logic relationships between DER units, developing the function of agents and creating additional agents.

References

1. H. Motallebi, M.A. Azgomi, Modeling and verification of hybrid dynamic systems using multisingular hybrid petri nets. Theor. Comput. Sci. **446**, 48–74 (2012)
2. H. Guo, G. Li, T. Li, Analysis of gearbox fault diagnosis system of wind turbine based on fuzzy petri nets, in *2012 9th International Conference on Fuzzy Systems and Knowledge Discovery* (2012), pp. 540–544
3. W. Yao, M. Chen, J. Matas, J.M. Guerrero, Z. Qian, Design and analysis of the droop control method for parallel inverters considering the impact of the complex impedance on the power sharing. IEEE Trans. Ind. Electr. **58**(2), 576–588 (2011)
4. J.M. Guerrero, J.C. Vasquez, J. Matas, L.G. de Vicuna, M. Castilla, Hierarchical control of droop-controlled ac and dc microgrids-a general approach toward standardization. IEEE Trans. Ind. Electr. **58**(1), 158–172 (2011)
5. B. Bahrani, M. Saeedifard, A. Karimi, A. Rufer, A multivariable design methodology for voltage control of a single-dg-unit microgrid. IEEE Trans. Ind. Inf. **9**(2), 589–599 (2013)
6. C. Dou, B. Liu, D.J. Hill, Hybrid control for high-penetration distribution grid based on operational mode conversion. IET Gen. Trans. & Distrib. **7**(7), 700–708 (2013)
7. K. Seethalekshmi, S.N. Singh, S.C. Srivastava, A synchrophasor assisted frequency and voltage stability based load shedding scheme for self-healing of power system. IEEE Trans. Smart Grid **2**(2), 221–230 (2011)
8. J. Lagorse, D. Paire, A. Miraoui, A multi-agent system for energy management of distributed power sources. Renew. Energy **35**(1), 174–182 (2010)
9. J. Lagorse, M.G. Simoes, A. Miraoui, A multiagent fuzzy-logic-based energy management of hybrid systems. IEEE Trans. Ind. Appl. **45**(6), 2123–2129 (2009)
10. C. Dou, B. Liu, J.M. Guerrero, Event-triggered hybrid control based on multi-agent system for microgrids. IET Gen. Trans. & Distrib. **8**(12), 1987–1997 (2014)
11. S. Wen, X. Yu, Z. Zeng, J. Wang, Event-triggering load frequency control for multiarea power systems with communication delays. IEEE Trans. Ind. Electr. **63**(2), 1308–1317 (2016)

Chapter 3
Multi-agent Based Hierarchical Hybrid Control for Smart Microgrid

This chapter studies the smart control issue for an autonomous microgrid in order to maintain the secure voltages as well as maximize economic and environmental benefits. A control scheme called as multi-agent based hierarchical hybrid control is proposed versus the hierarchical control requirements and hybrid dynamic behaviors of the microgrid. The control scheme is composed of an upper level energy management agent, several middle level coordinated control agents and many lower level unit control agents. The goals of smart control are achieved by designed control strategies. The simulations are given to demonstrate the effectiveness of proposed smart control for an autonomous microgrid.

3.1 The Multi-agent Based Hierarchical Hybrid Control Scheme

The framework of hierarchical multi-agent system is constructed as shown in Fig. 3.1, and the architecture of multi-agent based hierarchical hybrid control for an autonomous MG is shown as Fig. 3.2.

The architecture of the agents and the interaction among them are the two main aspects in a multi-agent system. In this chapter, there are three hierarchies of agents, and the architectures of different hierarchical agents and the interaction among them are designed as follows:

The lower level unit control agent is designed as a hybrid agent, which consists of both reactive and deliberative layer. The reactive layer that is defined as "perception and action" has priority to respond quickly to emergencies of environment. Such as, the reactive layer of renewable resource unit agent can perceive the sudden change of nature conditions so as to determine whether it should act immediately to ask for switch of operational mode. And the deliberative layer that is defined as "belief, desire

D. Yue et al., *Cooperative Optimal Control of Hybrid Energy Systems*,
https://doi.org/10.1007/978-981-33-6722-7_3

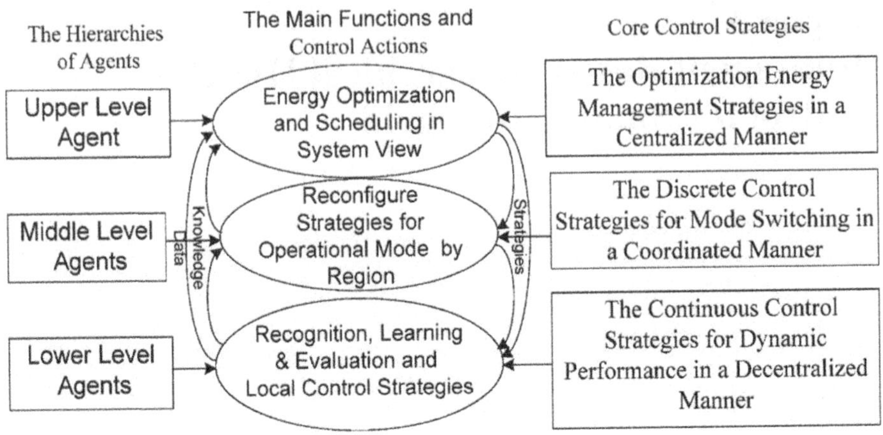

Fig. 3.1 Framework of hierarchical multi-agent system

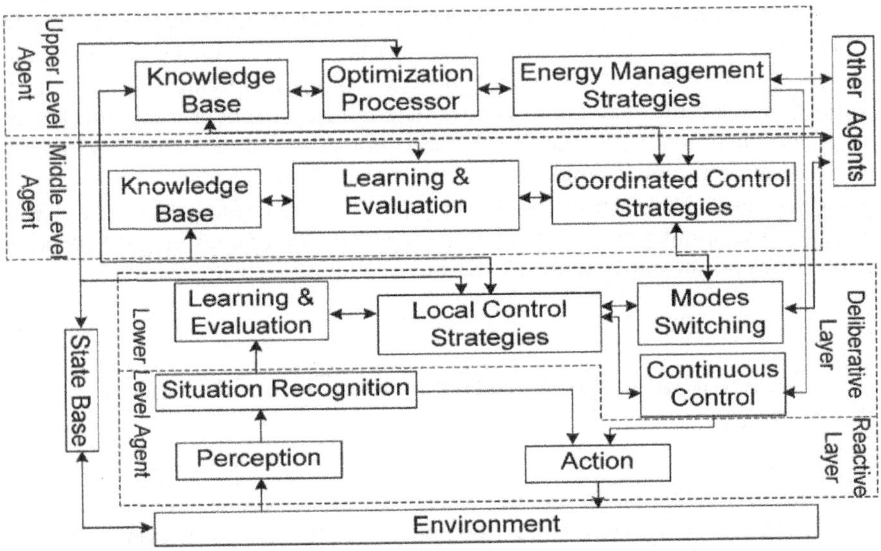

Fig. 3.2 Architecture of hierarchical multi-agent based on hybrid control system

and intention" has high intelligence of controlling or planning the behaviors of agent so as to achieve its desire or intention. In this chapter, versus the hybrid dynamic behaviors of distributed energy resource, the control strategies of the deliberative layer in unit agent are designed as hybrid control, which contain both discrete and continuous controls corresponding to mode switching and continuous running state respectively.

The middle level coordinated control agent is designed as a deliberative agent. Its intention is to coordinate switching of operational modes so as to maintain secure voltages. The reconfiguration plans of operational mode are carried out by the coordinated control strategies, which are determined based on the voltage security assessment by using knowledge data and state data.

The upper level energy management agent is also deliberative agent. It can plan the energy management strategies through the optimization process based on knowledge data and state data in order to manage the power assignment of whole autonomous MG.

In the designed multi-agent system, the interactions among agents include both direct agent-to-agent and indirect interactions. Form top to bottom, the power assignments from the upper level agent to the lower level the group of unit agents, and the coordinated control commands for mode switching from the middle level agent to the lower level the group of unit agents are direct interactions. Conversely, the interactions form the lower level agents to the both upper level and middle level agents are indirect interactions based on the environment. In other words, the lower level agents modify the state data of the environment so as to trigger the change of the control strategies of senior level agents. The interactions between unit agents include both direct and indirect interactions, such as when the cooperation request of a unit agent is responded by another unit agent, this interaction is direct, but when a unit agent modifies another unit agent's environment so as to trigger its reaction, this interaction is indirect.

The architecture sketch map of hierarchical multi-agent hybrid control system for an autonomous MG is shown in Fig. 3.3.

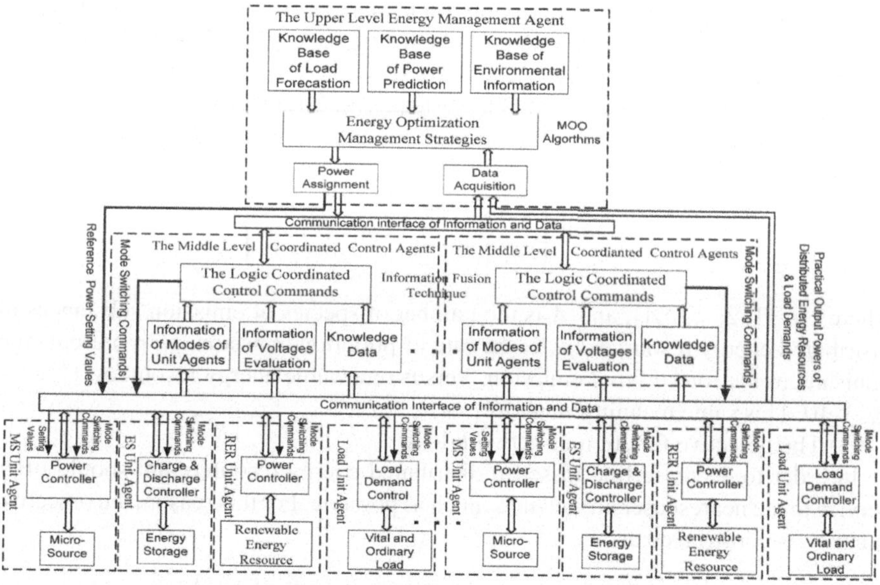

Fig. 3.3 The structure sketch map of hierarchical multi-agent based hybrid control for an autonomous MG

3.2 The Upper Level Energy Management Agent

The most important tasks of the upper level agent are to constitute energy management strategies, as well as according to the energy management strategies, to assign generating electricity task to each unit agent of distributed energy resource. Therefore, in this section, the researches are focused on how to establish the energy management strategies and how to implement the power assignment plans through interactive behaviors associated with other agents.

3.2.1 The Energy Management Strategies

(1) The Multi-Objective Functions: (1) The cost objective function is constructed as follows

$$O_1 = \min\left\{\sum_{i=1}^{n}\sum_{s=1}^{S_i}\phi_{\mathrm{is}}\left[h_i * F_{\mathrm{is}}\left(P_{\mathrm{DERis}}\right) + M_i\left(P_{\mathrm{DERis}}\right) + \lambda_{\mathrm{is}}C_{\mathrm{sti}}\right]\right\}, \tag{3.1}$$

where $i \in \{1, 2, \ldots, n\}$, and n is the number of distributed energy resources in MG. $s \in \{1, 2, \ldots, S_i\}$ represent operational mode of ith unit. P_{DERis} and $F_{\mathrm{is}}(P_{\mathrm{DERis}})$ denote power output and consumption characteristic function of ith unit in sth operational mode respectively. h_i is the price per unit fuel. Versus renewable energy resources, $h_i \equiv 0$. $M_{\mathrm{i}}(P_{\mathrm{DERis}})$ represents maintenance cost, which is deemed in proportion to P_{DERis}. C_{sti} is the starting cost of ith unit. $\lambda_{\mathrm{is}} \in [0, 1]$, when ith unit in sth operational mode is starting, $\lambda_{\mathrm{is}} = 1$, otherwise, $\lambda_{\mathrm{is}} = 0$. $\phi_{\mathrm{is}} \in [0, 1]$, when ith unit in sth operational mode is connected to system, $\phi_{\mathrm{is}} = 1$, otherwise, $\phi_{\mathrm{is}} = 0$.

(2) The objective function of pollution emission is constructed as follows

$$O_2 = \min\left\{\sum_{i=1}^{n}\sum_{s=1}^{S_i}\sum_{k=1}^{\Lambda}\phi_{\mathrm{is}}\rho_{\mathrm{is}}g_{\mathrm{is}}^{k}P_{\mathrm{DERis}}\right\}, \tag{3.2}$$

where $k \in \{1, 2, \ldots, \Lambda\}$, and Λ is the number of species of emission substances in distributed energy resources. ρ_{is} is the rate of pollution emission. g_{is}^{k} represents the punishment fee for kth kind emission. Versus renewable energy resources, $g_{\mathrm{is}}^{k} \equiv 0$. $\phi_{\mathrm{is}} \in [0, 1]$ is same meaning as above.

(3) The objective function of network loss

In order to decrease line losses, a distributed energy resource unit should offer service to the nearest user around it as much as possible. For this reason, the objective function is constructed as

$$O_3 = \min\left\{\sum_{i=1}^{n}\sum_{s=1}^{S_i}\sum_{j=1}^{m}\mu\phi_{is}d_{ij}R_{ij}\frac{P_{ijs}^2 + Q_{ijs}^2}{U_{ij}^2}\right\}, \tag{3.3}$$

where $j \in \{1, 2, \ldots, m\}$, and m is the number of load units. d_{ij} and R_{ij} represent the distance and resistance between ith unit and jth load unit respectively. μ denotes electrovalency. P_{ijs} and Q_{ijs} denote the transmitted power from ith unit agent to the jth load agent in sth operational mode. U_{ij} is the rating line voltage. In order to limit line losses, if $d_{ij} \geq d_{\text{enactment}}$, P_{ijs} and Q_{ijs} are denoted as zero. $\phi_{\text{is}} \in [0, 1]$ is same meaning as above.

(2) The Multi-Objective Optimization: The energy management problem is formulated as the following multi-objective optimization function satisfying with the constraint

$$\begin{aligned} &\min O = \gamma_1 O_1 + \gamma_2 O_2 + \gamma_3 O_3 \\ &\text{s.t.} \sum_{i=1}^{N_G} P_{Gi} + \sum_{i=1}^{N_{st}} \left[P_{st,di} - P_{st,ci}\right] = \sum_{i=1}^{m} P_{\text{load}i} + P_{\text{loss}}, \end{aligned} \tag{3.4}$$

where γ_1, γ_2 and γ_3 are the weight coefficients of each objective function respectively. $N_G + N_{st} = n$ is the number of distributed energy resource agents. $P_{\text{load}i}$ is the power demand of ith load unit agent. P_{loss} is the line loss of whole system. In (3.4), when a storage unit operates in discharging mode $P_{st,ci} = 0$; otherwise, $P_{st,di} = 0$.

Based on (3.4), by using the particle swarm optimization method, the energy management strategies can be determined, as well as the power assignment plans can be obtained.

3.2.2 The Interactive Behaviors Between Energy Management Agent with Other Agents

The power assignment plans are implemented through the interactive behaviors of energy management agent associated with other agents based on foundation for intelligent physical agents, agent communication language (FIPA-ACL) in Java agent development framework (JADE). FIPA-ACL messages are characterized by performative, conversation ID, content and receivers. The interactive behaviors based on FIPA-ACL messages are described in Table 3.1.

3.3 The Middle Level Coordinated Control Agents

In autonomous mode, the voltages of MG must be maintained by all distributed energy resources. As that the MG system usually runs under different conditions to meet the changes of load, environment and social, the intelligent reconfigure strategies are required to switch operational mode so as to meet different changes and maintain secure voltages. With respect to different architecture of MG, the recon-

Table 3.1 Interactive behaviors of the energy management agent associated with other agents

Agent	Behaviors	Receive message			Send message			
		Performative	Conversation ID	Content	Performative	Conversation ID	Content	Receivers
Load agent (LA)	LA reconfigure start	–	–	–	Query-ref	Reconfigure	Load demand	EMA
	LA reply	Accept proposal	Load control	Load demand	–	–	–	–
Energy management agent (EMA)	EMA reconfigure	Query-ref	Reconfigure	Load demand	Request	Reconfigure	Start new task	All DERAs
	EMA decision	DERAPropose	Reply	All agent information	Propose	EM strategies	Power assignments	All DERAs
	EMA reply	–	–	–	Propose	Reconfigure	Load demand	LA
Distributed energy resource agent (DERA)	DERA propose	EMARequest	Reconfigure	Start new task	Propose	Reply	All agent information	EMA
	DERA reply	Accept proposal	Local control	Power output	–	–	–	–

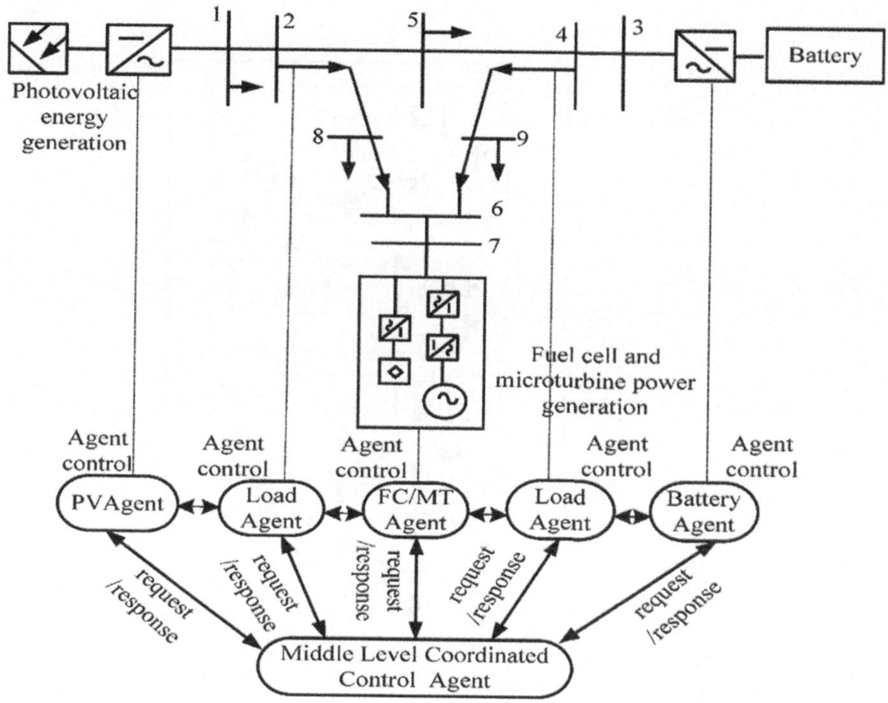

Fig. 3.4 The diagram of the investigated autonomous MG

figure strategies of operational mode should be very different. In this chapter, an example of autonomous MG is shown in Fig. 3.4, which contains a renewable PV source, a fuel microsource, a storage battery and a group of loads. Here, the switching of operational mode is carried out by the coordinated control strategies between unit agents in Sect. 3.3.1 and the logic coordinated control commands between middle level and lower level agents in Sect. 3.3.2.

3.3.1 Coordinated Control Strategies of Operational Mode Between Different Unit Agentsitle

The sketch map of the reconfigure strategies is shown in Fig. 3.5, where the mode switching process in respective unit agent is described based on Petri-net, and corresponding descriptions of places and transitions are given in Tables 3.2, 3.3, 3.4 and 3.5 respectively.

On Petri-net model, the mode switching process of unit agent depends on not only event-driven discrete behavior of itself, but also interactive behaviors associated with other agents. Such as storage unit agent usually ask for switching its charging

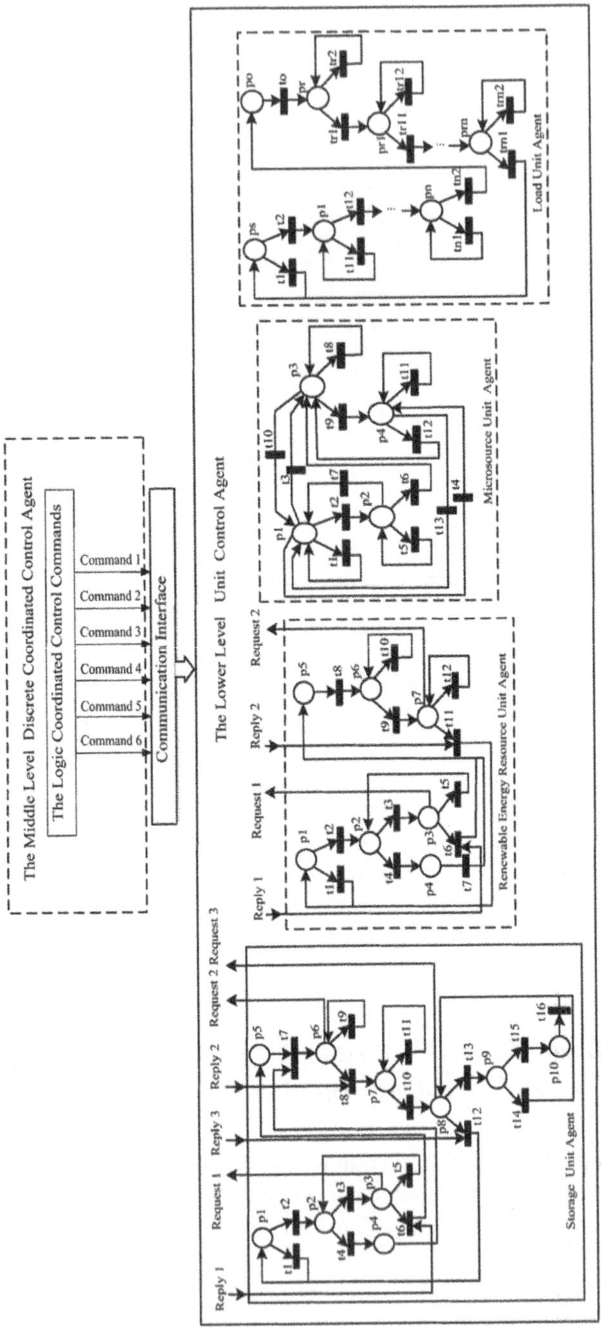

Fig. 3.5 The coordinated control strategies for operating mode

Table 3.2 Description of places and transitions in storage unit agent

Places/Transitions	Description
p1	Discharging state
p2	Discharging state with insufficient state of charge (SOC)
p3	Discharging state with insufficient SOC while re-questing other agents for stopping discharging
p4	Stopping state with the lowest critical SOC
p5	Stopping state with insufficient SOC
p6	Stopping state while requesting other agents for charging
p7	Charging state
p8	Charging state with sufficient SOC while requesting other agents for discharging
p9	Charging state with sufficient SOC
p10	Stopping state with the highest critical SOC
t1	SOC is sufficient
t2	SOC is insufficient
t3	SOC is not the lowest critical value
t4	SOC is the lowest critical value
t5	Other agents don't reply
t6	Other agents reply
t7	Storage unit agent prepares charging
t8	Other agents reply
t9	Other agents don't reply
t10	SOC is sufficient
t11	SOC is insufficient
t12	Other agents reply
t13	Other agents don't reply
t14	SOC is not the highest critical value
t15	SOC is the highest critical value
t16	Storage unit agent request other agents for discharging

or discharging state through interactive behaviors with other agents. With respect to renewable source unit, because of its uncertain power output, its power supply need be supported by other agents through interactive behaviors. When one unit agent need change mode by interactive behaviors, it can send a request to other agents. If other agents reply it, its mode switching can be implemented. The interactive behaviors regarding mode switching between two agents are implemented based on FIPA-ACL messages that are described in Table 3.6.

Table 3.3 Description of places and transitions in renewable resource agent

Places/Transitions	Description
p1	MPPT state at the normal condition
p2	Output power decreasing state
p3	Output power decreasing state while requesting other agents for support of power supply
p4	Stopping operating state
p5	Low voltage operating state
p6	Output power increasing state
p7	MPPT state while requesting other agents for stopping support of power supply
t1	Natural energy is sufficient
t2	Natural energy is insufficient
t3	Natural energy is not the lowest critical value
t4	Natural energy is the lowest critical value
t5	Other agents don't reply
t6	Other agents reply
t7	Natural energy satisfies threshold value condition again
t8	Natural energy is increasing
t9	Output power capacity is sufficient
t10	Output power capacity is insufficient
t11	Other agents reply
t12	Other agents don't reply

3.3.2 *The Logic Coordinated Control Commands Between Middle Level and Lower Level Agents*

The logic coordinated control commands will be determined based on information fusion technique of following characteristic parameters.

(1) *Characteristic Parameters Regarding Voltage Security:* Here, the voltage drop of nodes acts as characteristic parameter of voltage security and can be formulated as follows

$$C_{ih}^{l}(V) = \frac{V_i^l - V_{i0}}{V_{i0}}, \quad (i \in 1, 2, \dots, \upsilon_h), \tag{3.5}$$

where υ_h is the number of nodes in hth middle level agent colony. V_{i0} is rating voltage of ith node. V_i^l is the transient voltage of ith node at lth timing.

(2) *The Security Assessment Indexes Based on Information Fusion Technique of D-S Evidence Theory:* In order to determine security assessment indexes, here, we propose the information fusion technique of D-S evidence theory.

Table 3.4 Description of places and transitions in microresource agent

Places/Transitions	Description
p1	Rating operating state at the normal condition
p2	Low voltage operating state
p3	Hot standby state
p4	Cold standby state
t1	Maintain running
t2	Receive control command for switching into low voltage operating state
t3	Receive control command for stopping running and switching into hot standby state
t4	Receive control command for stopping running and switching into cold standby state
t5	Maintain low voltage operating
t6	Receive control command for stopping running and switching into hot standby state
t7	Receive control command for returning into rating operating state
t8	Maintain hot standby state
t9	Receive control command for switching into cold standby state from hot standby
t10	Receive control command for returning into rating operating state
t11	Maintain cold standby state
t12	Receive control command for switching into hot standby state from cold standby
t13	Receive control command for returning into rating operating state from cold standby

Firstly, we construct the basic probability functions based on the characteristic parameters as follows

$$m_{ih}^{l}(V) = C_{ih}^{l}(V)/\left[C_{ih}^{l}(V) + \left(1 - \gamma_{ihv}^{l}\right)(1 - \alpha_{ihv}\beta_{ihv})\right], \tag{3.6}$$

where $m_{ih}^{l}(V)$ denotes basic probability assignments of voltages on ith node at lth timing in hth local region. γ_{ihv}^{l} are weight coefficients of voltages. These weight coefficients are limited to between 0 and 1. α_{ihv} and β_{ihv} are maximum voltages and relative voltages among all timings on ith node respectively.

According to the timing sequences $l \in \{1, 2, \ldots, q\}$ and all node $i \in \{1, 2, \ldots, v_h\}$, the basic probability assignments $m_{ih}^{l}(V)$ are fused based on the fusion criterion of D-S evidence theory. The fusion result is integrated security assessment index of voltage in hth local region, which is denoted as $\bar{m}_h(V)$.

(3) *The Logic Coordinated Control Commands:* When the security assessment index approaches its threshold value, it implies that the voltage is insecure in this region. The threshold value can be enacted in critical working conditions in an offline

Table 3.5 Description of places and transitions in load agent

Places/Transitions	Description
ps	Load state at the normal condition
p1	The state at load shedding for the first time
pn	The state at the load shedding for the last time
po	The state without the ability of load shedding
pr	The readiness state for load restoration
pr1	The state at load restoration for the first time
prn	The state at load restoration for the last time
t1	Load voltage are normal
t2	Load voltage overstep their threshold values
t11	Load voltage return normal values
t12	Load voltage still overstep their threshold values
tn1	Load voltage return normal values
tn2	Load voltage still overstep their threshold values
to	Load agent is preparing for load restoration
tr1	Load restored condition is satisfied for the highest-level load
tr2	Load restored condition is unsatisfied
tr11	Load restored condition is satisfied again
tr12	Load restored condition is unsatisfied
trn1	Load restored condition is satisfied for the lowest-level load
trn2	Load restored condition is unsatisfied

manner in advance. The threshold value is defined as $\varepsilon_{h,\text{ threshold}} > 0$. According to whether the threshold is violated, we denote

$$
\begin{aligned}
&F_h = \bar{m}_h(V) - \varepsilon_{h,\text{ threshold}},\\
&\text{if } |\bar{m}_h(V)| \geq \varepsilon_{h,\text{ threshold}} \text{ and } \bar{m}_h(V) > 0,\\
&F_h = \bar{m}_h(V) + \varepsilon_{h,\text{ threshold}},\\
&\text{if } |\bar{m}_h(V)| \geq \varepsilon_{h,\text{ threshold}} \text{ and } \bar{m}_h(V) < 0.
\end{aligned}
$$

In light of the size of F_h, we map it into 6-tuple fuzzy set as {NB, NM, NS, PS, PM, PB}.

Corresponding to the architecture of MG as shown in Fig. 3.4, if original modes of unit agents are storage unit agent: P8, renewable unit agent: P1, fuel microsource unit agent: P2, load unit agent: Ps, then the original mode of the group of unit agents is defined as M(8,1,2,s). In this case, the coordinated voltage control commands as follows

No command, if $|\bar{m}_h(V)| < \varepsilon_{h,\text{ threshold}}$;
Command 1: M(8,1,2,s) to M(1,1,2,s), if F_h is PS;
Command 2: M(8,1,2,s) to M(1,1,1,s), if F_h is PM;
Command 3: M(8,1,2,s) to M(1,1,1,(1,2,...,n)), if F_h is PB;

Table 3.6 Interactive behaviors between two unit agents

Agent	Behaviors	Receive message			Send message			
		Performative	Conversation ID	Content	Performative	Conversation ID	Content	Receivers
Unit agent I(UAI)	UAI reconfigure start	–	–	–	Request	Reconfigure	Mode switch	UAII
	UAI reply	Accept proposal	Switch control	Mode switch	Inform	Accept proposal	Mode switch	UAII
Unit agent II(UAII)	UAII propose	UAIIRequest	Reconfigure	Mode switch	Propose	Reply	Agent information	UAI
	UAII inform	UAIIInform	Accept proposal	Mode switch	–	–	–	–

Command 4: M(8,1,2,s) to M(10,1,2,s), if F_h is NS;
Command 5: M(8,1,2,s) to M(10,1,3,s), if F_h is NM;
Command 6: M(8,1,2,s) to M(10,5,3,s), if F_h is NB.

The interactive behaviors regarding logic coordinated control commands between middle level and lower level agents are similar to Table 3.6.

3.4 The Lower Level Unit Control Agents

The design of continuous control strategies for the unit agents is highly dependent on their operational modes and dynamic characteristics, which is very different issue. However, in this section, we present a generality design idea of continuous control strategies for various distributed generation unit agents. The continuous control strategies of distributed generation units are usually divided into two kinds: grid-following control and grid-forming control. The grid-following control is usually selected as the control strategies of the non-dispatchable resources, which mainly includes MPPT control for the renewable energy resource units, V-Q control for the DC resource units and P-Q control. The grid-forming control is generally employed as the control strategies of the dispatchable resources, which mainly includes f-V control and load sharing based on droop characteristics. In an autonomous MG, at least one or more dispatchable distributed generation units actively participate in grid stabilization and voltage regulation, usually f-V control are employed in such unit as shown in Fig. 3.6. And other non-dispatchable units ordinarily use P-Q control based on the pre-specified reference values as shown in Fig. 3.7.

Beside the control strategies need be designed for the continuous control, another important issue is how to select appropriate control algorithm for the controller to deal with robust stabilization problem at multi-mode switching scenarios. As an optimal means, the H_∞ robust control method based on multiply Lyapunovs can effectively solve the multi-mode stabilization problem. The method had been introduced in detail in the author's previous contribution. Taking consideration of the chapter length, here we will not repeat it.

Fig. 3.6 f-V control strategy

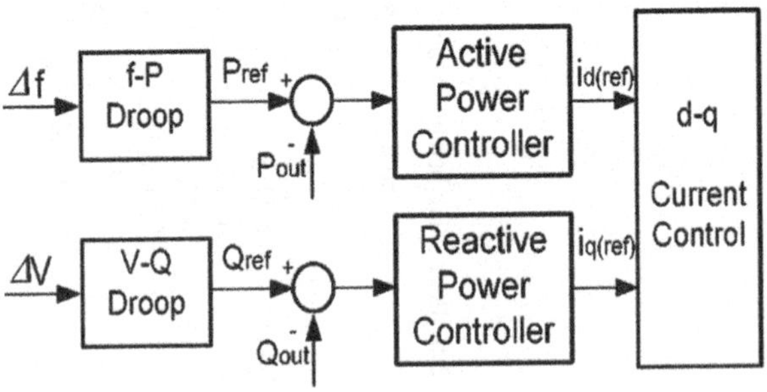

Fig. 3.7 P-Q control strategy

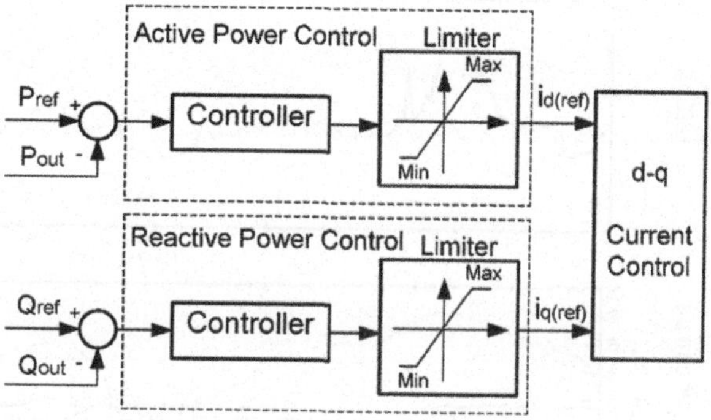

3.5 Simulation Studies

In the researched autonomous MG as shown in Fig. 3.4, the single-master operation (SMO) approach is used, where the battery unit acts as the "master-VSC" in f-V control mode, and the FC/MT, PV units is dynamically controlled in P-Q mode.

The Case 1: The Load Following Performance: Three distributed generation unit agents offer power supply for the loads as shown in Fig. 3.8a. The load following performance under the proposed hybrid control is shown in Fig. 3.8b and c.

Figure 3.8a is load curve from 8.00 to 24.00 during a day. And Fig. 3.8b shows the optimal active power outputs of three distributed generation unit agents. From Fig. 3.8b, it can be seen that PV unit agent runs in MMPT mode as far as possible. Only after 17.00, due to there is no solar energy, the PV unit agent stops operating. During all of the simulation time, besides the mode of no operation, the battery unit agent is switched between both discharging and charging modes to match disequilibrium powers. And during most of the time, the FC/MT unit agent operates in low voltage mode to reduce operational cost as far as possible. The above power assignments among unit agents indicate that the proposed hybrid control can guarantee the operating cost of the system as small as possible. Figure 3.8c displays the highest and lowest voltage performance of the autonomous MG. In the Fig. 3.8c, it can be seen that during all of the simulation time, the voltage are controlled between 0.98 p.u. and 1.02 p.u., and the fluctuation value is limited in the range of of ±2% of rating value. It indicates that the hybrid control contributes to maintain better voltages and load following performance. Figure 3.8d displays the multi-objective optimization performance under the load demand.

The Case 2: The Performance Under Load Disturbance: The purpose of the case 2 is to verify the autonomous MG's control performance under a lager load disturbance. Figure 3.9 shows the load curve, the optimal active power outputs of three distributed generation unit agents and voltage responses respectively.

When t = 10 h, a transient load leads to about 100% of increasing on the load. This moment, the voltages of MG drop to 0.92 p.u. In order to maintain secure voltages, the middle level coordinated control agent sends control commands to switches

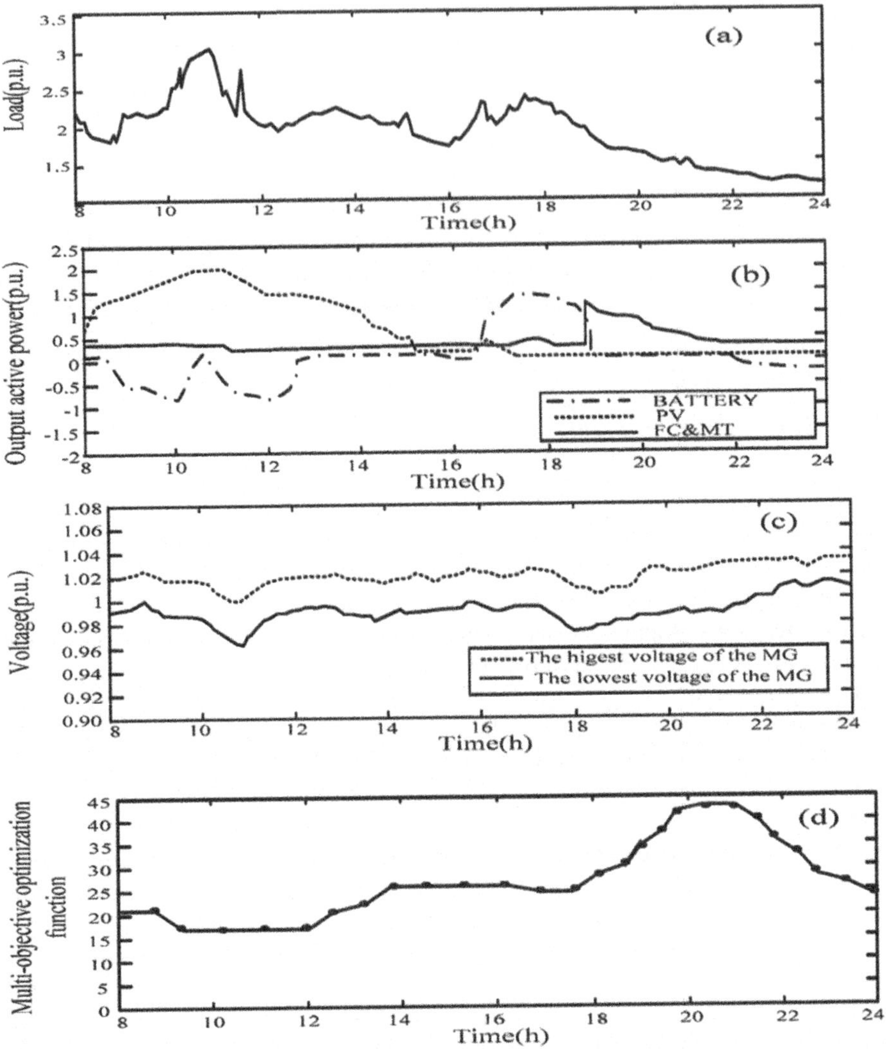

Fig. 3.8 Load following performance in the case 1

FC&MT unit agent to rating operation state, and to switch battery unit agent to discharging state. When loads return to normal state, the operating situations of distributed generation unit agents are similar to the case 1.

The Case 3: The Performance Under Unplanned Islanding: It is well known that MG can run in both grid-connected and islanded modes of operation. An unplanned disconnection from main grid usually is regarded as the most severe test to the islanded MG. The purpose of the case 3 is to investigate MG's control performances during unplanned islanding. The unplanned islanding occurs at t = 10 h.

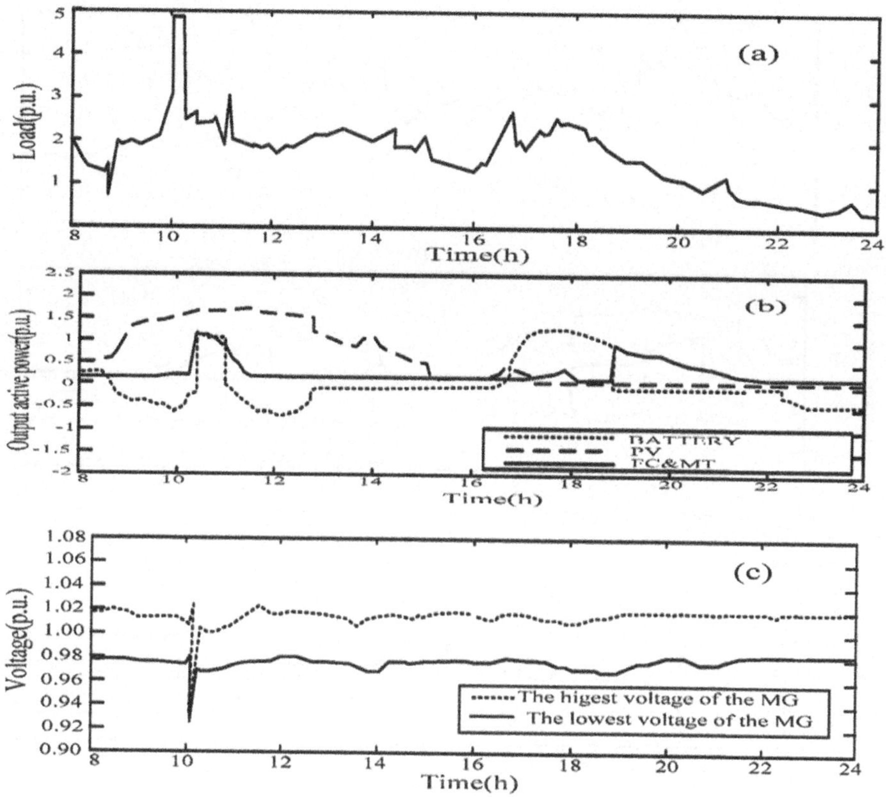

Fig. 3.9 The responses under a large load disturbance in the case 2

Figure 3.10 shows the load curve, the optimal active power outputs of three distributed generation unit agents and voltage responses respectively at the unplanned islanding scenario. From Fig. 3.10, at islanding moment, voltages drop to 0.91 p.u., in order to maintain secure voltages, the middle level coordinated control agent switches the FC&MT unit agent to rating operation state, and switches battery unit agent to discharging state. Compared with the case 2, both FC&MT and battery unit agents keep the switched states longer period of time in the case 3. The voltages are finally controlled between 0.95 p.u. and 1.02 p.u.

3.6 Conclusion

In this chapter, a main contribution is that a multi-agent based hierarchical hybrid control is proposed for autonomous MG. In particular, in the hybrid control scheme, the intelligent reconfigure strategies of operational mode are established via Petri

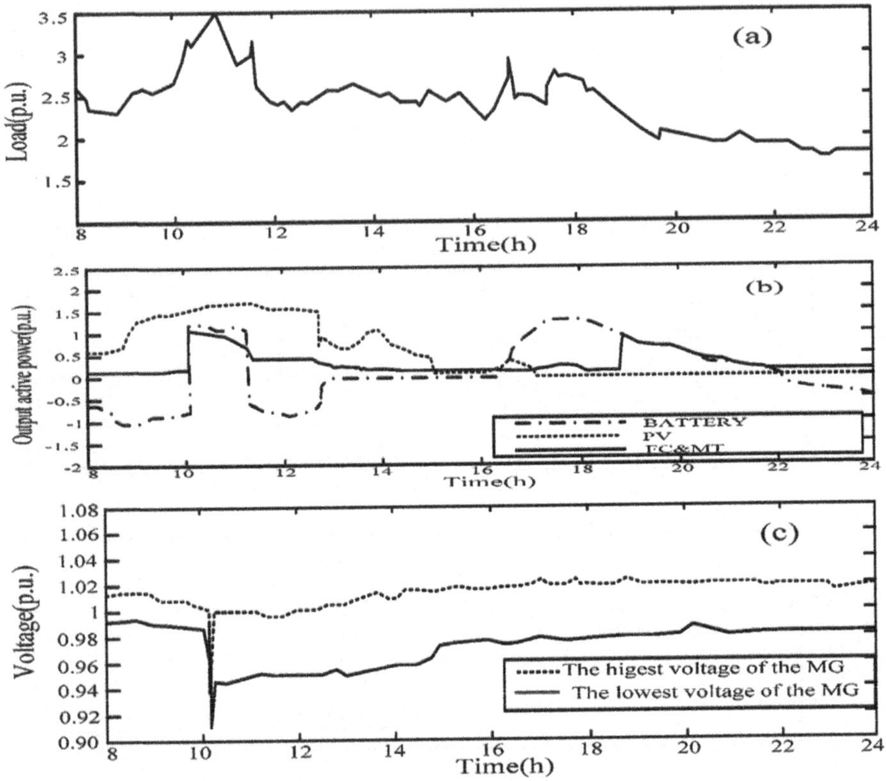

Fig. 3.10 The responses under unplanned islanding in the case 3

nets and the information fusion technique, which is rather innovative. Simulation results show that the proposed hybrid control can maintain secure voltages, as well as maximize economical benefits for an autonomous MG.

Chapter 4
Two-Stage Optimal Operation Strategy of Isolated Power System with TSK Fuzzy Identification of Supply-Security

Due to the uncertainty of intermittent energy and system load, it is a big challenge to optimally operate an isolated power system. This chapter proposes a two-stage optimal operation strategy with a Takagi–Sugeno–Kang (TSK) fuzzy system to address the supply-security under uncertainty circumstance. For proper analysis of the uncertainty characteristics, adjustable uncertainty parameters of intermittent energy resource and system load are taken as fuzzy sets, with consideration of the robustness of these uncertainty parameters on isolated power system, it creates supply-security identification model with TSK fuzzy approach under RBF neural network, and deduces optimal weight values with a recursive least square (RLS) method. For properly avoiding potential risks, security index is classified into several degrees, each degree of risk can switch a different operation model, which can ensure the supply-security of an isolated power system. For properly solving the optimization model, gradient descent based multi-objective cultural differential evolution (GD-MOCDE) is employed to minimize economic cost and emission rate simultaneously. With simulations on isolated regional network, the obtained results reveal that the proposed method can be a viable alternative for optimal operation in isolated power systems.

4.1 Takagi–Sugeno–Kang Fuzzy System with Recursive Least Square for Identification

Generally, a TSK fuzzy system can be considered as a typical nonlinear and dynamic system [1], consisting of several rules that can be expressed as follows:

$$\begin{aligned} &\text{Rule } i: \text{IF}(x_1 \in A_{i1})\text{AND...AND}(x_N \in A_{iN}) \\ &\text{THEN } \ y_i = f_i(x_1, \ldots, x_N) \ \ i = 1, 2 \ldots, N_R \end{aligned} \tag{4.1}$$

D. Yue et al., *Cooperative Optimal Control of Hybrid Energy Systems*,
https://doi.org/10.1007/978-981-33-6722-7_4

The membership functions are employed to produce membership degree with a Gaussian function as follows:

$$\mu_{ij} = exp\left[-\frac{(x_j - m_{ij})^2}{b_{ij}^2}\right] \tag{4.2}$$

For each rule i, its firing degree τ_i can be considered as "AND" operators for all membership $\mu_{ij}(j = 1, 2, \ldots, N)$, which can obtain:

$$\tau_i = \prod_{j=1,2..,N} \mu_{ij} = exp\left[-\sum_{j=1,2,\ldots,N} \frac{(x_j - m_{ij})^2}{b_{ij}^2}\right] \tag{4.3}$$

Combined with a weighted average method, it can obtain an output of this fuzzy model as follows:

$$y = \frac{\sum_{i=1}^{N_R} \tau_i y_i}{\sum_{i=1}^{N_R} \tau_i} \tag{4.4}$$

Since the output function can be expressed with a nonlinear style, RBF neural network is employed to approximate it with consideration of two reasons: (1) The RBF has good universal approximation ability; (2) RBF networks are more stable because each individual RBF unit operates only on selected input patterns [2].

$$f_i(X) = \sum_{j=1}^{N} w_{ij}\phi(||X - c_j||) + d_i \tag{4.5}$$

where $X = (x_1, x_2, \ldots, x_N)$, $\phi()$ denotes an RBF function, which can be presented with inverse multi-quadratics as:

$$\phi(||X - c_j||) = \frac{1}{\sqrt{||X - c_j||_2^2 + \kappa_j^2}} \tag{4.6}$$

Since system load can be a dynamic process with on-line sampling data, a recursive model can be taken into consideration here. For all recent training sample data, it needs to minimize the global objective as follows:

$$J(\theta_1, \theta_2, \ldots, \theta_N) = \sum_{n=k-p+1}^{k} \left(y_n - \sum_{j=1}^{N} \theta_j \phi(||X_n - c_j||)\right)^2 \tag{4.7}$$

where $X_n = [x_{n-j+1}\ y_{n-j+2}\ \cdots\ y_j]^T$ is an input vector of the nth sample data. For simplicity, the above equation can be rewritten as:

$$J(\boldsymbol{\Theta}) = ||Y_n - \boldsymbol{\Phi\Theta}||_2^2 \tag{4.8}$$

where the weight vector $\boldsymbol{\Theta} = [\theta_1\ \theta_2\ \ldots\ \theta_N]^T$, output vector $Y_n = [y_{n-p+1}\ y_{n-p+2}\ \cdots\ y_n]^T$, network function vector $\boldsymbol{\Phi} = [\phi]_{nj}$. To minimize this global objective, the RLS is utilized with several recursive equations as follows:

$$\begin{cases} \boldsymbol{\Theta}(n) = \boldsymbol{\Theta}(n-1) + \alpha(n)G(n) \\ \alpha(n) = Y(n) - \boldsymbol{\Phi}(n)^T \boldsymbol{\Theta}(n-1) \\ G(n) = Q(n-1)\boldsymbol{\Phi}(n)[\rho + \boldsymbol{\Phi}(n)^T Q(n-1)\boldsymbol{\Phi}(n)]^{-1} \\ Q(n) = \rho^{-1}Q(n-1) - G(n)\boldsymbol{\Phi}(n)^T \rho^{-1} Q(n-1) \end{cases} \tag{4.9}$$

The initial conditions of the above parameters can be set as follows:

$$\begin{cases} \boldsymbol{\Theta}(0) = \mathbf{0} \\ X(n) = X_n = 0,\ n = -p, -p+1, \ldots, -1 \\ Y(n) = y_n = 0,\ n = -p, -p+1, \ldots, -1 \\ Q(0) = \delta I, \delta \in R \end{cases} \tag{4.10}$$

Since RLS can be considered as a rolling optimization approach, it can revise the weight value $\boldsymbol{\Theta}$ as time goes, which can finally achieve a global optima when the time step ends.

4.2 Problem Formulation: Risk Degree Based Isolated Power System with Different Switching Models

4.2.1 Uncertainty Analysis of Intermittent Energy Resources and System Load

Since intermittent energy resources and system load have great uncertainty, it can bring a potential risk for hybrid energy system stability. The intermittent power output and system load can be expressed as follows:

$$\begin{cases} P_{Ij} = \overline{P_{Ij}} + r_{Ij}\widetilde{P_{Ij}} \\ P_{load} = \overline{P_{load}} + r_{load}\widetilde{P_{load}} \\ \widetilde{P_{Ij}} \in [\widetilde{P_{Ij,min}}, \widetilde{P_{Ij,max}}] \\ \widetilde{P_{load}} \in [\widetilde{P_{load,min}}, \widetilde{P_{load,max}}] \\ r_{Ij}, r_{load} \in [0, 1] \end{cases} \tag{4.11}$$

For further analysis on uncertainty, the uncertainty interval can be divided into several small intervals as:

$$[\widetilde{P_{Ij,min}}, \widetilde{P_{Ij,max}}] = \bigcup_{k=1}^{K_j}[\widetilde{P^{[k]}_{Ij,min}}, \widetilde{P^{[k]}_{Ij,max}}] \tag{4.12}$$

$$[\widetilde{P_{load,min}}, \widetilde{P_{load,max}}] = \bigcup_{k=1}^{K_{load}}[\widetilde{P^{[k]}_{load,min}}, \widetilde{P^{[k]}_{load,max}}] \tag{4.13}$$

For simplicity, suppose that $K_j = K_{load} = K$, it can be labeled as:

$$\begin{cases} x_j = \widetilde{P_{Ij}}, \quad j = 1, 2 \ldots, J-1 \\ x_J = \widetilde{P_{load}} \\ A_{jk} = [\widetilde{P^{[k]}_{Ij,min}}, \widetilde{P^{[k]}_{Ij,max}}], \quad j = 1, 2 \ldots, J-1 \\ A_{Jk} = [\widetilde{P^{[k]}_{load,min}}, \widetilde{P^{[k]}_{load,max}}] \\ \theta_j = r_{Ij}, \quad j = 1, 2 \ldots, J-1 \\ \theta_J = -r_{load} \end{cases} \tag{4.14}$$

where $x = [x_1, x_2, \ldots, x_J]$, $A = [A_{jk}]_{J\times K}$, $\theta = [\theta_1, \theta_2, \ldots, \theta_J]$, those small intervals are equally divided, it can obtain

$$\begin{cases} \widetilde{P^{[k]}_{Ij,\min}} = P_{Ij,\min} + (k-1)\frac{P_{Ij,\max}-P_{Ij,\max}}{K} \\ \widetilde{P^{[k]}_{Ij,\max}} = P_{Ij,\min} + k\frac{P_{Ij,\max}-P_{Ij,\max}}{K} \\ \widetilde{P^{[k]}_{\text{load},\min}} = P_{\text{load},\min} + (k-1)\frac{P_{\text{load},\max}-P_{\text{load},\max}}{K} \\ \widetilde{P^{[k]}_{\text{load},\max}} = P_{\text{load},\min} + k\frac{P_{\text{load},\max}-P_{\text{load},\max}}{K} \end{cases} \tag{4.15}$$

4.2.2 Economic Dispatch Model with Uncertainty Degree in Isolated Power System

The economic issue is crucial in power systems. It consists of different kinds of economic cost, i.e. power generation cost and benefit, load cut-off cost, on/off cost, and simultaneously it also produces emission pollution from thermal units, which can be expressed as follows:
(1) Power generation cost: Generally, the power generation cost can be described with quadratic functions of power output. Here, the on/off state of each thermal units is also taken into consideration, then it can be expressed as:

$$C_{gen} = \sum_{k=1}^{N_c} H_{ck}(\alpha_{k0} + \alpha_{k1}P_{ck} + \alpha_{k2}P_{ck}^2) \tag{4.16}$$

(2) Emission volume: Since thermal units can produce emission pollutant during power generation, emission issue can also be taken into consideration, which can be presented as:

$$E = \sum_{k=1}^{N_c} H_{ck}(\gamma_{k0} + \gamma_{k1} P_{ck} + \gamma_{k2} P_{ck}^2) \tag{4.17}$$

(3) Load cut-off cost: Some system load can be controllable, it can be cut-off to ensure the system balance, but it needs to compensate consumers, which can generate cut-off cost as:

$$C_{cut} = \sum_{s \in S_0} \gamma_{cut,s} P_{con,s} \tag{4.18}$$

$$S_0 = \{s | H_{con,s} = 0\} \tag{4.19}$$

If it is on, $H_{con,s} = 1$, otherwise, $H_{con,s} = 0$.
(4) Switching on/off cost: The on/off operation of thermal units and energy storage can bring economic cost, the switching cost can be presented as:

$$C_{swi,thermal} = \sum_{k=1}^{N_c} \gamma_{swi,ck} |H_{ck}(t) - H_{ck}(t-1)| \tag{4.20}$$

$$C_{swi,store} = \sum_{k=1}^{N_b} \gamma_{swi,bl} H_{bl} |P_l^{store}| \tag{4.21}$$

$$C_{swi} = C_{swi,thermal} + C_{swi,store} \tag{4.22}$$

Its total cost F can be expressed with three parts as:

$$F = C_{gen} + C_{cut} + C_{swi} \tag{4.23}$$

(5) Power system transmission loss: In this chapter, the isolated power system consists of thermal units, intermittent energy sources and energy storage units. Since different energy resources are widely distributed, it exists transmission loss among them.

$$P_{tot} = P_{load} + P_{loss} = P_{fix} + \sum_{s=1}^{N_{con}} H_{con,s} P_{con,s} + P_{loss} \tag{4.24}$$

$$P_{loss} = \sum_{i,j \in \Omega} B_{ij} P_i P_j + \sum_{j \in \Omega} B_{0j} P_j + B_{00} \tag{4.25}$$

$$P_{tot} = \sum_{j=1}^{J-1} P_{Ij} + \sum_{k=1}^{N_c} H_{ck} P_{ck} + \sum_{l=1}^{N_b} H_{bl} P_{bl} \tag{4.26}$$

(6) Thermal power generation constraints: During thermal power generation, it needs to satisfy maximum and minimum output limits. With consideration of equipment management issue, each thermal unit can not be with on/off state permanently, its on/off time must be limited as:

$$\begin{cases} P_{ck,min} \leq P_{ck} \leq P_{ck,max} \\ (T^{on}_{ck,t-1} - T^{on}_{ck,min})(H_{ck}(t-1) - H_{ck}(t)) \geq 0 \\ (T^{off}_{ck,t-1} - T^{off}_{ck,min})(H_{ck}(t) - H_{ck}(t-1)) \geq 0 \end{cases} \tag{4.27}$$

(7) Energy storage constraints: Energy storage is a supplementary energy resource for intermittent energy, it has storage limits and charging/discharging limits, all these constraints can be presented as follows:

$$\begin{cases} V^{store}_l(t+1) = V^{store}_l(t) + P_{bl}(t) * \Delta T \\ P_{bl}(t) = \eta_l P^{store}_l(t) \\ V^{store}_{l,min} \leq V^{store}_l(t) \leq V^{store}_{l,max} \\ P^{store}_l(t) = P^{cha}_l(t), if\ P^{store}_l(t) \geq 0 \\ P^{store}_l(t) = -P^{dis}_l(t), if\ P^{store}_l(t) < 0 \\ 0 \leq P^{dis}_l(t) \leq P^{dis}_{l,max} \\ 0 \leq P^{cha}_l(t) \leq P^{cha}_{l,max} \\ V^{store}_l(0) = V^{store}_{l,initial} \end{cases} \tag{4.28}$$

4.3 Fuzzy System Based Two Stage Optimization Strategy

4.3.1 TSK Fuzzy System Identification of Supply-Security in Isolated Power System

The confidence degree of active power supply-security can be considered as an important issue in isolated power system, since it is mainly affected by both power supply and demand uncertainty, but the relationship between these is still uncertain. For properly dealing with this problem, a TSK fuzzy system is employed to build up the relationship, an RBF neural network approximates the nonlinear function, and the robustness constraint is also satisfied for avoiding potential risk. The confidence degree can be considered as the output of a fuzzy system, which is a function of the adjustable parameter θ_j and input variables with an RBF neural network approach as:

$$y = f(X) = \sum_{j=1}^{J} \theta_j \phi(||X - c_j||) \tag{4.29}$$

Combined with the introduced fuzzy system model in section II, the robustness constraint is also taken into consideration with an RLS approach, those recursive iterations can be updated with several procedures. For proper analysis on uncertainty, the system load balance can be rewritten as:

$$\sum_{j=1}^{J} \theta_j x_j = \overline{P_{load}} - \sum_{j=1}^{J-1} \overline{P_{Ij}} - \sum_{k=1}^{N_c} P_{ck} - \sum_{l=1}^{L} P_l^{store} \tag{4.30}$$

Moreover, system robustness can also be taken into consideration for avoiding the worst case, which has been referred to in [3]. Then, the following constraint should be satisfied:

$$\sum_{j=1}^{J-1} \theta_j - \theta_J \leq \delta \tag{4.31}$$

where $\delta \in [0, J]$ represents the uncertainty degree. Here, it satisfies:

$$\delta \geq \sqrt{-2Jln\xi} \tag{4.32}$$

where $1 - \xi \in (0, 1]$ denotes the probability of satisfying the constraints requirement. Since the robustness constraint is added into the fuzzy system model, it can be converted into a constrained problem. With consideration of the correlation among sequences, taking forgetting factors also into consideration, the Lagrange function can be presented as follows:

$$L(\boldsymbol{\Theta}_n) = ||Y_n^{'} - \boldsymbol{\Phi}_n^{'}\boldsymbol{\Theta}_n||_2^2 + \lambda(I_0\boldsymbol{\Theta}_n - (\delta - \delta^{'})ee^T)^2 \tag{4.33}$$

where $0 << \zeta < 1$ represents the forgetting factor, $I_0 = [1\ 1\ \cdots\ 1\ -1], 0 \leq \delta^{'} \leq \delta$, $e = [1\ 0\ \cdots\ 0\ 0]$. Here, $X_n^{'} = [\zeta^{j-1} x_{n-j+1}\ \zeta^{p-2} x_{n-j+2}\ \cdots\ x_n]^T$, and the matrix $\boldsymbol{\Phi}_n^{'}$ can also be updated as:

$$\begin{pmatrix} \phi_{11}^{'} & \phi_{12}^{'} & \cdots & \phi_{1J}^{'} \\ \phi_{21}^{'} & \phi_{22}^{'} & \cdots & \phi_{2J}^{'} \\ \vdots & \vdots & \vdots & \vdots \\ \phi_{p1}^{'} & \phi_{p2}^{'} & \cdots & \phi_{pJ}^{'} \end{pmatrix} \tag{4.34}$$

The RBF of the neural network can be updated with $\phi_{nj}^{'}$, which can be rewritten as:

$$\phi_{nj}^{'} = \phi(||X_n^{'} - c_j||) = \frac{1}{\sqrt{||X_n^{'} - c_j||_2^2 + \kappa_j^2}} \tag{4.35}$$

The relationship between $X_n^{'}$ and $X_{(n)}$ can be expressed as:

$$X_n^{'} = \boldsymbol{B}X_n \tag{4.36}$$

where $\boldsymbol{B}$ can be defined as:

$$\begin{pmatrix} \zeta^{j-1} & 0 & 0 & 0 \\ 0 & \zeta^{j-2} & \cdots & 0 \\ \vdots & \vdots & \vdots & \vdots \\ 0 & 0 & \cdots & 1 \end{pmatrix} \tag{4.37}$$

Define an exponentially-weighted covariance matrix $\boldsymbol{U}_n$ and vector $\boldsymbol{V}_n$, then it can be expressed as:

$$\begin{cases} \boldsymbol{U}_n = X_n^{'}(X_n^{'})^T = \zeta^2\boldsymbol{U}_{n-1} + x_n x_n^T \\ \boldsymbol{V}_n = X_n^{'}Y_n^{'} = \zeta\boldsymbol{V}_{n-1} + y_n x_n \end{cases} \tag{4.38}$$

For optimizing the above Lagrange function, a relaxation approach is utilized with Lagrange multipliers. $\boldsymbol{\Theta}_n$ can be calculated as:

$$\boldsymbol{\Theta}_n = (\boldsymbol{U}_n + \lambda I_0^T I_0)^{-1}(\boldsymbol{V}_n + \lambda(\delta - \delta^{'})I_0^T) \tag{4.39}$$

To decrease the cumulative error caused by iterations, it needs to update $\boldsymbol{\Theta}$ with experienced information:

$$I_0\boldsymbol{\Theta}_n = \delta - \delta^{'} \tag{4.40}$$

With those above iterations, the weight vector $\boldsymbol{\Theta}_n$ can be deduced with several samples, which also means that the nonlinear relationship between the confidence degree of the system balance and the uncertainty sets can be approximated.

4.3.2 Switching Mechanisms Under Confidence Degree of Supply-Security

For proper description of supply-security as well as convenience for collecting output value from samples, the confidence degree is defined as:

$$Security_{con} = Prob\left(\sum_{i\in N_{node}} \beta_i |P_{tot,i} - P_{load,i} - P_{loss,i}| < \epsilon\right) \tag{4.41}$$

where N_{node} represents the number of nodes in isolated power system, β_i is state parameter of ith node, $P_{tot,i}$, $P_{load,i}$ and $P_{loss,i}$ denote total power output, load demand and transmission loss respectively. The confidence degree is in the mainly range [0, 1], with safe when it approaches 1. If the confidence degree cannot be properly satisfied, it needs to take some measures to improve it, energy storage can be turned on to supplement it, and even cut off some controllable load in the worst case. Since the probability of real-world supply security is unknown, the confidence degree can be easily obtained from samples, confidence degree of each sample can be collected by frequency of same events, and then it can be taken as the output value of TSK fuzzy model. Moreover, within total power generation, thermal power bears 80% of system load to ensure stable power supply. The confidence degree can be classified at three levels: Excellent, good and bad, and each level will switch on different adjusting scheme. An excellent level of confidence degree is in the range [0.9, 1], which means that the system balance has been completely satisfied, then its goal is merely to minimize economic cost. A good level of confidence degree is in the range [0.8, 0.9), which indicates that the energy storage should be turned on to decrease the uncertainty for ensuring system balance. A bad level of confidence degree is mainly in the range [0, 0.8), then some controllable load can also be cut off to the keep system balance. For each time period, the confidence degree must achieve an excellent level, then it can be optimized. From both a security and economic view, the adjusting procedures can be presented in Algorithm 4.1.

4.3.3 GD-MOCDE Approach for Optimizing the Economic Dispatch Problem of Isolated Power

Combined with above procedures, the potential risk has been properly avoided and all the on/off or cut-off states of all power generators and controllable load becomes a certain state, the remaining problem can be considered as a classical economic dispatch problem. Without loss of generality, the optimal operation model can be converted as follows:

$$\begin{cases} minF_1, minF_2 - E \\ P_{ck,min} \leq P_{ck} \leq P_{ck,max} \\ V_l^{store}(t+1) = V_l^{store}(t) + P_{bl}(t) * \Delta T \\ V_{l,min}^{store} \leq V_l^{store}(t) \leq V_{l,max}^{store} \\ P_l^{store}(t) = P_l^{cha}(t), if\ P_l^{store}(t) \geq 0 \\ P_l^{store}(t) = -P_l^{dis}(t), if\ P_l^{store}(t) < 0 \\ 0 \leq P_l^{dis}(t) \leq P_{l,max}^{dis} \\ 0 \leq P_l^{cha}(t) \leq P_{l,max}^{cha} \\ V_l^{store}(0) = V_{l,initial}^{store} \\ H_{ck} = 1, H_{bl} = 1, H_{con,s} = 1, \\ k = 1, 2 \ldots \tau_c, l = 1, 2 \ldots \tau_{store}, s = 1, 2 \ldots \tau_{con} \end{cases} \tag{4.42}$$

Here, the GD-MOCDE algorithm is utilized to solve the above problem, the output of the thermal unit and the charging/discharging output of the energy storage units are taken as decision variables. In the framework of the cultural algorithm, a differential evolution strategy promotes the evolution of the population space. In the differential evolution, the mutation operator of DE/rand/1/bin can be expressed as:

Algorithm 4.1

```
procedure S(w)itched adjusting scheme under different levels of confidence degree
    Evaluate confidence degree y
Case 1:
    if y ≥ 0.9 then
        Goto optimize ED problem
    end if
Case 2:
    if 0.8 ≤ y < 0.9 then
        Turn on energy storage
        For l = 1 : L
        H_l = 1
        δ = δ − 0.1
        Evaluate confidence degree y
        if y ≥ 0.9 then
            Goto Case 1
        end if
    end if
Case 3:
    if y < 0.8 then
        Turn on energy storage
        For l = 1 : L
        H_l = 1
        δ = δ − 0.1
        Evaluate confidence degree y
        if y < 0.9 then
            switch off controllable load
            For s = 1 : N_con
            H_con,s = 0
            δ = δ − 0.1
            Evaluate confidence degree y
            if y ≥ 0.9 then
                Goto Case 1
            end if
        end if
    end if
end procedure
```

$$R_{r,G+1} = Q_{r,G} + \beta[(Q_{r1,G} - Q_{r2,G}) + (Q_{r3,G} - Q_{r4,G})] \tag{4.43}$$

where $Q_{r,G}$, $Q_{r1,G}$, $Q_{r2,G}$, $Q_{r3,G}$ and $Q_{r4,G}$ are selected from the current population set and it also satisfies $r1 \neq r2 \neq r3 \neq r4 \neq r$, G denotes the generation index, $R_{r,G+1}$ is a trial vector, and β is a mutation parameter. For improving the optimization efficiency, the gradient decent mechanism can be obtained with two spaces Ω^+ and Ω^-, which can be presented as:

$$\begin{cases} \Omega^+ = d \in R^m | \bigtriangledown g(z)^T d > 0 \\ \Omega^- = d \in R^m | \bigtriangledown g(z)^T d < 0 \end{cases} \tag{4.44}$$

where $g(z) = [g_1(z), g_2(z), \ldots, g_m(z)]^T$ represents the objective function vector, $z = [z_1, z_2, \ldots, z_n]$ denotes the variable, d is an arbitrary vector, and $\nabla g(z)$ can be considered as a Jacobi matrix as follows:

$$\begin{pmatrix} \frac{\partial g_1(z)}{\partial z_1} & \frac{\partial g_1(z)}{\partial z_2} & \cdots & \frac{\partial g_1(z)}{\partial z_n} \\ \frac{\partial g_2(z)}{\partial z_1} & \frac{\partial g_2(z)}{\partial z_2} & \cdots & \frac{\partial g_2(z)}{\partial z_n} \\ \vdots & \vdots & \vdots & \vdots \\ \frac{\partial g_m(z)}{\partial z_1} & \frac{\partial g_m(z)}{\partial z_2} & \cdots & \frac{\partial g_m(z)}{\partial z_n} \end{pmatrix} \tag{4.45}$$

It also means that the deviation between Q_{G+1} and Q_G can be described as:

$$Q_{G+1} - Q_G = -\Upsilon_G \sum_{i \in m} \chi_i \frac{\nabla g_i(z)}{|| \nabla g_i(z)||} \tag{4.46}$$

With consideration of a discrete version, if it is a bi-objective optimization problem, and it can be converted as:

$$Q^j_{G+1} = Q^j_G + \psi_1(Q_{r1,G} - Q_{r2,G}) + \psi_2(Q_{r3,G} - Q_{r4,G}) \tag{4.47}$$

where ψ_1 and ψ_2 are two control parameters, which can be presented as:

$$\begin{cases} \psi_1 = -\frac{\Upsilon_G \chi_1 sgn(g_1(Q_{r1,G}) - g_1(Q_{r2,G}))}{(Q_{r1,G}-Q_{r2,G})^2 \sqrt{\sum_{j \in n} \frac{1}{(Q_{r1,G}-Q_{r2,G})^2}}} \\ \psi_2 = -\frac{\Upsilon_G \chi_2 sgn(g_2(Q_{r3,G}) - g_2(Q_{r4,G}))}{(Q_{r3,G}-Q_{r4,G})^2 \sqrt{\sum_{j \in n} \frac{1}{(Q_{r3,G}-Q_{r4,G})^2}}} \\ \Upsilon_G = \Upsilon_0[(G_{max} - G + 1)/G_{max}]^p \end{cases} \tag{4.48}$$

where Υ_G, χ_i, ψ_1 and ψ_2 are control parameters, G_{max} denotes the maximum generation number. By replacing the classic mutation operator with the above gradient descent operator, the convergence speed can be improved. Moreover, since there are some inequality and equality constraints in the model, an embedded constraint-handling technique is employed to deal with them especially those equality constraints, the details can be found in literature [4]. The whole procedure of the GD-MOCDE for economic dispatch can be presented in Algorithm 4.2. In the algorithm, $Size$ represents the current size of the archive set, $|B|$ means the size of the archive set, $violation$ denotes the total violation of the equality constraints.

Algorithm 4.2

procedure G(D)-MOCDE algorithm for economic dispatch in isolated power system
  Initialize population set Q and archive set B
  $G = 0$, $B = \emptyset$, $Size = 0$
  while $G < G_{max}$ **do**
    Check total constraint violation
    if $violation > \epsilon$ **then**
      Embedded constraint handling strategy
    end if
    GD based mutation operator
    Crossover operator
    Selection operator
    Store non-dominated solutions in archive set B
    $Size = Size + 1$
    if $Size > |B|$ **then**
      Truncate archive set B
    end if
    $G = G + 1$
  end while
end procedure

4.4 Case Study

The isolated power system consists of 4 wind farms, 3 photovoltaic fields, 10 thermal units and 4 energy storage units, its structure has been shown in Fig. 4.1, related details can be found in [5, 6]. The system load includes a fixed load and a controllable load, which can be cut off when necessary. The output of wind power and photovoltaic power can be obtained with wind speed and illumination intensity prediction shown in Tables 4.1 and 4.2, system load interval and controllable load are shown in Fig. 4.2. The outputs of wind power, solar power and system load can be considered as fuzzy sets, which can be calculated by dividing the output interval into four parts. Here, four typical periods are selected for verifying the efficiency of both fuzzy identification and optimal operation. It includes 00:00–02:00, 06:00–08:00, 10:00–12:00 and 19:00–21:00, which represent the key periods in one day.

4.4.1 Fuzzy System Identification with Intermittent Energy Resources and System Load

The fuzzy system has $4^4 * 4^3 * 4 = 65536$ possible fuzzy sets and the output value can be calculated with a membership function. Combined with the sampling data, the relationship between the uncertainty variables and supply-security can be identified, and then optimal schemes can be properly applied. A static model can be converted into a dynamic one due to its on-line identification approach. With the consideration of constraints robustness, the lower bound of the uncertainty budget can be calculated

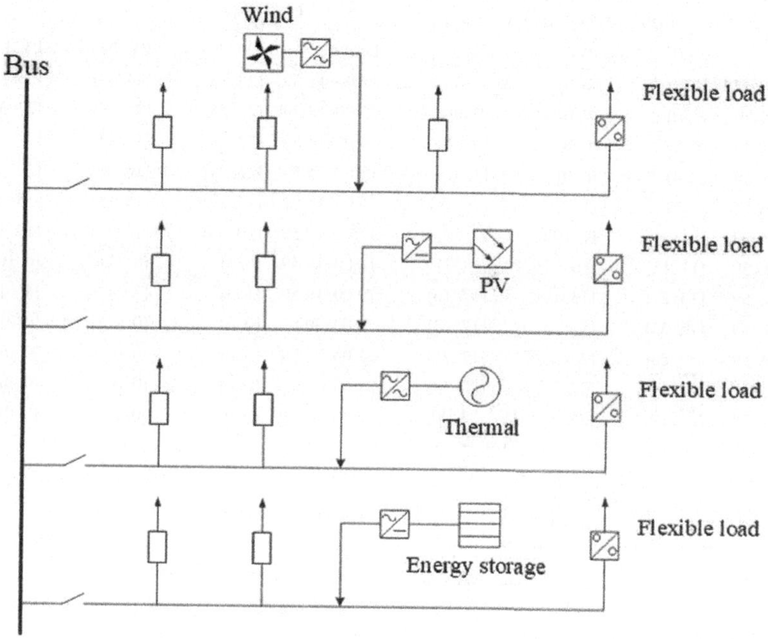

Fig. 4.1 The structure of isolated power system

Table 4.1 The output interval of wind power generation

Period	Wind 1	Wind 2	Wind 3	Wind 4	Period	Wind 1	Wind 2	Wind 3	Wind 4
00:00–00:59	[32, 45]	[30, 42]	[30, 40]	[25, 34]	12:00–12:59	[16, 22]	[15, 21]	[12, 16]	[13, 19]
01:00–01:59	[35, 45]	[35, 41]	[32, 38]	[29, 35]	13:00–13:59	[20, 26]	[20, 26]	[17, 23]	[17, 23]
02:00–02:59	[35, 44]	[34, 40]	[30, 36]	[25, 43]	14:00–14:59	[25, 31]	[22, 30]	[22, 28]	[21, 27]
03:00–03:59	[29, 35]	[27, 35]	[23, 29]	[18, 24]	15:00–15:59	[30, 38]	[28, 36]	[27, 35]	[25, 33]
04:00–04:59	[20, 28]	[20, 26]	[16, 24]	[12, 18]	16:00–16:59	[26, 34]	[24, 32]	[24, 30]	[22, 28]
05:00–05:59	[15, 21]	[13, 19]	[12, 18]	[10, 18]	17.00–17.59	[24, 30]	[22, 26]	[20, 26]	[19, 25]
06:00–06:59	[18, 26]	[15, 23]	[13, 20]	[13, 20]	18:00–18:59	[22, 28]	[19, 25]	[18, 24]	[17, 23]
07:00–07:59	[22, 28]	[19, 25]	[17, 23]	[14, 22]	19:00–19:59	[15, 20]	[17, 23]	[15, 21]	[15, 21]
08:00–08:59	[22, 30]	[22, 28]	[20, 24]	[17, 23]	20:00–20:59	[22, 28]	[23, 29]	[20, 26]	[19, 25]
09:00–09:59	[20, 26]	[18, 24]	[15, 23]	[15, 21]	21:00–21:59	[25, 32]	[28, 34]	[25, 33]	[23, 29]
10:00–10:59	[17, 23]	[15, 19]	[12, 18]	[12, 18]	22:00–22:59	[31, 39]	[27, 35]	[25, 33]	[23, 30]
11:00–11:59	[17, 23]	[15, 21]	[12, 18]	[12, 16]	23:00–23:59	[33, 43]	[32, 40]	[30, 38]	[27, 35]

as 6.0697, which also means that the uncertainty budget can only range between [6.0697, 8]. The uncertainty budget for 24 periods are listed in Table 4.3, because the uncertainty is mainly caused by intermittent power generation and system load, the uncertainty budget can be large in some periods where huge intermittent power generation or system load requirement occurs. For each time period, the potential

Table 4.2 The output interval of PV fields

Period	PV 1	PV 2	PV 3	Period	PV 1	PV 2	PV 3
00:00–00:59	[0, 0]	[0, 0]	[0, 0]	12:00–12:59	[28, 36]	[24, 32]	[26, 34]
01:00–01:59	[0, 0]	[0, 0]	[0, 0]	13:00–13:59	[25, 35]	[23, 29]	[27, 33]
02:00–02:59	[0, 0]	[0, 0]	[0, 0]	14:00–14:59	[23, 29]	[20, 24]	[23, 29]
03:00–03:59	[2, 4]	[0, 0]	[1, 3]	15:00–15:59	[20, 24]	[16, 20]	[20, 24]
04:00–04:59	[4, 6]	[1, 3]	[2, 6]	16:00–16:59	[15, 19]	[14, 18]	[15, 21]
05:00–05:59	[8, 12]	[6, 10]	[7, 11]	17:00–17:59	[10, 14]	[11, 15]	[10, 14]
06:00–06:59	[11, 15]	[10, 14]	[8, 12]	18:00–18:59	[6, 8]	[8, 12]	[6, 10]
07:00–07:59	[15, 21]	[13, 19]	[12, 16]	19:00–19:59	[1, 3]	[3, 5]	[4, 6]
08:00–08:59	[16, 22]	[17, 23]	[17, 23]	20:00–20:59	[0, 0]	[0, 0]	[0, 2]
09:00–09:59	[20, 26]	[20, 26]	[17, 23]	21:00–21:59	[0, 0]	[0, 0]	[0, 0]
10:00–10:59	[23, 29]	[22, 28]	[20, 24]	22:00–22:59	[0, 0]	[0, 0]	[0, 0]
11:00–11:59	[23, 29]	[25, 31]	[24, 30]	23:00–23:59	[0, 0]	[0, 0]	[0, 0]

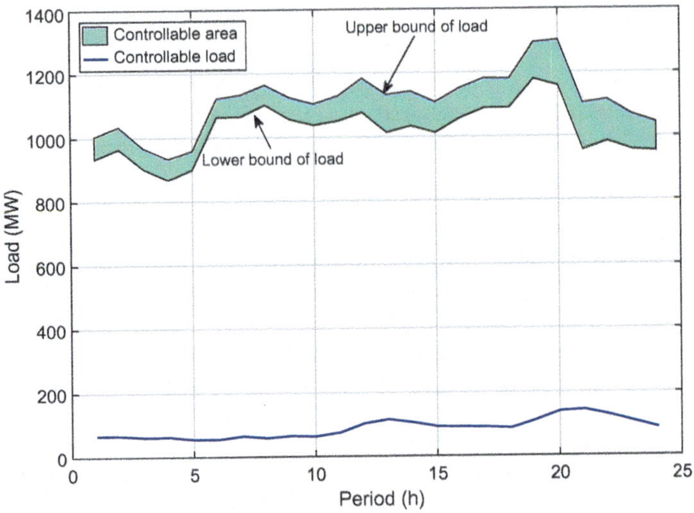

Fig. 4.2 Controllable load and system load with lower and upper bounds

Table 4.3 The uncertainty budget of 24 time periods

Period	1	2	3	4	5	6	7	8	9	10	11	12
δ	7.342	7.123	6.775	6.642	6.5321	7.221	7.327	7.452	7.4631	7.462	7.4853	7.514
Period	13	14	15	16	17	18	19	20	21	22	23	24
δ	7.631	7.642	7.454	7.535	7.498	7.346	7.245	7.524	7.511	7.201	7.132	7.265

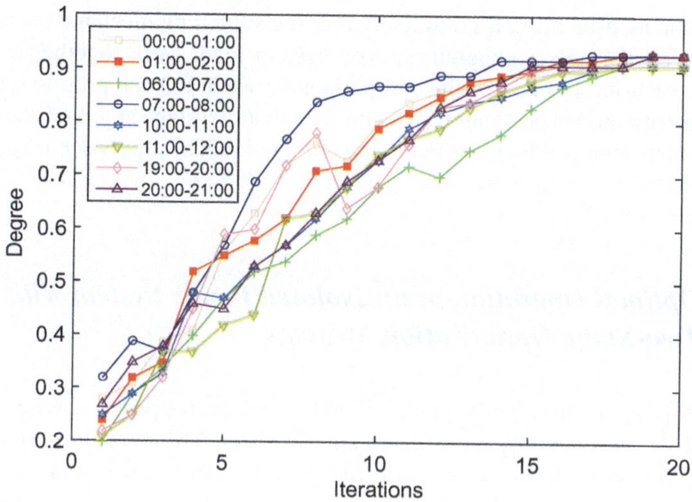

Fig. 4.3 The convergence analysis of supply-security in some typical periods

degree of supply-security can be evaluated with the TSK fuzzy approach, and then two-stage optimization strategy is utilized to optimize the isolated power system with comparison with RO. The convergence process of TSK identification in some typical periods are shown in Fig. 4.3. It can be seen that identification in different typical periods converges in less than 20 iterations to reach an excellent degree (larger

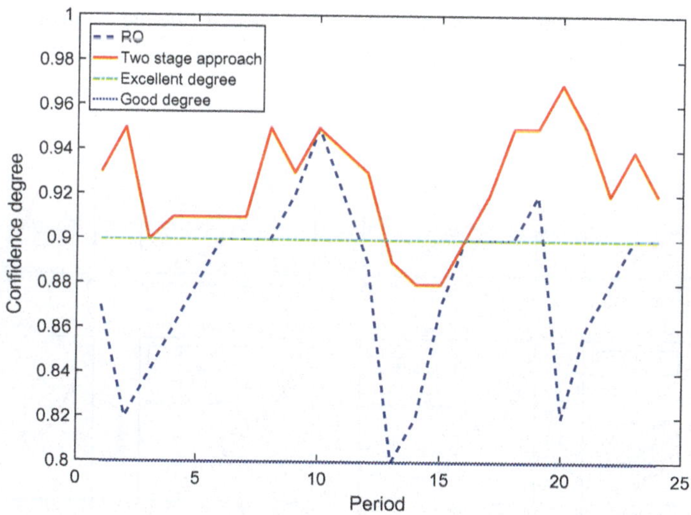

Fig. 4.4 The confidence degree between RO and two stage approach

than 0.9), the identification is completed after the switching mechanism, which also means that all the power generators, even system load, work together to improve the supply-security. In Fig. 4.4, the comparison between RO and proposed two stage optimization method is presented, it can be seen that two stage approach can improve the confidence degree, which can avoid the potential risk of supply security to certain extent.

4.4.2 *Optimal Operation of an Isolated Power System with a Two-Stage Optimization Strategy*

Once the supply-security has been identified for each period, optimal schemes can be calculated with a switching mechanism and GD-MOCDE algorithm. Since 00:00–02:00, 06:00–08:00, 10:00–12:00 and 19:00–21:00 can be taken a four typical periods for the system load in one day, the analysis below focus on the results of these four periods. Since economic cost and emission rate can be taken as two objectives, 20 Pareto schemes are obtained by GD-MOCDE and multi-objective differential evolution (MODE) [7]. The results in some typical periods have been presented in Fig. 4.5. It can be clearly observed that those Pareto optimal schemes obtained by MODE are disorder, and the GD-MOCDE has both better convergence ability and diversity distribution. For better analysis on optimization results, the 10th scheme of those Pareto optimal schemes is taken as a compromise scheme, which has been labeled in Fig. 4.5. According to different confidence degree, different switching mechanisms have been turned on, the on/off state of all power generators and energy storage units in 24 h are listed in Table 4.4. It can be seen that Unit 1 and Unit 2 are almost turned off in the whole day, energy storage 2 is always turned off, the controllable load is

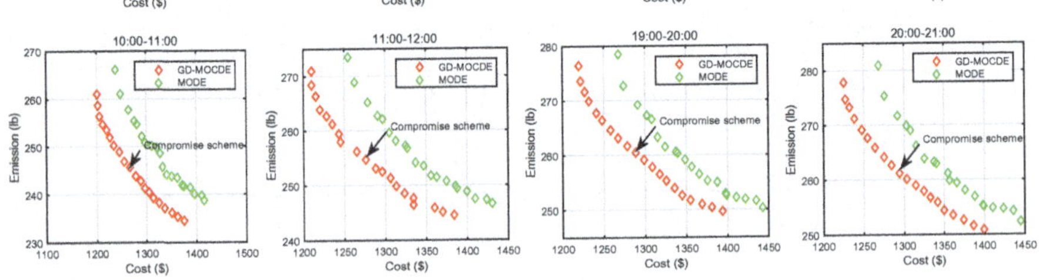

Fig. 4.5 Pareto optimal schemes by GD-MOCDE and MODE in some typical periods

Table 4.4 The on/off state of power generators and system load

Period	Unit1	Unit2	Unit3	Unit4	Unit5	Unit6	Unit7	Unit8	Unit9	Unit10	Storage1	Storage2	Storage3	Storage4	Load
1	Off	Off	On	On	On	On	On	On	On	On	On	Off	Off	Off	On
2	Off	Off	On	On	On	On	On	On	On	On	On	Off	Off	On	On
3	Off	Off	On	On	On	On	On	On	On	On	Off	Off	Off	On	On
4	Off	Off	Off	On	On	On	On	On	On	On	Off	Off	Off	Off	On
5	Off	Off	On	On	On	On	On	On	On	On	Off	Off	On	On	On
6	Off	Off	On	On	On	On	On	On	On	On	Off	Off	Off	Off	On
7	Off	Off	On	On	On	On	On	On	On	On	On	Off	Off	Off	On
8	Off	Off	On	On	On	On	On	Off	On	On	On	Off	Off	Off	On
9	Off	Off	On	On	On	On	On	Off	On	On	Off	Off	Off	Off	On
10	Off	Off	On	On	On	On	On	Off	On	On	Off	Off	Off	Off	On
11	On	Off	On	On	On	On	On	Off	On	On	On	Off	Off	Off	On
12	Off	Off	On	On	On	On	On	Off	On	On	On	Off	On	Off	On
13	Off	Off	On	On	On	On	On	Off	On	On	On	Off	On	Off	On
14	Off	Off	On	On	On	On	On	On	On	On	On	Off	On	Off	On
15	Off	Off	On	On	On	On	On	On	On	On	On	Off	On	Off	On
16	Off	Off	On	On	On	On	On	On	On	On	On	Off	On	Off	On
17	Off	Off	On	On	On	On	On	On	On	On	Off	Off	Off	Off	On
18	Off	Off	On	On	On	On	On	On	On	On	Off	Off	Off	Off	On
19	Off	Off	On	On	On	On	On	On	On	On	Off	Off	Off	Off	On
20	Off	Off	On	On	On	On	On	On	On	On	Off	Off	Off	Off	On
21	Off	Off	On	On	On	On	On	On	On	On	Off	Off	Off	Off	On
22	Off	Off	On	On	On	On	On	On	On	On	On	Off	On	Off	On
23	Off	Off	On	On	On	On	On	On	On	On	Off	Off	On	Off	On
24	On	On	On	On	On	On	On	On	On	On	Off	Off	Off	Off	On

Table 4.5 The economic cost and emission of some typical periods

Periods	RO		Two stage approach	
	Cost	Emission	Cost	Emission
00:00–01:00	1780.168	223.142	1535.836	226.8745
01:00–02:00	1776.541	224.348	1526.278	224.077
06:00–07:00	1796.878	237.766	1651.548	235.0249
07:00–08:00	1903.875	239.416	1795.792	240.6667
10:00–11:00	2026.398	248.735	1966.264	245.9567
11:00–12:00	2036.938	254.179	1975.633	254.7754
19:00–20:00	2047.687	259.274	1987.835	260.6285
20:00–21:00	2054.649	261.367	1993.399	261.2676
Total	45733.04	5756.712	43130.62	5766.276

not turned off, and thermal power is still taken as the main power source. Further analysis is taken on those typical periods, the obtained economic cost and emission rate are presented in Table 4.5, in which the proposed two stage approach achieves better results than RO. All the evaluation indexes are presented in Table 4.6, fuel cost, on/off cost of thermal units, charging/discharging cost, cut-off load cost, total cost, total emission, average confidence degree and average transmission loss are listed, it can be revealed that the proposed two stage approach has better results than RO. Since energy storage is seldom used, the switching cost can be expensive. In those typical periods, the outputs of all power generators and energy storage units are shown in Fig. 4.6, where U1, U2,...,U10 represent Unit 1, Unit 2,..., Unit 10, S1, S2, S3 and S4 represent energy storage 1, energy storage 2, energy storage 3 and energy storage 4. With above analysis, optimal operation with supply-security identification is a complicated problem. The system model can be uncertain, multi-objective and complex-coupled, hence this chapter proposes a two stage optimization

Table 4.6 The comparison of total economic cost ($), emission rate (lb) and transmission loss (MW) with RO

	RO	Two stage approach
Fuel cost	31623.432	30134.26
On/off cost of thermal unit	5571.96	4812.65
Charging/discharging cost	8537.652	8183.712
Cut-off load cost	0	0
Total cost	45733.044	43130.622
Total emission	5756.712	5766.276
Average confidence degree	0.852	0.915
Average transmission loss	78	66

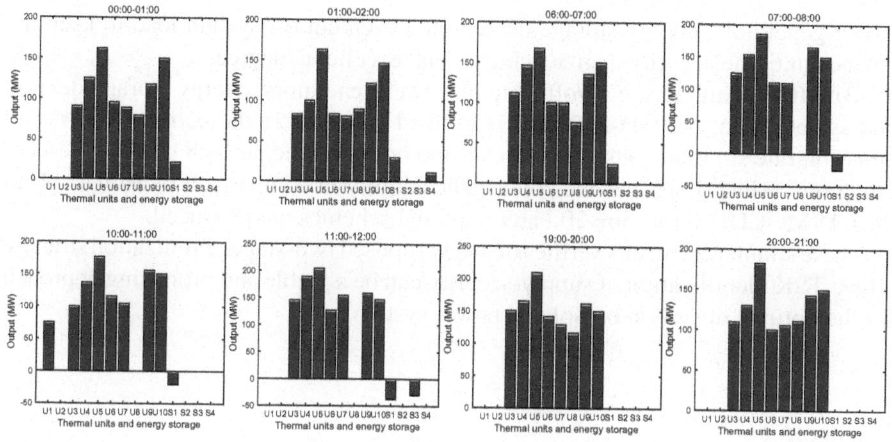

Fig. 4.6 The output of thermal units and energy storage in some typical periods

approach with supply-security identification. The TSK fuzzy approach can properly identify the relationship between uncertainty parameters and supply-security, and a two-stage optimization strategy can switch on/off power generators to improve the security degree according to three levels of confidence degree, then optimizes the economic dispatch problem with the known on/off state of the power generators and energy storage units with the GD-MOCDE algorithm, those obtained results can reveal that the proposed method can be an effective way for solving isolated power system problem.

4.5 Conclusion

The uncertainty and complexity of the optimal operation in isolated power system pose a great challenge optimization. This chapter proposes a two-stage optimization strategy with a TSK fuzzy identification of supply-security. According to the obtained results, the merits can be concluded as follows:

(1) Supply-security can be considered as an important issue in isolated power system, and it is mainly affected by the uncertainty of power generation and system load. For better dealing with uncertainty problem and decreasing potential risk, this chapter proposes a TSK fuzzy model for identifying the relationship between uncertainty parameters and confidence degree of supply security with TSK fuzzy model under RBF neural network.

(2) Due to the complexity coupled characteristics of the system model, a two-stage optimization strategy is proposed to optimize the operation model. Combined with different confidence degree of supply security, a switching mechanism is made before economic dispatch to ensure the security degree. It adjusts the on/off state of

power generators, energy storage devices and even cut off system load to keep the balance until the security degree achieves an "excellent" degree.

(3) After obtaining the on/off state of power generators, energy storage devices and system load, the GD-MOCDE is utilized to optimize the economic cost and emission rate simultaneously. It searches the optimal scheme with gradient descent directions, which improves the convergence ability in comparison to MODE. With the GD-MOCDE algorithm, 20 Pareto optimal schemes are produced.

These simulation results verifies that the proposed two-stage optimization strategy with a TSK identification of supply-security can be a viable and promising approach for the optimal operation of isolated power system.

References

1. C. Juang, P. Chang, Designing fuzzy-rule-based systems using continuous ant-colony optimization. IEEE Trans. Fuzzy Syst. **18**(1), 138–149 (2010)
2. X. Meng, P. Rozycki, J. Qiao, B.M. Wilamowski, Nonlinear system modeling using RBF networks for industrial application. IEEE Trans. Ind. Inform. **14**(3), 931–940 (2018)
3. C. Peng, P. Xie, L. Pan, R. Yu, Flexible robust optimization dispatch for hybrid wind/photovoltaic/hydro/thermal power system. IEEE Trans. Smart Grid **7**(2), 751–762 (2016)
4. H. Qin, J. Zhou, Y. Lu, Y. Wang, Y. Zhang, Multi-objective differential evolution with adaptive cauchy mutation for short-term multi-objective optimal hydro-thermal scheduling. Energy Convers. Manag. **51**(4), 788–794 (2010)
5. H. Zhang, D. Yue, X. Xie, Distributed model predictive control for hybrid energy resource system with large-scale decomposition coordination approach. IEEE Access **4**, 9332–9344 (2016)
6. J. Aghaei, T. Niknam, R. Azizipanah-Abarghooee, J.M. Arroyo, Scenario-based dynamic economic emission dispatch considering load and wind power uncertainties. Int. J. Electr. Power Energy Syst. **47**, 351–367 (2013)
7. W. Qian, A. Li, Adaptive differential evolution algorithm for multiobjective optimization problems. Appl. Math. Comput. **201**(1–2), 431–440 (2008)

Part II
Multi-objective Optimization for Optimal Operation of Hybrid Energy System

Chapter 5
Probabilistic PBI Approach for Risk-Based Optimal Operation of Hybrid Energy Systems

The stochastic nature of intermittent energy resources has brought significant challenges to the optimal operation of the hybrid energy systems. This chapter proposes a probabilistic multiobjective evolutionary algorithm based on decomposition (MOEA/D) method with two-step risk-based decision-making strategy to tackle this problem. A scenario-based technique is first utilized to generate a stochastic model of the hybrid energy system. Those scenarios divide the feasible domain into several regions. Then, based on the MOEA/D framework, a probabilistic penalty-based boundary intersection (PBI) with gradient descent differential evolution (GDDE) algorithm is proposed to search the optimal scheme from these regions under different uncertainty budgets. To ensure reliable and low risk operation of the hybrid energy system, the Markov inequality is employed to deduce a proper interval of the uncertainty budget. Further, a fuzzy grid technique is proposed to choose the best scheme for real-world applications. The experimental results confirm that the probabilistic adjustable parameters can properly control the uncertainty budget and lower the risk probability. Further, it is also shown that the proposed MOEA/D-GDDE can significantly enhance the optimization efficiency.

5.1 Problem Formulation of the Stochastic Hybrid Energy System

5.1.1 *Intermittent Power Generation with Uncertainty Budget*

To properly handle the uncertainty issue of intermittent energy resources, the intermittent power output P_{Ijt} can be described as follows with adjustable intervals [1]

D. Yue et al., *Cooperative Optimal Control of Hybrid Energy Systems*,
https://doi.org/10.1007/978-981-33-6722-7_5

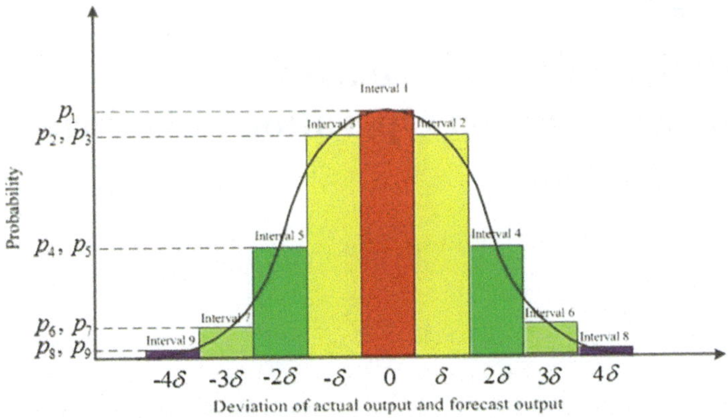

Fig. 5.1 Division of output deviations between actual output and forecasted output

$$\begin{cases} P_{Ijt} \in \left[\bar{P}_{Ijt} + \gamma_{Ijt}\tilde{P}_{Ijt}^{\min}, \bar{P}_{Ijt} + \gamma_{Ijt}\tilde{P}_{Ijt}^{\max}\right] \\ \gamma_{Ijt} \in [0, 1] \end{cases} \quad (5.1)$$

where P_{Ijt} is the forecasted output of the intermittent power, $\tilde{P}_{Ijt}^{\min}$ and $\tilde{P}_{Ijt}^{\max}$ are the lower and upper limits of deviation, and γ_{Ijt} is the adjustable parameter. Since the power generation forecasting of intermittent energy resources is described with probabilistic intervals, the probability of each obtained intervals is also taken into consideration. To properly analyze the uncertainty budget, the interval in (5.1) can be divided into several levels with different adjustable parameters γ_{Ijt}, which satisfies $\gamma_{Ijt} \in \{0, 1/4, 1/2, 3/4, 1\}$ and $\tilde{P}_{wj,t}^{\min} = -\tilde{P}_{wj,t}^{\max}$. The deviation of actual output and forecasted output and its probability is illustrated in Fig. 5.1.

The probability of each interval can be calculated as follows:

$$\begin{cases} p_1 = \text{Prob}\left(\bar{P}_{Ijt} - \delta/2 \le P_{Ijt} \le \bar{P}_{Ijt} + \delta/2\right) \\ p_2 = \left(\text{Prob}\left(\bar{P}_{Ijt} - 3\delta/2 \le P_{Ijt} \le \bar{P}_{Ijt} + 3\delta/2\right) - p_1\right)/2 \\ p_4 = \left(\text{Prob}\left(\bar{P}_{Ijt} - 5\delta/2 \le P_{Ijt} \le \bar{P}_{Ijt} + 5\delta/2\right)\right. \\ \quad \left. -2p_2 - p_1\right)/2 \\ p_6 = \left(\text{Prob}\left(\bar{P}_{Ijt} - 7\delta/2 \le P_{Ijt} \le \bar{P}_{Ijt} + 7\delta/2\right)\right. \\ \quad \left. -2p_4 - 2p_2 - p_1\right)/2 \\ p_8 = \left(\text{Prob}\left(\bar{P}_{Ijt} - 9\delta/2 \le P_{Ijt} \le \bar{P}_{Ijt} + 9\delta/2\right)\right. \\ \quad \left. -2p_6 - 2p_4 - 2p_2 - p_1\right)/2 \end{cases} \quad (5.2)$$

where δ represents the deviation unit, Prob(•) is the probability of the interval, $p2 = p3$, $p4 = p5$, $p6 = p7$, and $p8 = p9$. The uncertainty budget Δ_t is utilized to control the uncertainty degree of intermittent energy resources, which is allocated for intermittent power generation as [1]:

$$\sum_{j=1}^{N_I} \gamma_{Ijt} \leq \Delta_t \tag{5.3}$$

where N_I is the number of intermittent energy resources, and the uncertainty budget Δ_t is in the range $[0, N_I]$. It can be satisfied with adjusting those parameters, which determines the amplitude of output disturbance in each intermittent energy resource. If the adjustable parameters are continuous, the probability of (5.3) can be formulated as

$$P\left(\sum_{j=1}^{N_I} \gamma_{Ijt} \leq \Delta_t\right) = \iint f\left(\gamma_{I1t}, \gamma_{I2t}, \ldots, \gamma_{IN_It}\right) \sum_{j=1}^{N_I} \gamma_{Ijt} \leq \Delta_t \; d\gamma_{I1t} d\gamma_{I2t} \cdots d\gamma_{IN_It} \tag{5.4}$$

where f(•) represents probability density function (PDF) of the intermittent power generation. The uncertainty budget can be adjusted to control the potential risk caused by the power generation uncertainty, and different combinations of adjustable parameters γ_{Ijt} can achieve certain uncertainty budget.

5.1.2 Problem Formulation

A hybrid energy system may consist of energy storage (ES), thermal power, and intermittent power (mainly wind power and photovoltaic power), and all energy resources cooperate together to achieve the minimum economic cost and pollutant emissions. The economic cost is mainly caused by the operation cost of ES and fuel cost of thermal power generations. Since scenario-based approach can improve dispatch performance while guaranteeing a quantifiable risk level [2], which can be more suitable for optimal operation especially with considering potential risk, scenario-based approach is utilized instead of the Monte Carlo method. On the basis of generated scenarios, the economic cost can be expressed as follows:

$$\begin{cases} \min F_1 = \sum_{s \in N_s} \Pr(s) \left(f_{\mathrm{ES}}(s) + f_{\mathrm{The}}^{(1)}(s)\right) \\ f_{\mathrm{ES}} = \sum_{t=1}^{T} \sum_{l \in N_l} c_{\mathrm{ops},l} P_{l,t}^{B} \\ f_{\mathrm{The}}^{(1)}(s) = \sum_{t=1}^{T} \sum_{i \in N_c} \left[a_i + b_i P_{\mathrm{cits}} + c_i P_{\mathrm{cits}}^2 \right. \\ \left. + \left|d_i \sin\left(e_i \left(P_{ci,\min} - P_{\mathrm{cits}}\right)\right)\right|\right] \end{cases} \tag{5.5}$$

where N_s represents the total scenario number, $Pr(s)$ is the probability of scenario s, T is the length of the operation period, Nc is the number of thermal units, Nl is the number of $ES, a_i, b_i, c_i, d_i, and e_i$ are the coefficients of fuel cost for thermal power generation, P_{cits} and $P_{ci,min}$ are the output and minimum output of thermal unit, $c_{ops,l}$ is the cost efficient of the lth ES, $P_{l,t}^{B}$ denotes the charging or discharging output of ES, respectively. Further, the pollutant emissions from thermal units should be minimized. Similarly, pollutant emission can be formulated as

$$\begin{cases} \min F_2 = \sum_{s\in N_s} \Pr(s) f_{\text{The}}^{(2)}(s) \\ f_{\text{The}}^{(2)}(s) = \sum_{t=1}^{T} \sum_{i\in N_c} \left(\alpha_{1i} + \alpha_{2i} P_{\text{cits}} + \alpha_{3i} P_{\text{cits}}^2\right. \\ \left. +\alpha_{4i} \exp\left(\alpha_{5i} P_{\text{cits}}\right)\right) \end{cases} \tag{5.6}$$

where $\alpha_{1i}, \alpha_{2i}, \alpha_{3i}, \alpha_{4i}, \; and \; \alpha_{5i}$ are the coefficients of emission rate for thermal power generation.

5.1.3 Constraints

(1) System Load Balance: The balance between power generation load demand must be satisfied

$$\sum_{i\in N_c} P_{\text{cits}} + \sum_{j\in N_w} P_{\text{wjts}} + \sum_{k\in N_p} P_{pkts} + \sum_{l\in N_b} P_{l,t}^{B} = P_{D,t} \tag{5.7}$$

where P_{wjts} and P_{pkts} describe power output of wind power and solar power, $P_{l,t}^{B}$ denotes charge or discharge output of battery ES, $N_w, N_p, \; and \; N_b$ are the index sets of wind farms, PV arrays and batteries, $P_{D,t}$ is system load demand and transmission loss.

(2) Power Generation Limits: The output of each power generator must be within feasible interval

$$\begin{cases} P_{ci,\min} \le P_{\text{cits}} \le P_{ci,\max}, & i = 1, 2, \ldots, N_c \\ \text{Prob}\left(P_{wj,\min} \le P_{wjts} \le P_{wj,\max}\right) = \rho_w, & j = 1, 2, \ldots, N_w \\ \text{Prob}\left(P_{pk,\min} \le P_{pkts} \le P_{pk,\max}\right) = \rho_p, & k = 1, 2, \ldots, N_p \end{cases} \tag{5.8}$$

where $P_{ci,max}$ is the maximum output of thermal power, $P_{wj,min}$ and $P_{ci,max}$ are the minimum and maximum output of wind power, $P_{pk,min}$ and $P_{pk,min}$ are the minimum and maximum output of solar power, ρ_w and ρ_p are the required probability of wind power and PV power generation.

(3) Ramp Rate Limits: During the power generation process, power output can be adjusted within limited condition due to the power generation capacity

$$\begin{cases} \mathrm{DR}_{ci} \leq P_{\mathrm{cits}} - P_{ci,t-1,s} \leq \mathrm{UR}_{ci}, & i = 1, 2, \ldots, N_c \\ & t = 1, 2, \ldots, T \\ \mathrm{DR}_{wj} \leq P_{wjts} - P_{wj,t-1,s} \leq \mathrm{UR}_{wj}, & j = 1, 2, \ldots, N_w \\ & t = 1, 2, \ldots, T \\ \mathrm{DR}_{pk} \leq P_{pkts} - P_{pk,t-1,s} \leq \mathrm{UR}_{pk}, & k = 1, 2, \ldots, N_p \\ & t = 1, 2, \ldots, T \end{cases} \tag{5.9}$$

where DR_{ci} and UR_{ci} are the down and up ramp rate limits of thermal power generator, DR_{wi} and UR_{wi} are the down and up ramp rate limits of wind power, DR_{pk} and UR_{pk} are the down and up ramp rate limits of solar power.

(4) Wind Speed and Its PDF: The wind power generation is mainly related to the wind speed, suppose that the wind speed follows the Weibull distribution function, the distribution function of wind power can also be deduced [3]

$$P_{wjts} = \begin{cases} 0, & v_j < v_{j,\,\mathrm{in}} \text{ or } v_j \geq v_{j,\,\mathrm{out}} \\ P_{wj,\max} * \frac{v_j - v_{j,\,\mathrm{in}}}{v_{j,\,\mathrm{rate}} - v_{j,\,\mathrm{in}}}, & v_{j,\,\mathrm{in}} \leq v_j < v_{j,\,\mathrm{rate}} \\ P_{wj,\max}, & v_{j,\,\mathrm{rate}} \leq v_j < v_{j,\,\mathrm{out}} \end{cases} \tag{5.10}$$

$$\begin{aligned} F\left(P_{\mathrm{wjts}}\right) = & 1 - \exp\left\{-\left[\left(1 + \frac{v_{\mathrm{rate}} - v_{\mathrm{in}}}{v_{\mathrm{in}} P_{\mathrm{rate}}} P_{\mathrm{wits}}\right) \frac{v_{\mathrm{in}}}{c}\right]^k\right\} \\ & + \exp\left[-\left(v_{\mathrm{out}}/c\right)^k\right], 0 \leq P_{\mathrm{wits}} < P_{\mathrm{rate}} \end{aligned} \tag{5.11}$$

where v_j represents wind speed, $v_{j,\,\mathrm{in}}$, $v_{j,\,\mathrm{rate}}$, and $v_{j,\,\mathrm{out}}$ denote the cut-in, rated, and cut-out wind speeds, respectively. k and c are scaling parameters.

(5) PDF of Photovoltaic Power: Since photovoltaic power can also be taken as intermittent energy resource, it can be described in probabilistic forms. Generally, the PDF of photovoltaic power output η_j can be presented with beta distribution as follows:

$$f\left(\eta_j\right) = \frac{1}{B(\alpha, \beta)} \eta_j^{\alpha-1} \left(1 - \eta_j\right)^{\beta-1}, 0 \leq \eta_j \leq 1 \tag{5.12}$$

where $B(\alpha, \beta)$ represents beta function with two parameters.

(6) Battery Energy Storage System: Battery energy storage system (BESS) is also taken into consideration to complement the intermittent energy resources, and its energy management needs to satisfy

$$\begin{cases} V_{l,t+1}^B = V_{l,t}^B + \eta_l P_{l,t}^B * \Delta t \\ V_{l,\min}^B \le V_{l,t}^B \le V_{l,\max}^B \\ P_{l,t}^B = P_{l,t}^{\text{dis}}, \text{ if } P_{l,t}^B \ge 0 \\ P_{l,t}^B = -P_{l,t}^{\text{cha}}, \text{ if } P_{l,t}^B < 0 \\ 0 \le P_{l,t}^{\text{dis}} \le P_{l,\max}^{\text{dis}} \\ 0 \le P_{l,t}^{\text{cha}} \le P_{l,\max}^{\text{cha}} \end{cases} \tag{5.13}$$

where ${}_{I,t}^{dis}P$ and $P_{I,t}^{cha}$ are the output of discharging and charging state, ${}_{I,max}^{dis}P$ and $P_{I,max}^{cha}$ are the maximum discharging and charging output in the lth battery at the tth time period. The state of charge (SOC) is also taken into consideration, $V_{I,t}^B$ is the storage of the lth battery at the tth time period, $V_{I,min}^B$ and $V_{I,max}^B$ are the minimum and maximum storage of the lth battery, $\eta_l \in (0, 1]$ represents the efficiency of SOC.

(7) Minimum ON/OFF Time Constraints:

$$\begin{cases} \left[T_{i,t-1}^R - T_{i,\min}^R\right]\left[\tau_{i,t-1} - \tau_{i,t}\right] \ge 0 \\ \left[T_{i,t-1}^S - T_{i,\min}^S\right]\left[\tau_{i,t} - \tau_{i,t-1}\right] \ge 0 \end{cases} \tag{5.14}$$

where $T_{I,t-1}^R 1$, $T_{i,t-1}^S$ denote the continuous online and offline time of the unit until period $t-1$, $\tau_{i,t}$ is a binary decision variable for online state of thermal unit at period t.

(8) Spinning Reserve Constraint:

$$\begin{aligned} &\textstyle\sum_{i \in N_c} \left(P_{ci,\max} - P_{ci,t}\right) + \sum_{l \in N_l} \left(P_{l,\max}^{\text{dis}} - P_{l,t}^{\text{dis}}\right) \\ &\ge \textstyle\sum_{j \in N_I} \gamma_{Ijt} \left(\tilde{P}_{Ijt}^{\max} - \tilde{P}_{Ijt}^{\min}\right) \end{aligned} \tag{5.15}$$

Considering the stability of a hybrid energy resource system, it requires more additional power to prevent the disturbance caused by intermittent power uncertainty.

(9) Uncertainty Budget: As it is presented in (5.3), the uncertainty budget Δ_t is taken as rough constraint limit, the summation of adjustable parameters cannot exceed this limit. With consideration of required reliability of the hybrid energy system, it needs decision-making strategy to deduce the minimum deviation with utopia uncertainty budget.

5.2 MOEA/D with Penalty-Based Boundary Intersection Approach

MOEA/D, originally proposed by Zhang and Li [4], mainly decomposes an MOP into several scalar optimization subproblems, and each subproblem coordinates its neighborhoods to seek the optimal solution. Generally, MOEA/D can be regarded as an improved framework of cMODE proposed in [5]. Generally, an MOP can be stated as follows:

$$\begin{cases} \min\ F(x) = (f_1(x), f_2(x), \ldots, f_m(x))^T \\ \text{s.t. } h_j(x) \le 0, x \in R^n, j = 1, 2, \ldots, J \end{cases} \quad (5.16)$$

Since $H_j(x)$ are continuous functions, (5.16) can be considered as a continuous MOP. The decomposition approach involves weight vector $\lambda^i = \left(\lambda_1^i, \lambda_2^i, \ldots, \lambda_m^i\right)^T$ for the ith subproblem $\left(\sum_{j=1}^{m} \lambda_j^i = 1, \lambda_j^i \ge 0\right)$. Suppose that $z^* = \left(z_1^*, z_2^*, \ldots, z_m^*\right)^T$ is a utopian point, then the PBI approach can decompose (5.10) into several subproblems as follows:

$$\begin{aligned} \min g^{pbi}\left(x \mid \lambda^i, z^*\right) &= d_1^i + \beta d_2^i \\ d_1^i &= \left\|(F(x) - z^*)^T \lambda^i\right\| / \left\|\lambda^i\right\| \\ d_2^i &= \left\|F(x) - z^* - d_1^i \lambda^i\right\| \\ \text{s.t. } h_j(x) \le 0, x \in R^n&, j = 1, 2, \ldots, J \end{aligned} \quad (5.17)$$

where d_1^i represents the distance between z^* and projection of F(x) in the ith subproblem, β is the preset penalty parameter, and d_2^i denotes the distance between F(x) and direction line in the ith subproblem. In comparison with the Tchebycheff approach, the PBI approach has two advantages: (1) with the same weight vectors in more than two objective problem, the optimal solutions by PBI has more uniform distribution than those obtained by Tchebycheff approach and (2) if optimal solution x dominates another solution y, it is possible that $g^{pbi}\left(x \mid \lambda^i, z^*\right) = g^{pbi}\left(y \mid \lambda^i, z^*\right)$ when x dominates y, the attribute is however, rare for other boundary intersection aggregation functions, it can properly improve the diversity of Pareto optimal front [4].

Since the feasible domain is divided into several levels according to different uncertainty budgets with probability distribution, it searches optimal solutions as well as considers the probability of obtained scheme. The PBI method can be extended to solve stochastic optimization (SO) problem with probabilistic feasible region, it exists optimal solution in each feasible region. With consideration of the probabilistic distribution, PBI for probabilistic optimization problem can be expressed as

$$\begin{cases} \min g^{pbi}\left(x \mid \lambda^i, z^*\right) = \sum_{s=1}^{S} \Pr\left(\Delta_t^{(s)}\right)\left(d_1^{i(s)} + \beta d_2^{i(s)}\right) \\ +\mu\|\Theta\|_2 \\ d_1^{i(s)} = \left(F\left(x^{(s)}\right) - z^*\right)^T \lambda^i / \left\|\lambda^i\right\| \\ d_2^{i(s)} = \left\|F\left(x^{(s)}\right) - z^* - \left(d_1^{i(s)} / \left\|\lambda^i\right\|\right)\lambda^i\right\| \\ x^{(s)} \in \Omega, s = 1, 2, \ldots, S \end{cases} \quad (5.18)$$

where s is scenario index, and S is the total scenario number, $\Delta_t^{(s)}$ is the uncertainty budget of scenario s at the tth time period, $d_1^{(s)}$ denotes the distance between projection point and z^* and $d_2^{(s)}$ denotes the distance between initial point and projection point, μ is discount factor, it can be considered as a regularization parameter, which mainly controls the scale of scenario vector. $x^{(s)}$ is simulated value of scenario s. Since scenarios can increase computational complexity, the number of scenarios can-

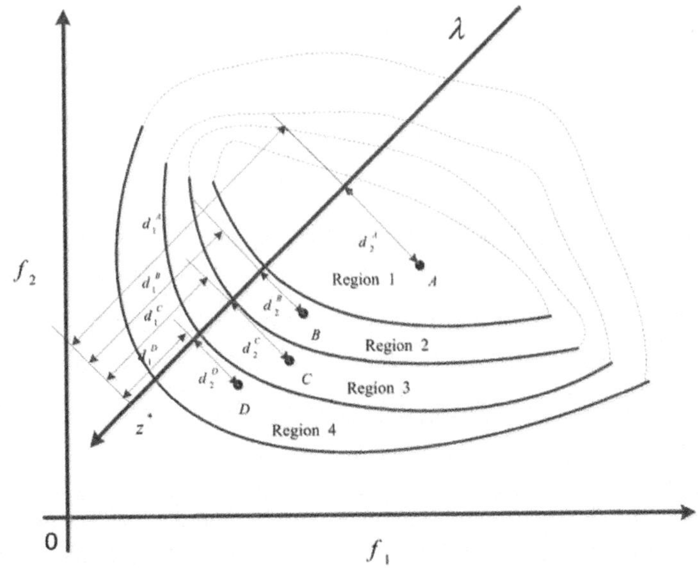

Fig. 5.2 Probabilistic PBI method with different scenarios

not exceed certain degree, regularization operator $\|\Theta\|_2$ can be employed to control the scale of scenarios, and where $\Theta = \left[\Pr\left(\Delta_t^{(1)}\right), \Pr\left(\Delta_t^{(2)}\right), \ldots, \Pr\left(\Delta_t^{(S)}\right)\right]^T$.

In probabilistic PBI method, those generated scenarios are scattered into the $f_1 - f_2$ space, which has been classified into different regions. Each region has several scenarios with certain probabilistic characteristics, to search the optimal solution of the stochastic problem, expected value replaces the objective in (5.17), as is shown in Fig. 5.2. The probabilistic characteristics can be obtained with PDF of wind power and PV power generation, which can be expressed with uncertainty budget. With above MOEA/D framework, this chapter utilizes differential evolution (DE) to solve above scalar subproblems with different weights due to its simple yet powerful search ability in comparison to other heuristic optimization algorithms, DE procedure is taken with mutation operator of DE/rand/1/bin, which is generally demonstrated as

$$\begin{aligned} V_{r,G+1} &= X_{r,G} + \gamma * \left[\left(X_{r1,G} - X_{r2,G}\right) + \left(X_{r3,G} - X_{r4,G}\right)\right] \\ & r1 \neq r2 \neq r3 \neq r4 \neq r \end{aligned} \tag{5.19}$$

where $V_{r,G}$ is the parameter vector for the $G + 1$th generation, γ is the mutation parameter, which is range in [0, 2]. $X_{r1,G}$, $X_{r2,G}$, $X_{r3,G}$, and $X_{r4,G}$ are randomly selected individual in the archive set, which mainly stores the nondominated solutions in each generation. For improving the search ability of DE, gradient decent-based DE procedure is taken as it is shown in [6]. The improved mutation operator can be improved as follows:

$$X_{G+1}^{j} = X_{r,G}^{j} + \gamma_1^j \left(X_{r1,G} - X_{r2,G}\right) + \gamma_2^j \left(X_{r3,G} - X_{r4,G}\right) \\ r1 \neq r2 \neq r3 \neq r4 \neq r \tag{5.20}$$

$$\begin{cases} \gamma_1^j = \dfrac{-\eta_G \lambda_1 * \operatorname{sgn}\left(f_1\left(X_{r1,G}\right) - f_1\left(X_{r2,G}\right)\right)}{\left(X_{r1,G}^j - X_{r2,G}^j\right)^2 \sqrt{\sum_{j=1}^{n} \frac{1}{\left(x_{r1,G}^j - X_{r,C,G}^j\right)^2}}} \\ \gamma_2^j = \dfrac{-\eta_G \lambda_2 * \operatorname{sgn}\left(f_2\left(X_{r3,G}\right) - f_2\left(X_{r4,G}\right)\right)}{\left(X_{r3,G}^j - X_{r4,G}^j\right)^2 \sqrt{\sum_{j=1}^{n} \frac{1}{\left(x_{r3,G}^j - X_{r4,G}^j\right)^2}}} \\ \eta_G = \eta_0 \left[\left(g_{\max} - G + 1\right) / g_{\max}\right]^p \end{cases} \tag{5.21}$$

where η_G is the scaling parameter at the Gth generation, and η_0 is the initial scaling parameter, G_{max} is the maximum generation, λ_i is weighted parameter in interval [0, 1], γ_1^i and γ_2^i are the mutation parameters. The gradient decent method searches the optimal solution along the shortest direction, which speeds up the search ability of DE. The weights of those subsystems can also be properly set, which can be seen in [7].

5.3 Fuzzy Decision-Making Method for the Probabilistic Optimal Problem

5.3.1 *Fuzzy Decision-Making Approach*

Due to the uncertainty of intermittent power introduced into the hybrid energy system, each optimal scheme from nondominated solutions contains different risk levels after multiobjective optimization. Hence, the best scheme should be a tradeoff among different objectives, and at the same time it also has the lowest risk level. Once, those Pareto-optimal solutions are obtained with above optimization method, it can be assumed that probability of the optimal solutions $X^* = \left[X_1^*, X_2^*, \ldots, X_{N_A}^*\right]$ can be expressed as follows:

$$\operatorname{Prob}\left(X_i^*\right) = \prod_{s=1}^{N_s} \operatorname{Pr}^{(i)}(s), \; (i = 1, 2, \ldots, N_A) \tag{5.22}$$

where $Pr^{(i)}(s)$ is the probability of the sth scenario in the ith optimal solution. With consideration of stochastic problem and multiple objectives, some remarks can be defined as follows.

Remark 5.1 Suppose the reference uncertainty budget at the tth time period $\Delta_t^{(\alpha)*}$, the uncertainty deviation between reference uncertainty budget and the resultant uncertainty budget can be defined as the uncertainty metric

$$\mathrm{Unc}(\eta) = \left| \sum_{t=1}^{T} \Delta_t^{(\eta)} - \sum_{t=1}^{T} \Delta_t^{(\eta)*} \right| \tag{5.23}$$

where $\Delta_t^{(\alpha)}$ is the Δ_t of the ηth situation, η is the number of the Pareto-optimal set. When uncertainty metric is large, it means that those uncertainty budget settings are not close to realworld application. Hence, the best solution should have less deviation to ensure the practicality. Combined with the evaluation value, the decision-making method can be used to choose the best optimal solution for applications. First, best uncertainty set η^* should be selected with the smallest uncertainty deviation, it can be obtained by

$$\eta^* = \arg\min_{\eta \in \Omega_\eta} \left| \sum_{t=1}^{T} \Delta_t^{(\eta)} - \sum_{t=1}^{T} \Delta_t^{(\eta)*} \right| \tag{5.24}$$

where η is the index set of uncertainty set. Once the optimal uncertainty set $\Delta_t^{(\eta*)}$ is obtained, the best Pareto-optimal front can be selected with above uncertainty deviation.

Remark 5.2 To properly evaluate each optimal solution in the archive set, the $f_1 - f_2$ space is first divided into several small grids, and width on f_1 direction of each box is $\delta_1 = \left(f_{1,\max} - f_{1,\min}\right)/N_A$, and the length on f2 direction of each box is $\delta_2 = \left(f_{2,\max} - f_{2,\min}\right)/N_A$. For two given optimal solutions X_i^*, $X_j^*(j \neq i)$, there exists an evaluation index $Eval_{ij}(m)(m = 1, 2, \ldots, M)$ and K_1, $K_2 \in Z^+$, which represent the location of objective value in $f_1 - f_2$ coordinate axis. Here $M = 2$ for simplicity, for $m = 1$, if $K_1\delta_1 \leq \left| f_1\left(X_i^*\right) - f_1\left(X_j^*\right) \right| \leq (K_1 + 1)\,\delta_1$, $\mathrm{Eval}_{ij}(m) = K_1\delta_1$; for $m = 2$, if $K_2\delta_2 \leq \mid f_2\left(X_i^*\right) - f_2\left(X_j^*\right) \mid \leq (K_2 + 1)\,\delta_2$, then $\mathrm{Eval}_{ij}(m) = K_2\delta_2$. The evaluation value between X_i^* and X_j^* can be expressed as

$$\mathrm{Eval}_{ij} = \sum_{m=1}^{M} \mathrm{Eval}_{ij}(m) \tag{5.25}$$

With consideration of the archive set, the evaluation value of X_i^* can be obtained as follows:

$$\mathrm{Eval}_i = \sum_{j=1, j \neq i}^{N_A} \mathrm{Eval}_{ij} \tag{5.26}$$

The optimal index can be obtained

$$i^* = \arg \max_{i=1,2,\ldots,N_A} (\text{Eval}_i) \tag{5.27}$$

5.3.2 *Probabilistic Risk Evaluation*

On the other side, the uncertainty budget $\varDelta_t$ also should be properly set, it mainly relates to the reliability of the hybrid energy system. The spinning reserve and system load balance can ensure the reliability of power system, since BESS can provide complimentary power, BESS can also be considered to assure the safety or reliability issue. Combine formulation (5.7) and (5.15) together can obtain

$$\begin{aligned} &\sum\nolimits_{i\in N_c} P_{ci\max} + \sum\nolimits_{l\in N_b} P_{l,\max}^{\text{dis}} + \sum\nolimits_{j\in N_w} P_{wjt} + \sum\nolimits_{k\in N_p} P_{pkt} \\ &\geq P_{D,t} + \sum\nolimits_{j\in N_I} \gamma_{Ijt}\left(\tilde{P}_{Ijt}^{\max} - \tilde{P}_{Ijt}^{\min}\right) \end{aligned} \tag{5.28}$$

According to [1], the probability of (5.28) satisfy

$$\begin{aligned} &\Pr o\left\{\sum\nolimits_{i\in N_c} P_{ci\max} + \sum\nolimits_{l\in N_b} P_{l,\max}^{\text{dis}} + \sum\nolimits_{j\in N_w} P_{wjt} + \sum\nolimits_{k\in N_p} P_{pkt} \right.\\ &\quad \left. < P_{D,t} + \sum\nolimits_{j\in N_I} \gamma_{Ijt}\left(\tilde{P}_{Ijt}^{\max} - \tilde{P}_{Ijt}^{\min}\right)\right\} \\ &\leq \Pr o\left(\sum\nolimits_{j\in N_I} w_{Ijt}\gamma_{Ijt} \geq \varDelta_t\right) \end{aligned} \tag{5.29}$$

where

$$w_{Ijt} = \begin{cases} 1, & j \in R_t^{\#} \\ \dfrac{\tilde{P}_{Ijt}^{\max}}{\min\left\{\tilde{P}_{Igt}^{\max}\right\}}, & j \in N_I \backslash R_t^{\#}, \text{ where } g \in R_t^{\#} \cup \{m\} \end{cases} \tag{5.30}$$

$R_t^{\#}$ is the set of intermittent power with extreme output, suppose that those are independent random variables and follow distribution in (5.11) and (5.12), with the Markov inequality it can obtain

$$\Pr o\left(\sum_{j\in N_I} w_{Ijt}\gamma_{Ijt} \geq \varDelta_t\right) \leq \frac{E\left(\sum\nolimits_{j\in N_I} w_{Ijt}\gamma_{Ijt}\right)}{\varDelta_t} \tag{5.31}$$

For simplicity, denote two random variables $x = \sum\nolimits_{j\in N_w} w_{Ijt}\gamma_{Ijt}$ and $y = \sum\nolimits_{j\in N_p} w_{Ijt}\gamma_{Ijt}$, then it can be deduced as follows:

$$\Pr o\left(\sum_{j\in N_I} w_{Ijt}\gamma_{Ijt} \geq \Delta_t\right) \leq (N_w - \Delta_t)\, e^{-(v_{\text{in}}/c)^k} + e^{-\Delta_t^2/2(N_p+N_w)} \tag{5.32}$$

Then, the probability of constraint limit (5.28) satisfies

$$\begin{aligned}&\Pr o\left\{\sum\nolimits_{i\in N_c} P_{ci\max} + \sum\nolimits_{l\in N_b} P_{l,\max}^{\text{dis}} + \sum\nolimits_{j\in N_w} P_{wjt} + \sum\nolimits_{k\in N_p} P_{pkt}\right.\\ &\quad\left. < P_{D,t} + \sum\nolimits_{j\in N_I} \gamma_{Ijt}\left(\tilde{P}_{Ijt}^{\max} - \tilde{P}_{Ijt}^{\min}\right)\right\}\\ &\leq (N_w - \Delta_t)\, e^{-(v_{\text{in}}/c)^k} + e^{-\Delta_t^2/2(N_p+N_w)}\end{aligned} \tag{5.33}$$

Suppose that the probability of (5.28) should be controlled at least with probability of $1 - \delta$, and then the appropriate upper bound of uncertainty budget can be calculated

$$\kappa \geq \Delta_t \geq 0 \tag{5.34}$$

Generally, Δ_t can be set as large as possible if possible risk has been properly avoided, it also means that k is the permitted maximum value of uncertainty budget, and the optimal uncertainty set $\Delta_t^{(\eta*)}$ can be deduced, which also means that potential risk can be prevented if (5.34) is properly satisfied. Since it is difficult to deduce the analytical solution, k value is deduced in simulation.

5.4 Simulations

The simulation can be implemented with following procedures.

(1) Making analysis on the probabilistic characteristics of wind and PV power generation.
(2) With consideration of potential risk, it deduces proper uncertainty budget.
(3) Optimizing the hybrid energy system with MOEA/DGDDE approach.
(4) Decision-making on those obtained optimal solutions, and produce best optimal scheme for the hybrid energy system operation.

5.4.1 *Parameters Settings and Basic Data*

The hybrid energy system consists of wind power, solar power, thermal power, and ES, it includes five thermal units, two ESs, four wind farms, and three photovoltaic fields, the data resource can be found in [3]. The uncertainty domain of intermittent power output is divided into five levels, instead of setting different uncertainty budgets at different periods, uncertainty budget at each time period can be consid-

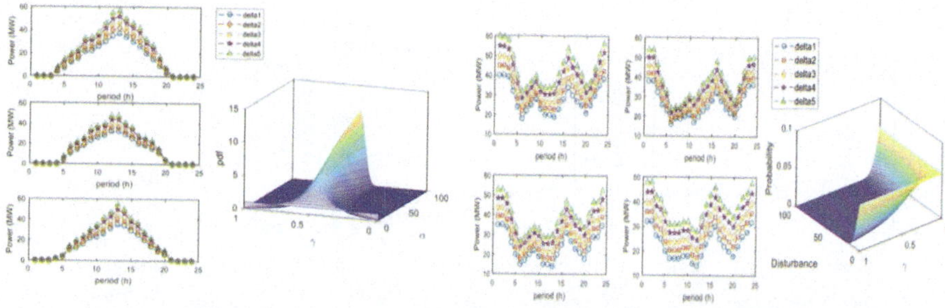

(a) (b)

Fig. 5.3 **a** PV and **b** wind power uncertainties with five levels and its PDF

ered as the same value. The PDF within uncertainty domain is illustrated in Fig. 5.3a and b respective. PV generators mainly work from 8:00 to 20:00 and its output achieves the maximum value at noon, while wind power fluctuates frequently, it mainly achieves the maximum output at 00:00 to 02:00 and 15:00 to 17:00. PV power follows Beta distribution (after normalization) and wind power follows density distribution of (5.11), where parameters are set as $\alpha = \beta = 2$, $c = v_{\text{in}} = 3\,\text{m/s}$, $v_{\text{rate}} = 13\,\text{m/s}$, $P_{\text{rate}} = 60\,\text{MW}$. Since the number of wind farms and PV cannot exceed 8, it satisfies $\left|\left[\left(\Delta_t^2\right)/\left(2\left(N_w + N_p\right)\right)\right]\right| < 1$, therefore the second term of (5.33) can be expressed by the first four terms of the Taylor series expansion, then (5.33) can be rewritten as

$$\begin{aligned} \Pr o & \left\{ \sum\nolimits_{i \in N_c} P_{ci\max} + \sum\nolimits_{l \in N_b} P_{l,\max}^{\text{dis}} + \sum\nolimits_{j \in N_w} P_{wjt} + \sum\nolimits_{k \in N_p} P_{pkt} \right. \\ & \left. < P_{D,t} + \sum\nolimits_{j \in N_l} \gamma_{Ijt} \left(\tilde{P}_{Ijt}^{\max} - \tilde{P}_{Ijt}^{\min} \right) \right\} \\ & \leq 1 + (N_w - \Delta_t)\, e^{-(v_{\text{in}}/c)^k} \quad \frac{5}{16} \frac{\Delta_t^2}{N_p + N_w} \end{aligned} \tag{5.35}$$

The parameter κ can also be calculated as

$$\begin{aligned} \kappa = & \sqrt{\frac{64}{25} N_I^2 e^{-2(v_{\text{in}}/c)k} + \frac{64}{25} N_I \left(1 + N_w e^{-(v_{\text{in}}/c)^k} - \delta\right)} \\ & - \frac{8}{5} N_I e^{-(v_{\text{in}}/c)^k} \end{aligned} \tag{5.36}$$

The parameters for population evolution can be set as follows: population size is set as 200, maximum generation size is 1000, the number of the Pareto optimal solutions is 20, the initial scaling parameter η_0 is set to 0.8, and the number of scenarios is 50, which is deduced as [8, 9]

$$\sum_{i=0}^{k} \binom{N}{i} \varepsilon^i (1 - \varepsilon)^{N-i} \leq \beta \tag{5.37}$$

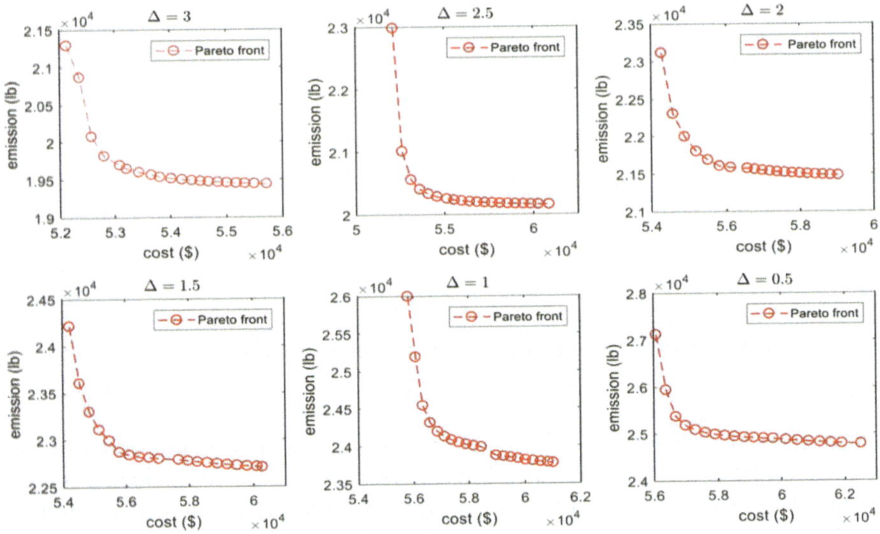

Fig. 5.4 Pareto fronts under different uncertainty budgets

where $\varepsilon \in (0, 1)$ is violation parameter (here it can be considered as scenario probability), β is confidence parameter, it is generally 10^{-6}.

5.4.2 *Results and Analysis*

Fifty scenarios are generated to simulate the stochastic process, the stochastic model of the hybrid energy system can be created, combined with MOEA/D approach, 20 Pareto-optimal schemes can be calculated with each uncertainty budget, which have been shown in Fig. 5.4. Economic cost, emission issue, and computational time are taken as metrics. In comparison to other representative MOEAs, including NSGA-II [10], MOEA/D-TPN [11], MOPSO [12], and MOHDE [13], the proposed MOEA/D-GDDE can obtain both lower cost and emission at certain time, which are listed in Table 5.1. It can be seen that the proposed MOEA/D-GDDE is superior to other alternatives on cost and emission objectives, since it integrates probabilistic analysis into optimization, computational time is merely better than a few of them. Here, three typical uncertainty budgets provide different uncertainty domain of intermittent power generations, it is also found that the results with larger uncertainty budget have lower/better economic cost and emission, it can be explained that large uncertainty budget provides large feasible domain for optimization method, and it further promotes search scale and find the global optima. Since the scheduling process of each time period can be quite similar for most of the time, uncertainty budget can be set as the same value.

Table 5.1 Comparison with other alternatives under different uncertainties

Algorithms	Uncertainty	Cost ($)	Emission (lb)	Time (s)
NSGA-II	$\Delta = 3$	54211	19823	660
	$\Delta = 2$	57229	22619	581
	$\Delta = 1$	59541	24766	543
MOEA/D-TPN	$\Delta = 3$	54358	19818	632
	$\Delta = 2$	57543	22153	558
	$\Delta = 1$	59712	24899	511
MOPSO	$\Delta = 3$	54206	19520	598
	$\Delta = 2$	57871	22763	532
	$\Delta = 1$	60012	25314	488
MOHDE	$\Delta = 3$	54206	19520	598
	$\Delta = 2$	56488	21590	512
	$\Delta = 1$	59012	24007	477
MOEA/D-GDDE	$\Delta = 3$	54006	19518	627
	$\Delta = 2$	56554	21573	553
	$\Delta = 1$	58971	23873	502

Without loss of generality, uncertainty budget can be set the same value at whole time period, three typical uncertainty budgets $\Delta = 3$, $\Delta = 2$, and $\Delta = 1$ are chosen to take further analysis of the optimization performance and scheduling process. After optimization with different uncertainty budgets, the adjustable parameters γ_{Ijt} can be obtained at each time period, which are shown in Table 5.2. Convergence process with different uncertainty budgets are illustrated in Fig. 5.5a–c, respectively, and the optimization process has a slower convergence as the uncertainty budget gets larger, it can also be found that both economic cost and emission rate achieve to converge within no more than 600 generations, it reflects that MOEA/D-GDDE can avoid premature problem to fall into local optima, and it also converges faster. Finally, those obtained optimal schemes with three typical uncertainty budgets $\Delta = 3$, $\Delta = 2$, and $\Delta = 1$ are shown in Figs. 5.6, 5.7, 5.8, 5.9, 5.10 and 5.11, where the output process of thermal units and charging and discharging process of ES are all illustrated. As shown in Figs. 5.6, 5.8, and 5.10, it can be found that the thermal unit with larger capacity bear more system load during the scheduling process, power output of five thermal units can be almost sorted with order $Unit5 > Unit4 > Unit3 > Unit2 > Unit1$.

According to the above results and analysis, it can be concluded that probabilistic PBI-based MOEA/D-GDDE method can deal with optimal operation of the hybrid energy system and deduce the According to the above results and analysis, it can be concluded that probabilistic PBI-based MOEA/D-GDDE method can deal with optimal operation of the hybrid energy system and deduce the best operation scheme by the proposed two step decision-making strategy.

Table 5.2 Optimal adjustable parameter γ_{Ijt} under different uncertainty budgets

Periods	$\Delta = 3$	$\Delta = 2$	$\Delta = 1$
00:00-00:59	{0, 0, 0}, {1, 1, 1, 0}	{0, 0, 0}, {1, 1, 0, 0}	{0, 0, 0}, {1, 0, 0, 0}
01:00-01:59	{0, 0, 0}, {1, 1, 1, 0}	{0, 0, 0}, {1, 1, 0, 0}	{0, 0, 0}, {1, 0, 0, 0}
02:00-02:59	{0, 0, 0}, {1, 1, 1, 0}	{0, 0, 0}, {1/2, 1/2, 1/2, 1/2}	{0, 0, 0}, {1/4, 1/4, 1/4, 1/4}
03:00-03:59	{0, 0, 0}, {3/4, 3/4, 3/4}	{0, 0, 0}, {1/2, 1/2, 1/2, 1/2}	{0, 0, 0}, {1/4, 1/4, 1/4, 1/4}
04:00-04:59	{0, 0, 0}, {3/4, 3/4, 3/4}	{0, 0, 0}, {1/2, 1/2, 1/2, 1/2}	{0, 0, 0}, {1/4, 1/4, 1/4, 1/4}
05:00-05:59	{1/2, 1/2, 0}, {1/2, 1/2, 1/2, 1/2}	{1/4, 1/4, 0}, {1/2, 1/2, 1/4, 1/4}	{0, 0, 0}, {1/4, 1/4, 1/4, 1/4}
06:00-06:59	{1/2, 1/2, 0}, {1/2, 1/2, 1/2, 1/2}	{1/4, 1/4, 0}, {1/2, 1/2, 1/4, 1/4}	{0, 0, 0}, {1/4, 1/4, 1/4, 1/4}
07:00-07:59	{1/2, 1/2, 0}, {1/2, 1/2, 1/2, 1/2}	{1/4, 1/4, 0}, {1/2, 1/2, 1/4, 1/4}	{0, 0, 0}, {1/4, 1/4, 1/4, 1/4}
08:00-08:59	{1/4, 1/4, 1/4}, {1/2, 1/4, 1/4, 1/4}	{1/4, 1/4, 1/4}, {1/2, 1/4, 1/4, 1/4}	{1/4, 0, 0}, {1/4, 1/4, 1/4, 0}
09:00-09:59	{1/4, 1/4, 1/4}, {1/2, 1/4, 1/4, 1/4}	{1/4, 1/4, 1/4}, {1/2, 1/4, 1/4, 1/4}	{1/4, 0, 0}, {1/4, 1/4, 1/4, 0}
10:00-10:59	{1/2, 1/2, 1/2}, {1/2, 1/2, 1/2, 0}	{1/4, 1/4, 1/4}, {1/2, 1/4, 1/4, 1/4}	{1/4, 1/4, 0}, {1/4, 1/4, 0, 0}
11:00-11:59	{1/2, 1/2, 1/2}, {1/2, 1/2, 1/2, 0}	{1/4, 1/4, 1/4}, {1/2, 1/4, 1/4, 1/4}	{1/4, 1/4, 0}, {1/4, 1/4, 0, 0}
12:00-12:59	{3/4, 3/4, 0}, {1/2, 1/2, 1/2, 0}	{1/2, 1/4, 1/4}, {1/4, 1/4, 1/4, 1/4}	{1/2, 1/4, 0}, {1/4, 0, 0, 0}
13:00-13:59	{3/4, 3/4, 0}, {1/2, 1/2, 1/2, 0}	{1/2, 1/4, 1/4}, {1/4, 1/4, 1/4, 1/4}	{1/2, 1/4, 0}, {1/4, 0, 0, 0}
14:00-14:59	{1/2, 1/2, 1/2}, {1/2, 1/2, 1/2, 0}	{1/2, 1/4, 1/4}, {1/4, 1/4, 1/4, 1/4}	{1/2, 1/4, 0}, {1/4, 0, 0, 0}
15:00-15:59	{1/2, 1/2, 1/2}, {1/2, 1/2, 1/2, 0}	{1/2, 1/4, 1/4}, {1/4, 1/4, 1/4, 1/4}	{1/2, 1/4, 0}, {1/4, 0, 0, 0}
16:00-16:59	{1/4, 1/4, 1/4}, {1/2, 1/4, 1/4, 1/4}	{1/4, 1/4, 0}, {1/2, 1/2, 1/4, 1/4}	{1/4, 1/4, 0}, {1/4, 1/4, 0, 0}
17:00-17:59	{1/4, 1/4, 1/4}, {1/2, 1/4, 1/4, 1/4}	{1/4, 1/4, 0}, {1/2, 1/2, 1/4, 1/4}	{1/4, 1/4, 0}, {1/4, 1/4, 0, 0}
18:00-18:59	{1/4, 1/4, 1/4}, {1/2, 1/4, 1/4, 1/4}	{1/4, 0, 0}, {1/2, 1/2, 1/2, 1/4}	{1/4, 0, 0}, {1/4, 1/4, 1/4, 0}
19:00-19:59	{1/4, 1/4, 1/4}, {1/2, 1/4, 1/4, 1/4}	{1/4, 0, 0}, {1/2, 1/2, 1/2, 1/4}	{1/4, 0, 0}, {1/4, 1/4, 1/4, 0}
20:00-20:59	{0, 0, 0}, {3/4, 3/4, 3/4, 3/4}	{0, 0, 0}, {1/2, 1/2, 1/2, 1/2}	{0, 0, 0}, {1/4, 1/4, 1/4, 1/4}
21:00-21:59	{0, 0, 0}, {3/4, 3/4, 3/4, 3/4}	{0, 0, 0}, {1/2, 1/2, 1/2, 1/2}	{0, 0, 0}, {1/4, 1/4, 1/4, 1/4}
22:00-22:59	{0, 0, 0}, {3/4, 3/4, 3/4, 3/4}	{0, 0, 0}, {1/2, 1/2, 1/2, 1/2}	{0, 0, 0}, {1/4, 1/4, 1/4, 1/4}
23:00-23:59	{0, 0, 0}, {3/4, 3/4, 3/4, 3/4}	{0, 0, 0}, {1/2, 1/2, 1/2, 1/2}	{0, 0, 0}, {1/4, 1/4, 1/4, 1/4}

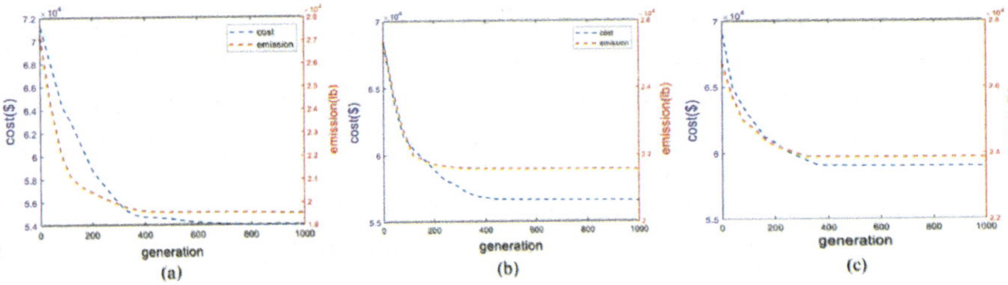

Fig. 5.5 Convergence of cost and emission with **a** $\Delta = 3$, **b** $\Delta = 2$, and **c** $\Delta = 1$

5.5 Conclusion

The increasing penetration of a large number of intermittent energy resources presents a new challenge for the optimal operation of the hybrid energy systems. To properly handle the uncertainty and to solve the probabilistic problems in the hybrid energy systems, this chapter mainly has several conclusions as follows.

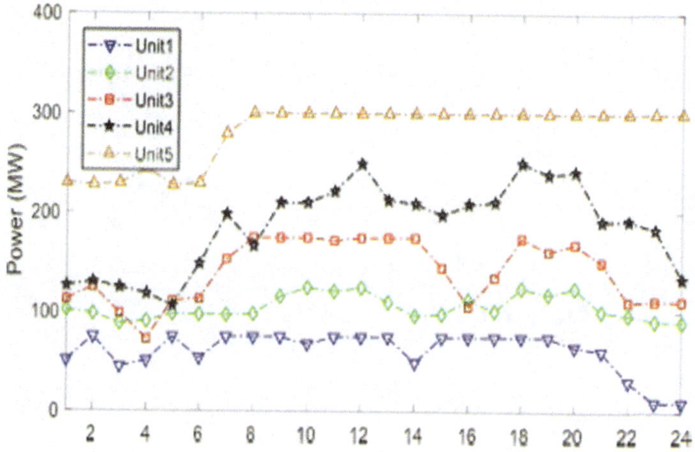

Fig. 5.6 Thermal output process with $\Delta = 3$

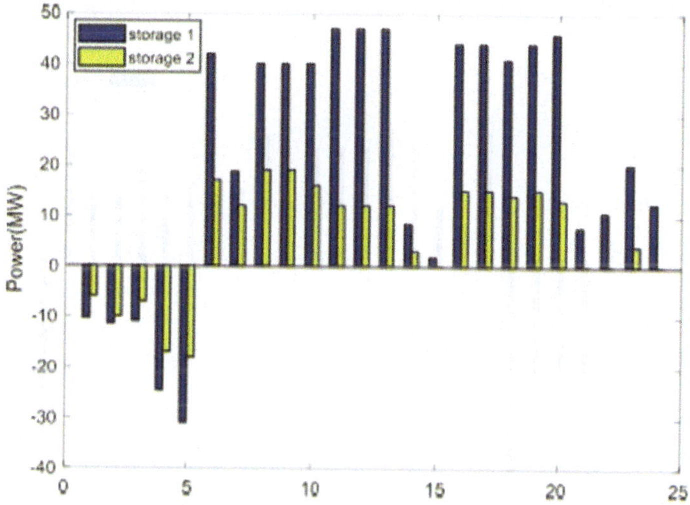

Fig. 5.7 Storage process of BES with $\Delta = 3$

1. Combined with uncertainty budget, scenarios-based approach can properly deal with stochastic problem in the hybrid energy system, adjustable parameters can decrease the conservation as well as decrease the potential risk.
2. On the basis of MOEA/D framework, gradient descent-based differential evolution can properly improve the optimization efficiency, the improved scaling parameter can be deduced to better fit the population evolution, which can further accelerate the convergence.

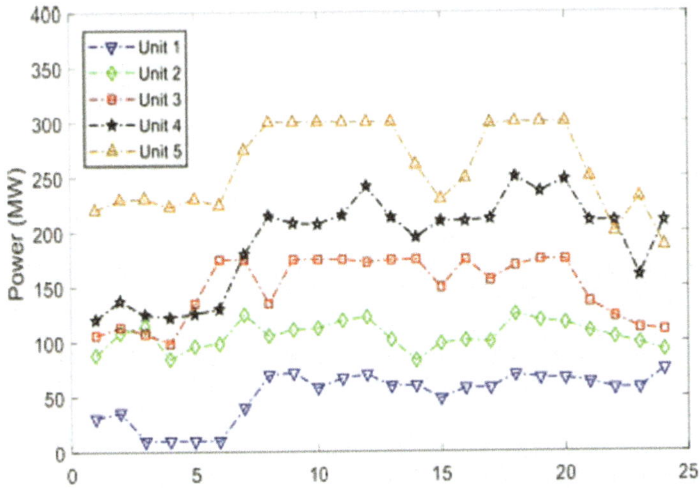

Fig. 5.8 Thermal output process with $\Delta = 3$

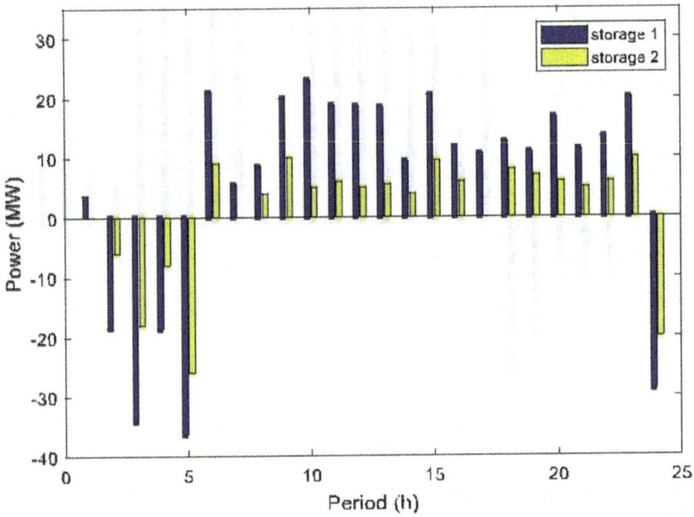

Fig. 5.9 Thermal output process with $\Delta = 3$

3. The two-step decision-making approach can deduce robust interval of uncertainty budget, which can avoid the potential risk in the hybrid energy system. Ultimately, best optimal scheme can be screened out from schemes set.

According to those simulation results, it reveals that the uncertainty budget can control the uncertainty of intermittent energy resources, MOEA/D-GDDE can improve the optimization efficiency and two-step decision-making strategy can ensure the robustness of the operation of the hybrid energy system.

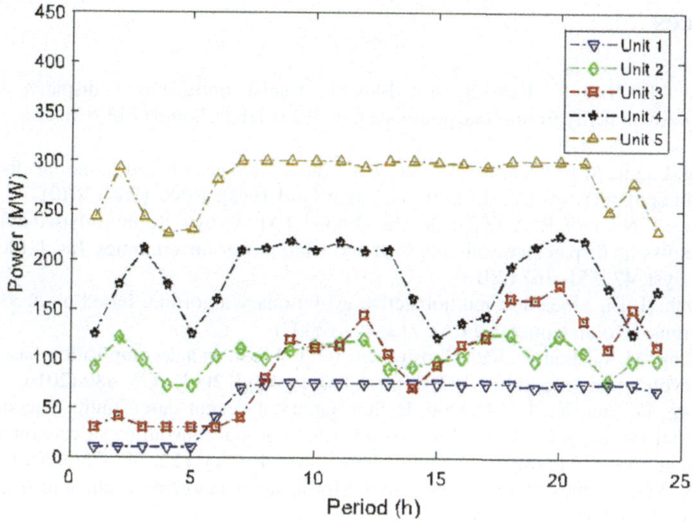

Fig. 5.10 Thermal output process with $\Delta = 3$

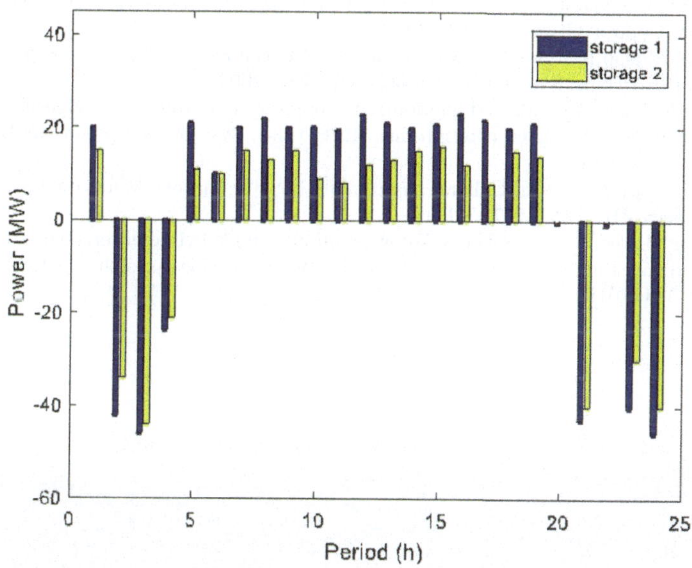

Fig. 5.11 Thermal output process with $\Delta = 3$

References

1. C. Peng, P. Xie, L. Pan, R. Yu, Flexible robust optimization dispatch for hybrid wind/photovoltaic/hydro/thermal power system. IEEE Trans. Smart Grid **7**(2), 751–762 (2016). March
2. H. Ming, L. Xie, M.C. Campi, S. Garatti, P.R. Kumar, Scenario-based economic dispatch with uncertain demand response. IEEE Trans. Smart Grid **10**(2), 1858–1868 (2019)
3. J. Aghaei, T. Niknam, R. Azizipanah-Abarghooee, J.M. Arroyo, Scenario-based dynamic economic emission dispatch considering load and wind power uncertainties. Int. J. Electr. Power Energy Syst. **47**, 351–367 (2013)
4. Q. Zhang, H. Li, Moea/d: a multiobjective evolutionary algorithm based on decomposition. IEEE Trans. Evolut. Comput. **11**(6), 712–731 (2007)
5. Q.L. Wang, M.A. Zhou, L. Jiao, Constrained subproblems in a decomposition-based multiobjective evolutionary algorithm. IET Control Theory Appl. **20**(3), 475–480 (2016)
6. H. Zhang, D. Yue, X. Xie, C. Dou, F. Sun, Gradient decent based multi-objective cultural differential evolution for short-term hydrothermal optimal scheduling of economic emission with integrating wind power and photovoltaic power. Energy **122**, 748–766 (2017)
7. Y. Qi, X. Ma, F. Liu, L. Jiao, J. Sun, J. Wu, Moea/d with adaptive weight adjustment. Evolut. Comput. **22**(2), 231–264 (2014)
8. S. Bittanti, M.C. Campi, M. Prandini, How many experiments are-needed to adapt? Lect. Notes Control Inf. Sci. 5–14 (2007)
9. M.C. Campi, S. Garatti, M. Prandini, The scenario approach for systems and control design. Ann. Rev. Control **33**(2), 381–389 (2008)
10. K. Deb, A. Pratap, S. Agarwal, T. Meyarivan, A fast and elitist multiobjective genetic algorithm: Nsga-ii. IEEE Trans. Evolut. Comput. **6**(2), 182–197 (2002)
11. S. Jiang, S. Yang, An improved multiobjective optimization evolutionary algorithm based on decomposition for complex pareto fronts. IEEE Trans. Syst., Man, Cybern. **46**(2), 421–437 (2016)
12. M. Daneshyari, G.G. Yen, Cultural-based multiobjective particle swarm optimization. Syst. Man Cybern. **41**(2), 553–567 (2011)
13. H. Zhang, D. Yue, X. Xie, S. Hu, S. Weng, Multi-elite guide hybrid differential evolution with simulated annealing technique for dynamic economic emission dispatch. Appl. Soft Comput. **34**, 312–323 (2015)

Chapter 6
Gradient Decent Based Multi-objective Optimization for Economic Emission of Hybrid Energy Systems

With the integration of wind power and photovoltaic power, optimal operation of hydrothermal power system becomes great challenge due to its non-convex, stochastic and complex-coupled constrained characteristics. This chapter extends short-term hydrothermal system optimal model into short-term hydrothermal optimal scheduling of economic emission while considering integrated intermittent energy resources (SHOSEE-IIER). For properly solving SHOSEE-IIER problem, a gradient decent based multi-objective cultural differential evolution (GD-MOCDE) is proposed to improve the optimal efficiency of SHOSEE-IIER combined with three designed knowledge structures, which mainly enhances search ability of differential evolution in the shortest way. With considering those complex-coupled and stochastic constraints, a heuristic constraint-handling measurement is utilized to tackle with them both in coarse and fine tuning way, and probability constraint-handling procedures are taken to properly handle those stochastic constraints combined with their probability density functions. Ultimately, those approaches are implemented on five test systems, which testify the optimization efficiency of proposed GD-MOCDE and constraint-handling efficiency for system load balance, water balance and stochastic constraint-handling measurements, those obtained results reveal that the proposed GD-MOCDE can properly solve the SHOSEE-IIER problem combined with those constraint-handling approaches.

6.1 Problem Formulation

The hybrid energy resource system mainly consists of hydro power, thermal power, wind power and photovoltaic power. The target of SHOSEE-IIER is to minimize the fuel cost and emission volume caused by thermal units, which requires proper assignment of power output by each power generator to decrease thermal output while

D. Yue et al., *Cooperative Optimal Control of Hybrid Energy Systems*,
https://doi.org/10.1007/978-981-33-6722-7_6

considering water balance constraint, system load balance constraint and various basic constraint limits of each power generator.

6.1.1 The Objective of Economic Cost

The fuel cost of hybrid energy resources is caused by thermal power, traditional formulation of fuel cost objective can't reflects the sharp increase of fuel cost in reality, it can be described as the objective of classical ED with considering valve point effects, it consists of fuel cost in N_c generating units for T time intervals, which can be presented as follows [1]:

$$\min F_1 = \sum_{t=1}^{T}\sum_{i=1}^{N_c}\left[a_i + b_i P_{cit} + c_i P_{cit}^2 + \left|d_i \sin\left(e_i\left(P_{ci,\min} - P_{cit}\right)\right)\right|\right] \tag{6.1}$$

Where F_1 denotes fuel cost objective of thermal power system, P_{cit} is the generated power in the ith thermal unit at tth time interval, T is the length of total dispatch period, N_c is the number of thermal units, a_i, b_i, c_i, d_i, e_i are the coefficients of the ith thermal unit, $P_{ci,\min}$ is the minimum output of the ith thermal unit.

6.1.2 The Objective of Emission Rate

With the increasing concern on the emission pollutant of thermal units, more and more focuses are taken on the emission rate of harmful gases. The nitric oxide can be taken as the representative harmful gas caused by thermal units [1], and the total emission rate of thermal power system can be presented as follows [2]:

$$\min F_2 = \sum_{t=1}^{T}\sum_{i=1}^{N_c}\left[\alpha_i + \beta_i P_{cit} + \gamma_i P_{cit}^2 + \zeta_i \exp\left(\lambda_i P_{cit}\right)\right] \tag{6.2}$$

Where F_2 denotes emission rate of thermal power system, and α_i, β_i, γ_i, ζ_i, λ_i are the coefficients of emission rate at the ith thermal unit.

6.1.3 Constraints

Since wind power and photovoltaic power are incorporated into hydro-thermal power system, stochastic characteristics of intermittent energy resources must be considered in the optimal scheduling while satisfying output limits of each power generator, ramp rate limits, water volume balance and system load balance constraints.

(1) System load balance constraint [3]:

$$\sum_{l=1}^{N_h} P_{hlt} + \sum_{i=1}^{N_c} P_{cit} + \sum_{j=1}^{N_w} P_{wjt} + \sum_{k=1}^{N_p} P_{pkt} = P_{Dt} + P_{Lt} \tag{6.3}$$

Where P_{hlt} is output of the lth hydro plant at tth time interval, P_{wjt} is the output of the jth wind farm at tth time interval, P_{pkt} is the output of the kth photovoltaic field at tth time interval, N_h, N_w and N_p are the number of hydro plants, wind farms and photovoltaic fields, P_{Dt} denotes the system load at tth time period, P_{Lt} is the transmission loss at tth time period, it is merely related to thermal output, and it can be expressed as follows:

$$P_{Lt} = \sum_{i=1}^{N_c} \sum_{j=1}^{N_c} P_{cit} B_{ij} P_{cjt} + \sum_{i=1}^{N_c} B_{0i} P_{cit} + B_{00} \tag{6.4}$$

Where B_{ij}, B_{0i} and B_{00} are loss coefficients at the ith thermal unit.

(2) Power generation limits:

$$\begin{cases} P_{ci,\min} \le P_{cit} \le P_{ci,\max}, i = 1, 2, \ldots, N_c, t = 1, 2, \ldots, T. \\ \text{Prob}\left(P_{wj,\min} \le P_{wjt} \le P_{wj,\max}\right) = \rho_w, j = 1, 2, \ldots, N_w, t = 1, 2, \ldots, T. \\ \text{Prob}\left(P_{pk,\min} \le P_{pkt} \le P_{pk,\max}\right) = \rho_p, k = 1, 2, \ldots, N_p, t = 1, 2, \ldots, T. \end{cases} \tag{6.5}$$

Where $P_{ci,\min}$, $P_{ci,\max}$ are the minimum and maximum outputs of the ith thermal unit, $P_{wj,\min}$, $P_{wj,\max}$ are the minimum and maximum outputs of the jth wind farm, $P_{pk,\min}$, $P_{pk,\max}$ are the minimum and maximum outputs of the kth photovoltaic field.

(3) Generating unit ramp rate limits:

$$\begin{cases} DR_{ci} \le P_{cit} - P_{ci,t-1} \le UR_{ci}, i = 1, 2, \ldots, N_c, t = 1, 2, \ldots, T. \\ DR_{wj} \le P_{wjt} - P_{wj,t-1} \le UR_{wj}, j = 1, 2, \ldots, N_w, t = 1, 2, \ldots, T. \\ DR_{pk} \le P_{pkt} - P_{pk,t-1} \le UR_{pk}, k = 1, 2, \ldots, N_p, t = 1, 2, \ldots, T. \end{cases} \tag{6.6}$$

Where UR_{ci}, DR_{ci} are the up-ramp and down-ramp limits of the ith thermal unit, UR_{wj}, DR_{wj} are the up-ramp and down-ramp limits of the jth wind farm, UR_{pk}, DR_{pk} are the up-ramp and down-ramp limits of the kth photovoltaic field.

(4) The constraint limits of hydropower plant:

Generally, hydro power output can be expressed with quadratic function of reservoir water head and water discharge, and water head can be also expressed with reservoir storage. Therefore, hydro power output can be in terms of reservoir storage and water discharge, which can be formulated as follows:

$$P_{hlt} = C_{1l} * V_{hlt}^2 + C_{2l} * Q_{hlt}^2 + C_{3l} * V_{hlt} * Q_{hlt} + C_{4l} * V_{hlt} + C_{5l} * Q_{hlt} + C_{6l} \tag{6.7}$$

Where C_{1l}, C_{2l}, C_{3l}, C_{4l}, C_{5l}, C_{6l} are the coefficients of the lth hydro plant, V_{hlt} is the reservoir storage of the lth hydro plant at tth time interval, Q_{hlt} is the water discharge of the lth hydro plant at tth time interval.

Simultaneously, topology of those hydro plants can also have great effect on the optimal scheduling of hydropower system, and water balance among water discharge, reservoir storage and water inflow should satisfy the following formulation:

$$\begin{gathered} V_{hlt} = V_{hl,t-1} + \left[I_{lt} - Q_{hlt} - S_{lt} + \sum_{k=1}^{N_l} \left(Q_{hl,t-\tau_l^k} + S_{l,t-\tau_l^k} \right) \right] \Delta t \\ l = 1, 2, \ldots, N_h; t = 1, 2, \ldots, T; 0 \le \tau_l^k \le T \end{gathered} \tag{6.8}$$

Where I_{lt} is inflow water at the lth hydro plant at tth time interval, S_{lt} is water spillage of the lth hydro plant at tth time period, τ_l^k is the transmission delay between the lth hydro plant and the kth upstream hydro plant at lth hydro plant.

For properly scheduling the hydro power generation in next day, initial storage and terminal storage of hydro plant are known as follows:

$$V_{hl0} = V_{l,\text{begin}}, \ V_{hlT} = V_{l,\text{end}} \quad (i = 1, 2, \ldots, N_h) \tag{6.9}$$

Where $V_{l,\text{begin}}$ is the initial reservoir storage at lth hydro plant, and $V_{l,\text{end}}$ is the terminal storage at lth hydro plant.

6.2 The Multi-objective Cultural Differential Evolution (MOCDE)

Based on the multi-objective differential evolution, culture algorithm can provide culture knowledge structures for population space evolution, several knowledge structures can provide elite experience, normative information and historical information, which can guide multi-objective differential evolution of population space to the global optimal solution.

6.2.1 The Principles of Differential Evolution

DE has been widely utilized due to its simple but powerful search ability in real-world application. It mainly consists of three evolution operators: mutation, crossover and selection, mutation and crossover operators are applied on the individual to yield trial vector, and selection operator determines whether the trial vector needs to be added into the population of next generation [4]. The mutation operator of DE/rand/1/bin strategy can be demonstrated as follows:

$$V_{r,G+1} = X_{r,G} + F * \left[\left(X_{r1,G} - X_{r2,G} \right) + \left(X_{r3,G} - X_{r4,G} \right) \right],$$
$$r1 \neq r2 \neq r3 \neq r4 \neq r \qquad (6.10)$$

Where $X_{r1,G}$, $X_{r2,G}$, $X_{r3,G}$, $X_{r4,G}$ are selected from non-dominated population, $r1, r2, r3, r4 \in \{1, 2, ..., NP\}$ are the integer index of the non-dominated population, NP is the size of non-dominated population, $V_{r,G+1}$ is the parameter vector for $G + 1$th generation, F is the mutation parameter, it is range in [0, 2]. The detail of crossover operator and selection operator can be seen in literature [4].

6.2.2 *The Culture Knowledge Structure for Multi-objective Differential Evolution*

The culture algorithm (CA) was proposed by Reynolds [5], it consists of belief space, population space and communication protocols, population space connects to belief space by those communications. The population space produce the offspring individual by Generate() function, and evaluate the fitness of individuals with Objective() function, and optimal individuals are selected by Select() function. The belief space accepts the experience of optimal individuals by Accept() function, elite experience of belief space is updated by update() function, and these elite experience can guide the population evolution in the population space. The framework of culture algorithm is presented in Fig. 6.1.

Belief space has various knowledge structures, situational knowledge, normative knowledge and historical knowledge structures have been proposed by Saleem [6].

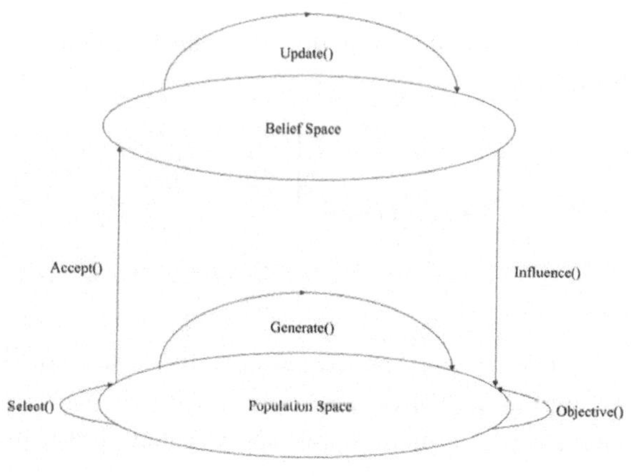

Fig. 6.1 The Framework of culture algorithm

The evolutionary algorithm can be integrated into the knowledge structure, and the efficiency of culture algorithm mainly depends on the knowledge structures. In this chapter, the situational knowledge, normative knowledge and historical knowledge structures are redesigned for multi-objective differential evolution for dynamic economic emission dispatch.

6.3 The Gradient Decent Based Multi-objective

In order to solve dynamic economic emission dispatch problem of hybrid energy resource system properly, some improvements have been proposed on the multi-objective culture differential evolution. The gradient decent based mutation operator is proposed to accelerate the search ability of differential evolution, which takes the shortest path to the optimal solution with gradient decent method [7–9]. Furthermore, several knowledge structures are redefined according to the problem formulation of dynamic economic emission dispatch of hybrid energy resource system, which can provide convenience for population space evolution. Due to those complex-coupled constraint in hybrid energy resource system, stochastic constraints of wind power and photovoltaic power are converted into deterministic constraints, and system load balance constraint is properly handled by a two-step coupled constraint-handling technique.

6.3.1 Gradient Decent Based Mutation Operator

It is known that the decent gradient represents the fastest direction of objective increasing or decreasing, which can provide fast search direction during the differential evolution. This chapter proposes a decent gradient based mutation operator, which improves the search ability of differential evolution [10].

For a given multi-objective function $f(x) = [f_1(x), f_2(x), \ldots, f_m(x)]^T, x \in R^n$, the gradient of function f(x) can be defined in terms of Jacobian matrix as follows:

$$\nabla_{m\times n} f(x) = \begin{bmatrix} \frac{\partial f_1(x)}{\partial x_1} & \frac{\partial f_2(x)}{\partial x_2} & \cdots & \frac{\partial f_1(x)}{\partial x_n} \\ \frac{\partial f_2(x)}{\partial x_1} & \frac{\partial f_2(x)}{\partial x_2} & \cdots & \frac{\partial f_2(x)}{\partial x_n} \\ \vdots & \vdots & \vdots & \vdots \\ \frac{\partial f_m(x)}{\partial x_1} & \frac{\partial f_m(x)}{\partial x_2} & \cdots & \frac{\partial f_m(x)}{\partial x_n} \end{bmatrix} = \begin{bmatrix} (\nabla f_1)^T \\ (\nabla f_2)^T \\ \vdots \\ (\nabla f_m)^T \end{bmatrix} \tag{6.11}$$

The ∇f_i represents the gradient direction, and $\frac{\partial f_i}{\partial x_j}$ $(i = 1, 2...m, j = 1, 2...n)$ is the partial derivative. For a given positive space H^+ and negative space H^-, the element $y \in R^n$ in these two spaces satisfy following conditions:

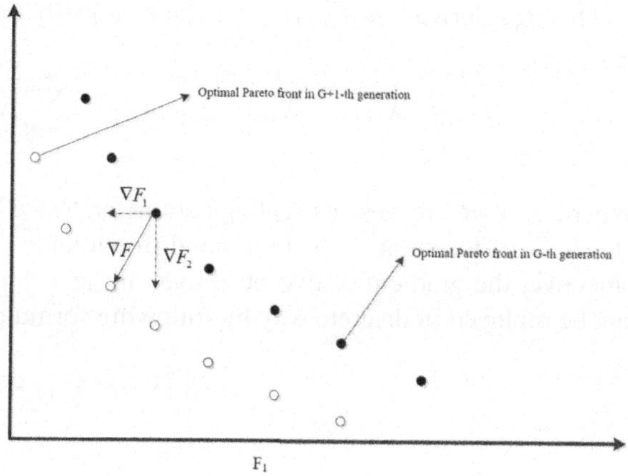

Fig. 6.2 The gradient direction of Pareto front during the evolution process

$$\begin{cases} H^+ = \left\{ y \in R^n \mid \nabla f(x)^T y > 0 \right\} \\ H^- = \left\{ y \in R^n \mid \nabla f(x)^T y < 0 \right\} \end{cases} \tag{6.12}$$

The positive direction leads the decision vector $y \in R^n$ to the direction that objective function decreases, and the negative directive leads it to the opposite direction. For the iterative vector X_{G+1}, $X_G(X_{G+1}$ is near to $X_G)$, the negative gradient needs to satisfy [10]:

$$F_i\left(X_{G+1}\right) - F_i\left(X_G\right) = \nabla F_i\left(X_{G+1} - X_G\right) < 0 \tag{6.13}$$

The above formulation describes the path that X_G moves to X_{G+1} on the i th objective direction. For simplicity, two objective optimization problem is taken for example, gradient direction of Pareto front changes as it is shown in Fig. 6.2. The optimal Pareto front in Gth generation converts to the Pareto front in the $G+1$th generation on the gradient direction, it can be seen that the gradient direction of multi-objective optimization can be represented by the weighted sum of gradient direction on each objective.

According to the results obtained in literature [10], the vector in the negative space can be taken:

$$y = -\sum_{i=1}^{m} \lambda_i \frac{\nabla f_i(x)}{\|\nabla f_i(x)\|} \tag{6.14}$$

Where λ_i is weighted parameter in interval [0, 1], $\|\cdot\|$ represents the Euclidian norm. The vector can ensure $\nabla F y < 0$, hence $y \in H^-$. The iterative vector in Formula (6.12) satisfy the following condition:

$$\nabla F\left(X_{G+1} - X_G\right) = \nabla F y \tag{6.15}$$

Then the derivation $X_{G+1} - X_G$ can be equally described as:

$$X_{G+1} - X_G = \eta_G y = -\eta_G \sum_{i=1}^{m} \lambda_i \frac{\nabla f_i(x)}{\|\nabla f_i(x)\|} \tag{6.16}$$

Where $\eta_G \in R^+$ represents scaling parameter. Since ∇f_i can't be obtained since the objective in formulation (6.1) is nondifferentiable. To avoid the derivative on the objective, the gradient of two objective on the j th dimension of decision variable can be replaced in discrete way by following formulation:

$$\nabla f_i\left(x_j\right) = \frac{f_i\left(x_j + \Delta x_j\right) - f_i\left(x_j\right)}{\Delta x_j} \tag{6.17}$$

Where

$$X_{G+1}^{j} = X_G^{j} + \gamma_1^{j}\left(X_{r_2,G} - X_{r_3,G}\right) + \gamma_2^{j}\left(X_{r_4,G} - X_{r_5,G}\right) \tag{6.18}$$

Where the parameters are set as follows:

$$\begin{aligned}
\gamma_1^{j} &= \frac{-\eta_G \lambda_1 \cdot \operatorname{sgn}\left(f_1\left(X_{r_2,G}\right) - f_1\left(X_{r_3,G}\right)\right)}{\left(X_{r_2,G}^{j} - X_{r_3,G}^{j}\right)^2 \sqrt{\sum_{j=1}^{n} \frac{1}{\left(X_{r_2,G}^{j} - X_{r_3,G}^{j}\right)^2}}} \\
\gamma_2^{j} &= \frac{-\eta_G \lambda_2 \cdot \operatorname{sgn}\left(f_2\left(X_{r_4,G}\right) - f_2\left(X_{r_5,G}\right)\right)}{\left(X_{r_4,G}^{j} - X_{r_5,G}^{j}\right)^2 \sqrt{\sum_{j=1}^{n} \frac{1}{\left(X_{r_4,G}^{j} - X_{r_5,G}^{j}\right)^2}}} \\
\eta_G &= \eta_0\left[\left(g_{\max} - G + 1\right)/g_{\max}\right]^p
\end{aligned} \tag{6.19}$$

According to the above obtained differential evolution, the search ability can be enhanced for seeking the shortest way to the optimal solution, the search direction can be obtained by summation of weighted two directions, which depend on the search direction of two objectives. Hence, this gradient decent method is proposed to improve the mutation operator for increasing the search ability of differential evolution.

6.3.2 *The Designed Knowledge Structure*

Due to high efficiency for dealing with single objective optimization problem, CA has been extended to solve multi-objective optimization problem. In this chapter, gradient decent based multi-objective differential evolution is embedded into the computational model provided by CA, and local random search (LRS) is also integrated into the historical knowledge structure of CA, which can avoid the premature problem caused by gradient decent based differential evolution at certain degree. In the population space, constraint limits of all those decision variables are stored in the normative knowledge structure, which can be interpreted as legal restraints in human society. Ultimately, the obtained elite experience during population space evolution is stored in the situational knowledge structure, which can guide the population space evolution.

6.3.2.1 Historical Knowledge Structure

Generally, historical knowledge is mainly utilized to find the patterns especially when the environment changes, and used to solve dynamic functions [11]. Due to great ability of gradient decent based differential evolution, it may tend to fall into the local optima, which suffers the premature problem as other evolutionary algorithm does. Here, LRS operation is used to overcome the premature problem by increasing the population diversity and exploit the promising area to get global optima.

The LRS presented in literature [12] improves the performance of DE for properly solving DED problem, DE is utilized to seek the available optimal area that global optima may be within, LRS operator is mainly used obtain the global optima from these local optimal solutions. The implementation of LRS can be generally taken as follows:

Step 1: Initialization. Combining with the differential evolution process, current optimal individual X_G is taken as initial individual X_{opt}, and set iteration step m as 0.

Step 2: Search scale setting. For a given real number $\theta \in (0, 1]$, the initial search scale S_0 can be defined as follows:

$$S_0 = \theta * (X_{\max} - X_{\min}) \tag{6.20}$$

Where $X_{\max}$, $X_{\min}$ are the maximum and minimum individual, and search scale needs to change adaptively for searching the global optima:

$$S_m = (1 - \rho) * S_{m-1} \tag{6.21}$$

Where ρ is the scaling parameter in the interval [0, 1).

Step 3: Local optima generating. In order to avoid falling into the local optima, N_l local optimal individuals need to be generated as follows:

$$X_m^n = X_{opt} + R_m * rnd(D, 1), \quad n = 1, 2, \ldots, N_l \tag{6.22}$$

Where $rnd(D, 1)$ is the D-dimensional uniform random number in the interval [-1, 1], if the value of local optimal individual violates the constraint limits, the constraint handling mechanism is taken to force it into the feasible area.
Step 4: Global optima selection. Dominance relationship can be obtained between the current optimal individual X_{opt} and those generated individuals. If $X_m^n > X_{opt}(X_m^n$ dominates $X_{opt})$, then set $X_{opt} = X_m^n$.
Step 5: Termination criterion. If $m < L_{\max}(L_{\max}$ is the maximum iteration number), then $m = m + 1$ and go to **Step 2**, otherwise, the LRS process is terminated, and the obtained X_{opt} can be taken as the optimal individual of LRS operation.

6.3.2.2 Normative Knowledge Structure

Due to those varieties of constraint limits in the hybrid energy resource system, those constraint limit information needs to be stored for convenience. The redefined normative knowledge structure mainly stores the interval information in hybrid energy resource system, and the constraints in problem formulation need to be converted into the interval formation. For properly controlling the constraint violation problem during the constraint handing process and population space evolution, all those variables that violates the constraint limits can be forced into the feasible domain. Generally, the following measurement can be taken:

$$x_i' = \begin{cases} l_i \ if \ \ x_i < l_i \\ x_i \ if \ \ l_i \le x_i \le u_i \\ u_i \ if \ \ x_i \ge u_i \end{cases} \tag{6.23}$$

Where l_i, u_i are the upper and lower bound, x_i represents the ith dimensional value of decision variable. Furthermore, ramp rate constraint can be converted into interval information combing with the output constraint limits, the output limit of each power generator can be presented as follows:

$$\begin{cases} \max\left(P_{ci,t-1} + DR_{ci}, P_{ci,\min}\right) \le P_{cit} \le \min\left(P_{ci,t-1} + UR_{ci}, P_{ci,\max}\right) \\ \max\left(P_{wj,t-1} + DR_{wj}, P_{wj,\min}\right) \le P_{wjt} \le \min\left(P_{wj,t-1} + UR_{wj}, P_{wj,\max}\right) \\ \max\left(P_{pk,t-1} + DR_{pk}, P_{pk,\min}\right) \le P_{pkt} \le \min\left(P_{pk,t-1} + UR_{pk}, P_{pk,\max}\right) \end{cases} \tag{6.24}$$

Where $P_{wj,\min}$, $P_{wj,\max}$, $P_{pk,\min}$, $P_{pk,\max}$ are the obtained low and upper bound after transforming stochastic constraint into deterministic constraint.

6.3.2.3 Situational Knowledge Structure

The situational knowledge structure mainly takes charge of storing the elite experience obtained in the population space evolution, and this elite experience can guide the population evolution by influence () protocol in return. The archive retention mechanism can be integrated into this knowledge structure, and the elite experience of non-dominated individual is stored. For a newly obtained individual, whether it can be added into the archive set needs to satisfy following norms:

(1) If the newly generated individual can dominate one individual in the archive set, delete all those individuals dominated by it, and add the new individual into the archive set and its elite experience is stored in the situational knowledge structure.
(2) If the newly generated individual can be dominated by the individual in the archive set, it can't be added into the archive set.
(3) If the newly generated individual can't dominate any individual in the archive set and also can't be dominated by any individual in the archive set, check the diversity distribution of Pareto front before it is added and after it is added, which can be found in literature [13]. If it has better diversity distribution after the newly generated individual is added, it can be added into the archive set and its elite experience can be stored in the situational knowledge structure. Otherwise, it can't be added into the archive set.

6.3.3 *Constraint Handling Technology*

Since intermittent energy resources have been jointed into the energy system, more complex constraints and great randomness can be taken into the optimal dispatch of hybrid energy resource system. In comparison to those traditional energy resources, the output limits of intermittent energy resources are uncertain, which brings great challenges to the dynamic economic emission dispatch of multiple energy resource system.

6.3.3.1 Constraint Handling for Probability Constraint Limits of Photovoltaic Power Output

Since the output process of wind power and photovoltaic power has strong randomness, output limits can be only demonstrated with probability terms as it is shown in Formula (6.5). However, it is difficult to deal with the probability constraint in the optimization process, it needs to convert the probability constraint into the deterministic constraint first. Here, the probability constraint can be properly handled as follows:

Firstly, the value of wind output needs to be normalized with $\eta_j = x_{pj}/x_{pjmax}$, it assumes that η_j follows the Beta distribution [14, 15], and its probability density function is described as:

$$f\left(\eta_j\right) = \frac{1}{B(\alpha, \beta)}\eta_j^{\alpha-1}\left(1-\eta_j\right)^{\beta-1}, \quad 0 \le \eta_j \le 1 \tag{6.25}$$

Then its distribution function can be obtained:

$$F\left(\xi_j\right) = \frac{\int_0^{\xi_j} \varpi_j^{\alpha-1}\left(1-\varpi_j\right)^{\beta-1} d\varpi_j}{B(\alpha,\beta)}, 0 < \xi_j < 1 \tag{6.26}$$

It equals 0 when $\xi_j < 0$, and equals 1 when $\xi_j \geq 1$. The $B(\cdot)$ represents the beta function with its two parameters α, β. Then probability constraint of output limits can be converted as [15]:

$$x_{pj\max} * F^{-1}(1-\rho) \leq x_{pj} \leq x_{pj\max} * F^{-1}(\rho) \tag{6.27}$$

6.3.3.2 Constraint Handling for Probability Constraint Limits of Wind Power Output

It is known that wind power generation is closely related to the wind speed, the relationship between wind power output and wind speed can be described as:

$$P_{wj} = \begin{cases} 0, & v_j < v_{j,in} \text{ or } v_j \geq v_{j,out} \\ P_{wj,\max} * \frac{v_j - v_{j,in}}{v_{j,rate} - v_{j,in}}, & v_{j,in} \leq v_j < v_{j,rate} \\ P_{wj,\max}, & v_{j,rate} \leq v_j < v_{j,out} \end{cases} \tag{6.28}$$

where $P_{wj,\max}$, $v_{j,rate}$ are the rated output (or maximum output) and rated wind speed at the j th wind turbine, $v_{j,in}$, $v_{j,out}$ are the cut-in and cut-out wind speed at the j th wind turbine. The wind output can not be zero when wind speed is between cut-in speed and cut-out speed, it has direct ratio relation with wind speed when wind speed is between cut-in speed and rated wind speed, and it works at the maximum output especially when wind speed ranges in the rated wind speed and cut-out speed. In literature [16], the probability density function and cumulative distribution function of wind speed can be described as follows:

$$\begin{cases} f\left(v_j\right) = (k/c)\left(v_j/c\right)^{k-1}\exp\left(-\left(v_j/c\right)^k\right), v_j \geq 0 \\ F_w\left(v_j\right) = 1 - \exp\left(-\left(v_j/c\right)^k\right), v_j \geq 0 \end{cases} \tag{6.29}$$

where k, c are the scaling parameters. According to the relationship presented in Formula (6.29), the cumulative distribution function of wind power output can be obtained as:

$$\begin{aligned} F_w\left(P_{wj}\right) =& 1 - \exp\left\{-\left[\left(1 + \frac{v_{j,rate} - v_{j,in}}{v_{j,in} P_{wjmax}} P_{wj}\right)\frac{v_{j,in}}{c}\right]^k\right\} \\ &+ \exp\left[-\left(v_{j,out}/c\right)^k\right], \quad 0 \leq P_{wj} < P_{wjmax} \end{aligned} \tag{6.30}$$

Combining with the method presented in Sect. 6.3.3.1, the deterministic interval can be obtained as:

$$F_w^{-1}(1-\rho) \le x_{wit} \le F_w^{-1}(\rho) \tag{6.31}$$

6.3.3.3 Constraint Handling Technique for Water Volume Balance

If cascaded hydro power is taken into consideration, water volume balance can be a headache during optimization process especially in hybrid energy operation system. Generally, it must be tackled properly before handling system load balance since other energy resource can't affect the hydro power output. Here, heuristic constraint handling method is taken to deal with water volume balance, it mainly controls the feasible domain by coarse adjustment and fine tuning technique, which can be presented in detail in literature [17].

6.3.3.4 Initial Priority Optimization Strategy on Constraint Handling for System Load Balance

Constraint handling efficiency has great influence on the optimization results especially when the hybrid energy resource system has series of complex-coupled constraints, the two-step constraint handling technique is used to tackle with this problem. Besides, since fuel cost and emission pollutant are mainly produced by thermal units, wind power and photovoltaic power need to be make full use to decrease the thermal output, which can produce operation scheme near to the optimal scheme and reduce the computational complexity of the whole population evolution. In this chapter, wind power and photovoltaic power are given relative high priority level at the initial step of each time period, the initial priority optimization strategy is presented before the constraint handling process [18].

Step 1: For simplicity, the output of thermal power, wind power and photovoltaic power is noted as $P_{\theta,t}$(it represents thermal power when $1 \le \theta \le N_c$, wind power when $N_c + 1 \le \theta \le N_c + N_w$, and photovoltaic power when $N_c + N_w \le \theta \le N_w + N_p$), and set $t = 1$, go to **Step 2.**

Step 2: Initialize the output of thermal power, wind power and photovoltaic power, and wind power and photovoltaic power is given priority as follows:

$$\begin{cases} P_{\theta,t} = P'_{c\theta,\min} + \text{rand}(0,1) * \left[P'_{c\theta,\max} - P'_{c\theta,\min}\right], \\ \theta = 1, 2, \ldots, N_c \\ P_{\theta,t} = P'_{w,\theta-N_c,\min} + 0.9 * \left[P'_{w,\theta-N_c,\max} - P'_{w,\theta-N_c,\min}\right], \\ \theta = N_c + 1, N_c + 2, \ldots, N_c + N_w \\ P_{\theta,t} = P'_{p,\theta-N_c-N_w,\min} + 0.9 * \left[P'_{p,\theta-N_c-N_w,\max} - P'_{p,\theta-N_c-N_w,\min}\right], \\ \theta = N_c + N_w + 1, N_c + N_w + 2, \ldots, N_c + N_w + N_p \end{cases} \tag{6.32}$$

Where rand(0, 1) is random number in interval (0,1], $P'_{c\theta,\min}$, $P'_{c\theta,\max}$, $P'_{w,\theta-N_c,\min}$, $P'_{w,\theta-N_c,\max}$ and $P'_{p,\theta-N_c-N_w,\min}$, $P'_{p,\theta-N_c-N_w,\max}$ are the obtained minimum and maximum output of thermal power, wind power and photovoltaic power, which can be obtained in normative knowledge in Sect. 6.3.2.2, and go to **Step 3.**

Step 3: Calculate the output deviation with Formula (6.32) If $\Delta P_t < \varepsilon_p(\varepsilon_p$ represents the permitted output accuracy), it represents that the system load balance is properly handled, and go to **Step 8**; Otherwise, a coupled coarse tuning method is implemented to tackle with it.

$$\Delta P_t = P_{D,t} + P_{loss,t} - \sum_{\theta=1}^{N_c+N_w+N_p} P_{\theta,t} \tag{6.33}$$

Set coarse iteration number $I_{coa} = 0$, and let all the power generator bears the output violation equally, and modify their output with Formula (6.33) and go to **Step 4:**

$$\begin{cases} \Delta P_{\text{ave},t} = \Delta P_t / \left(N_c + N_w + N_p\right) \\ P_{\theta,t} = P_{\theta,t} + \Delta P_{\text{ave},t} \end{cases} \tag{6.34}$$

Step 4: If the modified output exceeds the constraint limits, the measurements in the normative knowledge can force the output into the feasible domain. Set the iteration number of fine tuning $I_{fine} = 0$ and then go to **Step 5.**

Step 5: Calculate the output deviation with Formula (6.32) again. If $\Delta P_t < \varepsilon_p$ is satisfied, go to Step. Otherwise, choose a random unit number r, and modify the power output with Formula (6.34), go to **Step 6.**

$$P_{r,t} = P_{r,t} + \Delta P_{\text{ave},t} \tag{6.35}$$

Step 6: If the modified output exceeds the constraint limits, force it to the feasible domain with Formula (6.22). If $I_{fine} < I_{fine,max}(I_{fine,max}$ is the maximum iteration number of fine tuning), $I_{fine} = I_{fine} + 1$, and go to **Step 5.** Otherwise, go to **Step 7.**

Step 7: If $I_{coa} < I_{coa,max}(I_{coa,max}$ is the maximum iteration number of coarse adjusting), then $I_{coa} = I_{coa} + 1$ and go to **Step 3.** Otherwise, go to **Step 8.**

Step 8: If $t < T, t = t + 1$ and go to **Step 2.** Otherwise, system load balance constraint of each time period is properly handled after above procedures, and the constraint handling strategy is terminated.

6.3.3.5 Feasible Selection Mechanism for Constraint Handling Technique

Since dynamic economic emission dispatch of hybrid energy resource system consists of several constraint limits of photovoltaic power, wind power and thermal power, the constraint handling technique for these constraint limits plays an important role in the efficiency of dynamic economic emission dispatch of hybrid energy resource system. According to methods in above sections, these limits can be controlled in certain degree, but the modified individual still can be an infeasible one due to its finite iterations, it needs to ensure the feasible characteristics of the modified individual. Here, the total violation can be defined as:

$$\text{Totalviolate}\ (X_i) = \sum_{t=1}^{T} |\Delta P_t\ (X_i)| + \sum_{t=1}^{T} \left|\Delta W_{\text{water},t}\right| \tag{6.36}$$

Where the deviation $\Delta P_t\ (X_i) = \sum_{i=1}^{N_w} P_{wit} + \sum_{j=1}^{N_p} P_{pjt} + \sum_{k=1}^{N_c} P_{ckt} - L_t$, $\Delta W_{\text{water},t}$ is water balance violation, the modified individual can be taken as a feasible individual if $\text{Totalviolate}(X_i) < \varepsilon_p(\varepsilon_p$ is the permitted accuracy), and the modified individual can be taken as an infeasible individual if $\text{Totalviolate}(X_i) \geq \varepsilon_p$. In the selection operator, the feasible individual can be quite important, and generally several norms need to be followed [2]:

(1) If there is one feasible individual and one infeasible individual, select the feasible individual.
(2) If there are two infeasible individuals, select the individual that has less violation.
(3) If there are two feasible individuals, the selection mechanism can be implemented according to the situational knowledge in Sect. 6.3.2.3.

6.4 The Implementation of GD-MOCDE on Hybrid Energy Systems

Since the formulated model contains different power generators, the implementation of the proposed algorithm can be different from optimal dispatch of thermal units. The wind power and photovoltaic power has relative low output capacity and no relationship with fuel cost and emission pollutant, they can be treated priority to thermal power, which can also improve the optimization efficiency. Therefore, priority optimization strategy can be implemented before constraint handling and population evolution. The flowchart of gradient decent based multi-objective cultural differential evolution for solving optimal scheduling of hybrid energy resource system is shown in Fig. 6.3.

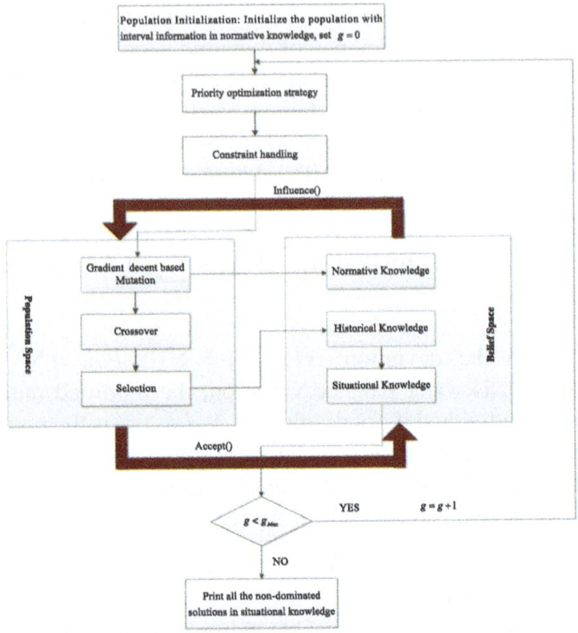

Fig. 6.3 The flowchart of gradient decent based multi-objective cultural differential evolution for solving optimal scheduling of hybrid energy resource system

6.5 Case Study

For testifying the efficiency of the proposed gradient decent based multi-objective cultural differential evolution, five test systems are designed in this case study. Test system 1 consists of 10 thermal units with considering transmission loss while integrating no wind power or photovoltaic power in 24 time periods. For testifying the efficiency of priority optimization strategy further, test system 2 is constructed with 30 thermal units with considering transmission loss in 24 time periods. Test system 3 consists of 4 thermal units, 1 wind farm and 1 photovoltaic field with considering the transmission loss between thermal units in 24 time periods. Test system 4 consists of four hydro plants and three thermal units with valve-point effect while considering transmission loss, but ramp rate of each power generator is not taken into consideration. Test system 5 presents a hybrid energy system consisted of four hydro plants, three thermal units, two wind farms and one photovoltaic fields, and all constraint limits are taken into consideration.

6.5.1 Test System 1

In this test system, all the data details can be found in literature [2], the output of each thermal unit at each time period is taken as the decision variable, each individual contains the output data of ten thermal units at 24 time periods with an interval of an hour. Here, the proposed GD-MOCDE is implemented on this test system in comparison with MODE, the obtained Pareto front is presented in Fig. 6.4, it consists of 30 representative non-dominated schemes, which can be seen in Table 6.1.

In comparison to MODE, the Pareto front obtain by the proposed GD-MOCDE has wide diversity distribution, all the non-dominated schemes are evenly distributed, and the maximum and minimum value of economic cost and emission rate are included. Moreover, the obtained non-dominated schemes can dominates that by MODE, which can also be seen in Table 6.1.

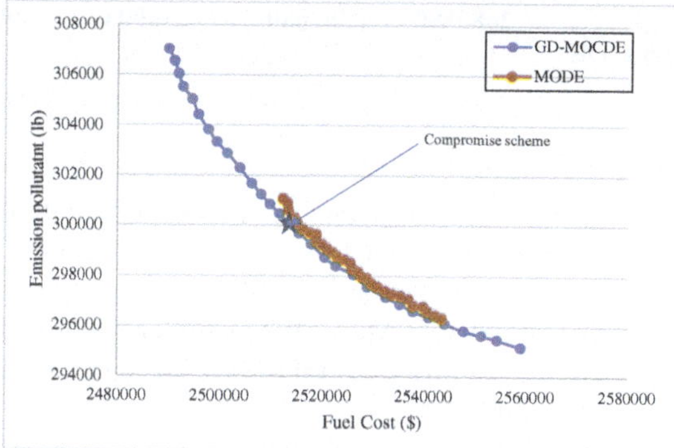

Fig. 6.4 The comparison of obtained schemes by GD-MOCDE and MODE for test system 1

Table 6.1 The obtained non-dominated schemes between GD-MOCDE and MODE for test system 1

Scheme	GD-MOCDE		MODE		Scheme	GD-MOCDE		MODE	
	Cost ($)	Emission (lb)	Cost ($)	Emission (lb)		Cost ($)	Emission (lh)	Cost ($)	Emission (lh)
1	2489890	307080	2512327	301130	16	2515458	299748	2525841	298344
2	2491028	306605	2513241	300963	17	2517832	299319	2526950	298219
3	2491822	306094	2513410	300671	18	2520478	298792	2527591	298046
4	2402678	305567	2514657	300362	19	2522612	298452	2528678	297976
5	2494439	305083	2515118	300156	20	2526021	298105	2529593	297754
6	2495685	304463	2516254	299927	21	2528746	297624	2530748	297634
7	2497590	303877	2517691	299764	22	2532389	297213	2532349	297467
8	2499230	303384	2518886	299698	23	2535183	296931	2533627	297336
9	2501296	302937	2518916	299434	24	2537752	296656	2535237	297278
10	2503713	302353	2520008	299295	25	2540856	296420	2536875	297147
11	2506041	301743	2520942	299161	26	2544095	296175	2537852	296888
12	2507926	301301	2521992	298996	27	2547820	295879	2539795	296844
13	2509660	300924	2522867	298836	28	2551325	295681	2540534	296675
14	2511560	300541	2524285	298703	29	2554396	295502	2542125	296515
15	2513263	300141	2525429	298580	30	2559089	295215	2543560	296387

Table 6.2 The comparison among optimization alternatives in literature [2, 19, 20]

Methods	Minimum cost		Minimum emission		Compromise result	
	Cost ($)	Emission (lb)	Cost ($)	Emission (lb)	Cost ($)	Emission (lb)
NSGA-II [2]	–	–	–	–	2522600	309940
RCGA [2]	2516800	317400	265630	304120	2525100	312460
IBFA [19]	2481773	327501	2614341	295883	2517116	299036
MAMODE [20]	2492451	315119	2581621	295244	2514113	302742
MODE	2512327	301130	2543560	296387	2525429	298580
GD-MOCDE	**2489890**	**307080**	**2559089**	**295215**	**2513263**	**300141**

In Table 6.1, 30 representative non-dominated schemes on Pareto front are presented, the economic cost is range from 2480000 $ to 2559000 $ and emission rate is between 295000 lb and 307000 lb. The obtained minimum economic cost and emission rate by GD-MOCDE is smaller than that of MODE. In comparison to other alternatives, the proposed GD-MOCDE can also have promising results, which can be seen in Table 6.2.

Since all the non-dominated schemes are evenly distributed, scheme (15) can be taken as the compromise scheme for simplicity. In comparison to other alternatives, scheme (15) by GD-MOCDE has the minimum economic cost and relative small emission rate, and its dispatch process is presented in Table 6.3, which can be taken for further analysis on the efficiency of the proposed GD-MOCDE. In Table 6.3, the output of 10 the thermal units at 24 time periods is presented with system load and transmission loss. It can be found that the obtained output of each thermal unit at each time period is controlled in the feasible constraint limits, and the transmission loss is range from 19 to 93 MW. Moreover, the transmission loss at each time period can't exceed the 5% of the system load, which also means that system load balance constraint is properly controlled.

According to the analysis on the obtained results, the proposed GD-MOCDE has better performance than that of MODE, which reveals that the gradient decent based mutation operator has more powerful search ability than DE. Furthermore, since culture algorithm is integrated into the differential evolution with three knowledge structures, which provides more convenience for population space evolution with constraint handling measurements.

The parameters setting on this test system can be presented as follows: The maximum generation number g_{max} is set to 1000, the size of evolutionary population and archive set are 50 and 30, the initial scaling parameter η_0 is set to 0.8, the permitted output accuracy is set to 0.01, total run time of the proposed GD-MOCDE is 127 s with 4.23 s to generate each scheme.

Table 6.3 The output (MW) details of compromise scheme obtained by GD-MOCDE for test system 1

Hours	P_1	P_2	P_3	P_4	P_5	P_6	P_7	P_8	P_9	P_{10}	Load	Loss
1	150	135	78.8	102.64	122.819	122.615	129.441	119.96	79.988	14.381	1036	19.644
2	150	135	80.719	119.347	172.141	122.786	129.539	119.983	80	22.928	1110	22.443
3	150	135.944	134.221	130.638	202.304	160	129.641	120	80	43.769	1258	28.517
4	150	174.864	182.467	180.638	223.275	160	130	119.979	80	40.513	1406	35.736
5	150	214.411	185.026	183.266	242.33	160	130	120	80	55	1480	40.033
6	200.039	219.709	236.219	233.266	242.806	160	129.906	120	80	54.987	1628	48.932
7	222.369	223.316	282.286	241.185	242.951	160	129.969	120	80	53.765	1702	53.841
8	231.064	274.325	283.617	258.289	243	160	130	120	80	55	1776	59.295
9	297.629	310.299	299.235	300	243	159.986	129.995	120	79.995	54.918	1924	71.057
10	335.804	338.087	340	300	243	160	130	120	80	54.68	2022	79.571
11	377.668	388.48	339.829	299.968	243	160	130	120	80	54.946	2106	87.891
12	402.803	411.799	340	300	242.987	160	129.888	120	80	55	2150	92.477
13	361.874	366.699	339.982	299.944	243	159.962	130	120	80	55	2072	84.461
14	300.405	309.374	300.266	299.976	242.893	160	130	120	79.989	52.193	1924	71.096
15	226.506	267.339	291.41	261.998	243	159.934	130	119.989	80	55	1776	59.176
16	164.252	222.342	211.573	233.852	243	160	130	120	80	33.203	1554	44.222
17	152.217	220.91	186.139	183.852	243	160	130	120	80	43.975	1480	40.093
18	225.945	222.472	209.02	231.826	243	160	130	120	80	55	1628	49.263
19	226.92	268.024	279.554	272.688	242.987	160	130	120	80	55	1776	59.173
20	304.747	317.342	337.171	299.768	243	159.974	130	119.989	79.922	55	1972	74.913
21	302.573	302.927	311.8	289.762	243	160	130	120	80	55	1924	71.062
22	222.573	222.927	231.8	239.762	235.796	160	129.595	120	80	34.713	1628	49.166
23	150	143.652	151.8	189.762	211.122	159.955	130	120	79.998	27.55	1332	31.839
24	150	135	105.349	139.762	171.46	158.427	129.534	119.925	80	19.799	1184	25.256

6.5.2 *Test System 2*

On the basis of test system 1, test system 2 extends 10 thermal units to 30 thermal units by tripling the 10 thermal units, all data details can also be found in literature [3]. Here, the proposed GD-MOCDE is implemented on this test system, and the obtained Pareto front is presented in Fig. 6.5, its representative non-dominated schemes are shown in Table 6.4.

In Fig. 6.5, it can be seen that the Pareto front consists of several non-dominated schemes, it has wider diversity distribution than that of MODE [13], which reveals that the maximum and minimum schemes are included in the Pareto front by GD-MOCDE. Moreover, the non-dominated schemes by MODE can be dominated by that of GD-MOCDE, which can be versified according to the data list shown in Table 6.4. The scheme by GD-MOCDE can dominate the scheme by MODE when their economic cost or emission rate is close to each other.

According to the comparison between GD-MOCDE and MODE, GD-MOCDE has more powerful search ability than MODE due to its efficient gradient decent based mutation operator. For further analysis on the efficiency of the proposed GD-MOCDE

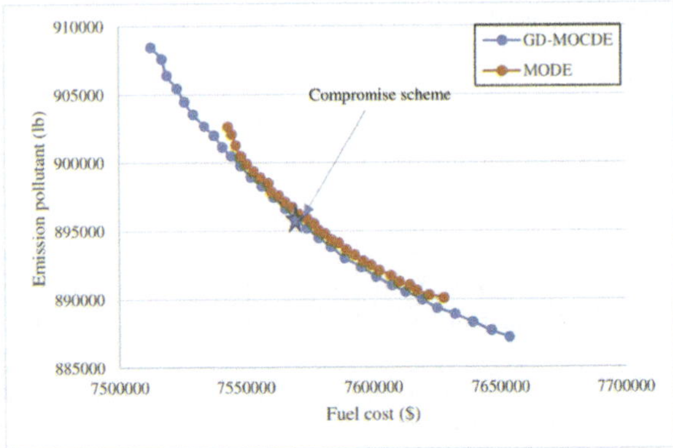

Fig. 6.5 The comparison of obtained schemes by GD-MOCDE and MODE for test system 2

Table 6.4 The obtained non-dominated schemes between GD-MOCDE and MODE for test system 2

Scheme	GD-MOCDE		MODE		Scheme	GD-MOCDE		MODE	
	Cost ($)	Emission (lb)	Cost ($)	Emission (lb)		Cost ($)	Emission (lb)	Cost ($)	Emission (lb)
1	7512698	908460	7543339	902682	16	7570424	895784	7578806	895074
2	7517057	907615	7544834	902102	17	7574099	895183	7581397	894805
3	7519031	906404	7546358	901297	18	7578714	894493	7583755	894362
4	7523032	905460	7548404	900480	19	7583391	893828	7586708	894118
5	7526022	904477	7550745	899914	20	7588798	893004	7589605	893632
6	7529408	903564	7553308	899391	21	7595174	892349	7592825	893249
7	7533904	902690	7556104	898905	22	7601177	891633	7596125	892823
8	7537738	902009	7559272	898515	23	7607476	891011	7599293	892509
9	7540961	901140	7560360	897886	24	7612856	890525	7602521	892117
10	7544545	900481	7563322	897586	25	7619351	889958	7607167	891745
11	7548101	899782	7565701	897142	26	7625108	889321	7610479	891286
12	7552086	898947	7568593	896755	27	7631962	888874	7614611	891036
13	7556468	898253	7571145	896269	28	7638948	888281	7617423	890654
14	7561210	897445	7574349	895908	29	7646313	887677	7622183	890320
15	7565745	896624	7576973	895544	30	7653408	887176	7627960	890081

on the larger scale of thermal system, scheme (16) is taken as the compromise scheme, which is also labeled in Fig. 6.5. The output process in scheme (16) is presented in Fig. 6.6, where the output of 30 thermal units can be seen at each time period.

Each bar represents the output of 30 thermal units at one time period, the bar of different colors presents the cumulative output of different thermal units. Since the transmission loss is taken into consideration, the transmission loss and system load at each time period is shown in Table 6.5. It can be seen that the transmission loss is range from 50 to 280 MW, and actually the obtained transmission loss doesn't exceed 5% of system load at each time period, which also reveals that the constraint handling technique is efficient on this DEED problem.

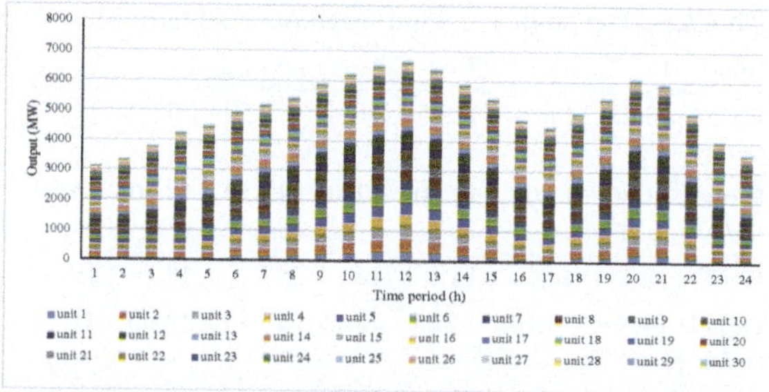

Fig. 6.6 The output (MW) process of compromise scheme obtained by GD-MOCDE for test system 2

Table 6.5 The system load and transmission loss for test system 2

Hour	1	2	3	4	5	6	7	8	9	10	11	12
Load (MW)	3108	3330	3774	4218	4440	4884	5106	5328	5772	6066	6318	6450
Loss (MW)	58.659	67.196	85.498	106.977	120.067	147.743	162.259	178.341	213.088	238.68	263.677	277.445
Hour	13	14	15	16	17	18	19	20	21	22	23	24
Load (MW)	6216	5772	5328	4662	4440	4884	5328	5916	5772	4884	3996	3552
Loss (MW)	253.331	213.332	178.392	133.499	120.04	147.752	177.779	225.169	213.343	147.645	95.631	75.743

In this DEED problem of larger-scale thermal system, the proposed GD-MOCDE also has excellent performance and keeps all the constraint limits under control. According to the comparison and analysis, GD-MOCDE can tackle with complex DEED problem due to its powerful search ability and constraint handling ability, which also reveals that the proposed GD-MOCDE can provide a promising way for complex system with multiple constraints or even stochastic constraints.

The parameter setting is presented as follows: The maximum generation number of GD-MOCDE and MODE is 2000, the size of evolutionary population and archive set are 50 and 30, the initial scaling parameter h0is set to 0.9, the permitted output accuracy is set to 0.05, and total run time of the proposed GD-MOCDE is 332s with 11.07s to generate each scheme.

6.5.3 *Test System 3*

The test system 3 consists of 7 thermal units combining with 2 wind farms and 1 photovoltaic field, and the transmission loss is also taken into consideration. In comparison to the DEED in test system 1, thermal unit 8 and thermal unit 9 are replaced by two wind farms, and thermal unit 10 is replaced by photovoltaic field. The wind speed data is adopted in literature [21], which is shown in Table 6.6. All

Table 6.6 The predicted wind speed on wind farm #1 and wind farm #2 for test system 3

Hour	1	2	3	4	5	6	7	8	9	10	11	12
Farm 1 (m/s)	13.25	14	12.75	11.90	12.50	13.90	11.80	12.75	12.90	12.20	15	13.25
Farm 2 (m/s)	11.80	12.00	12.20	12.40	12.50	14.00	15.00	14.50	13.00	13.75	13.40	13.40
Hour	13	14	15	16	17	18	19	20	21	22	23	24
Farm 1 (m/s)	14.30	14.10	14.25	11.75	13.75	12.60	11.50	11.90	14.50	16.00	12.70	13.00
Farm 2 (m/s)	12.80	12.25	11.40	11.50	11.00	11.25	11.10	11.00	11.45	11.80	11.75	12.25

Table 6.7 The corresponding output of wind farm #1 and wind farm #2 for test system 3

Hour	1	2	3	4	5	6	7	8	9	10	11	12
Farm 1 (MW)	99	108	93	82.8	90	106.8	81.6	93	94.8	86.4	120	99
Farm 2 (MW)	54.4	56	57.6	59.2	60	72	80	76	64	70	67.2	67.2
Hour	13	14	15	16	17	18	19	20	21	22	23	24
Farm 1 (MW)	111.6	109.2	111	81	105	91.2	78	82.8	114	120	92.4	96
Farm 2 (MW)	62.4	58	51.2	52	48	50	48.8	48	51.6	54.4	54	58

the optimization parameter settings are the same as that in test system 1, which can be seen in Sect. 6.5.1.

According to the wind speed predicted in Table 6.6, the output P_{rate} is set to 120 MW, wind speed v_{in}, v_{rate} and v_{out} are set as 5 m/s, 15 and 45 m/s, then the possible output of wind farm #1 and wind farm #2 can be calculated by formulation (27), which is presented in Table 6.7. Since power generation of wind turbine relays more on the wind speed, and the predicted power output can be taken as the maximum output. Here, it is assumed that the wind farm and photovoltaic power can be adjusted, the confidence interval needs to be taken into consideration due to the uncertainty of wind power.

With the probability density and cumulative distribution in- formation presented in Sect. 6.3.3.2, the output confidence interval of wind farm #1 and wind farm #2 can be properly obtained. If the degree of confidence ρ that the constraint limits are satisfied in Formula (6.5) is set to 0.8, the maximum output of wind farm #1 and wind farm #2 can be calculated as 103.4637 and 68.97582 MW. The parameter c is set to 15, and k is set to 2.2. Combining with the constraint limits of their output, the confidence interval of wind farm #1 and wind farm #2 are [47, 103.467] and [20, 68.976].

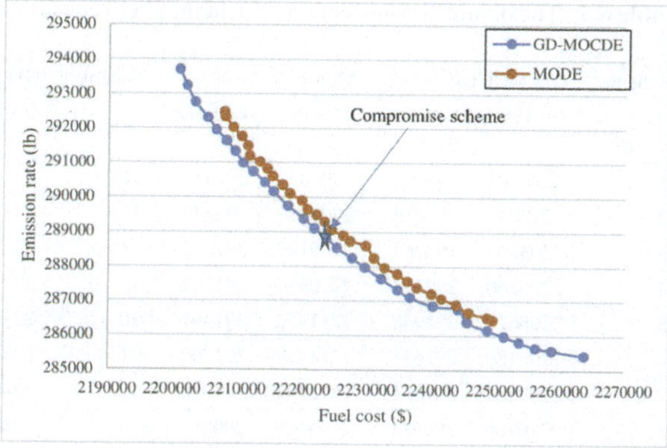

Fig. 6.7 The comparison of obtained schemes by GD-MOCDE and MODE for test system 3

Simultaneously, power generator is taken as the photovoltaic power generator, its probability density and cumulative distribution functions are presented in Sect. 6.3.3.1. According to the constraint handling method presented in Sect. 6.3.3.2, the confidence interval of photovoltaic output can be obtained. The parameter α and β are set to 2.0 and 1.0, the degree of confidence is set to 0.8. According to the calculation in Sect. 6.3.3.1, the confidence interval of photovoltaic power output is obtained as [10, 49.194].

On the basis of above analysis, the proposed GD-MOCDE is implemented on this test system, the parameter settings are taken as the same as that in test system 1, and the obtained Pareto front is presented in Fig. 6.7. For testifying the efficiency of the proposed GD-MOCDE, MODE is taken for comparison. It can be seen that the obtained Pareto front by GD-MOCDE is prior to that of MODE on both the convergence ability and diversity distribution, the non-dominated schemes on the Pareto front can dominate that of MODE.

The obtained 30 non-dominated schemes by GD-MOCDE and MODE are shown in Table 6.8. Since the number of thermal units in test system 3 decreases to 7, it can be seen that the total fuel cost and emission rate decreases in comparison to test system 1. Each obtained scheme by GD-MOCDE can dominate that of MODE on the identical fuel cost or emission rate, which means that the proposed GD-MOCDE has more powerful search ability than that of MODE for solving DEED of hybrid energy resource system.

For further analysis on the dispatching scheme obtained by GD-MOCDE, scheme (15) is taken as the compromise scheme, the output process of thermal units, wind farms and photovoltaic power generator is presented in Fig. 6.8. The output of different power generators is properly controlled in the feasible domain, and the ramp rate of each power generator is properly satisfied. It can be seen that two wind farms keep the maximum output under the confidence degree $\rho = 0.8$, and photovoltaic power generator also sustains the maximum output at the most time periods, which reveals that the renewable energy resources are made full use. Moreover, the transmission

Table 6.8 The obtained non-dominated schemes between GD-MOCDE and MODE for test system 3

Scheme	GD-MOCDE		MODE		Scheme	GD-MOCDE		MODE	
	Cost ($)	Emission (lb)	Cost ($)	Emission (lb)		Cost ($)	Emission (lb)	Cost ($)	Emission (lb)
1	2200904	293703	2207832	292481	16	2225285	288550	2224511	289066
2	2202041	293239	2207989	292342	17	2227651	288253	2226246	288904
3	2203305	292763	2209192	292024	18	2229635	287981	2227331	288742
4	2205260	292314	2210514	291762	19	2232181	287655	2229743	288605
5	2206621	291948	2211452	291491	20	2234701	287323	2231000	288254
6	2208186	291635	2211779	291202	21	2236676	287103	2232643	287971
7	2209525	291331	2213304	291030	22	2240295	286857	2234637	287792
8	2210766	290988	2214425	290824	23	2244159	286764	2236219	287579
9	2212286	290735	2215299	290606	24	2245663	286391	2237735	287397
10	2214109	290431	2216824	290359	25	2248779	286156	2240027	287210
11	2215468	290164	2217965	290104	26	2251362	285983	2241597	287066
12	2217679	289751	2219817	289901	27	2253721	285817	2243899	286893
13	2220142	289374	2220712	289655	28	2256267	285655	2245851	286663
14	2221791	289118	2222088	289498	29	2258783	285577	2248671	286525
15	2223621	288851	2223385	289302	30	2263760	285424	2249552	286454

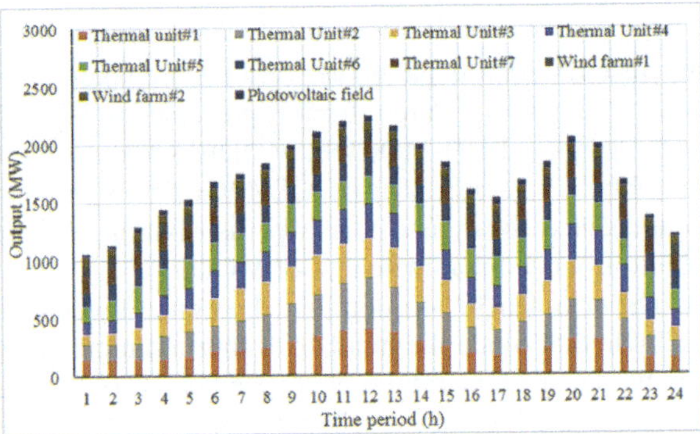

Fig. 6.8 The output (MW) details of compromise scheme under the degree of confidence $\rho = 0.8$ for test system 3

loss of each time period is also controlled under 5% in Table 6.9, which means that system load balance is properly handled.

According to above analysis on the DEED with several renewable energy resources, DEED problem become random while wind power and photovoltaic power integrating into the thermal power system, but if thermal cost, emission rate and transmission loss are taken for the main goal of DEED, these renewable energy resources need to be make full use, which provides a new challenge for solving DEED due to the uncertainty of wind power and photovoltaic power. This chapter takes the confidence interval to replace the power generation capacity of wind farm and photovoltaic power, which can be a relative feasible and reliable for solving this problem. Furthermore,

Table 6.9 The transmission loss under the degree of confidence $\rho = 0.8$ test system 3

Period (h)	1	2	3	4	5	6	7	8	9	10	11	12
Loss (MW)	19.545	22.305	28.511	36.035	40.363	49.322	54.315	59.718	71.506	80.206	88.624	93.241
Period (h)	13	14	15	16	17	18	19	20	21	22	23	24
Loss (MW)	85.156	71.414	59.792	44.586	40.295	49.391	59.542	75.452	71.678	49.677	32.12	25.166

the obtained results also reveal that the proposed GD-MOCDE can properly solve this DEED problem with its constraint handling technique for uncertainty in wind power and photovoltaic power.

6.5.4 *Test System 4*

The hydrothermal system mainly consists of four hydro plants coupled hydraulically and three thermal units, and the topology of four hydro plants (or reservoirs) is shown in Fig. 6.9. Reservoir 1 and reservoir 2 are located on the upstream of reservoir 3, which is also on the upstream of reservoir 4. The entire scheduling horizon is 24 h with each hour an interval, system load demand, water inflows, output limits and coefficients of hydro plants and thermal units can be seen from literature [22], and comparisons between proposed method and other alternatives are taken in this test system. The parameter settings are presented as follows: The population size is 100, the size of archive set is 30, evolutionary generation is 2000, the initial scaling parameter η_0 is set to 0.8, the total permitted accuracy is set to 0.1, coarse adjustment number is 15, fine tuning number is 10, the permitted water balance and output violation is 0.01.

30 non-dominated schemes are produced by the proposed GD-MOCDE in comparison to MODE [23], which can be seen in Fig. 6.10. It shows that the Pareto optimal schemes by GD-MOCDE dominates that of MODE, and also reveals that

Fig. 6.9 Hydraulic relationships among the four hydro-plants

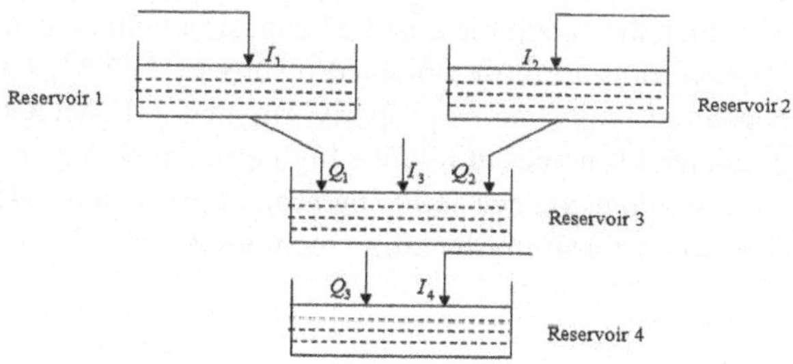

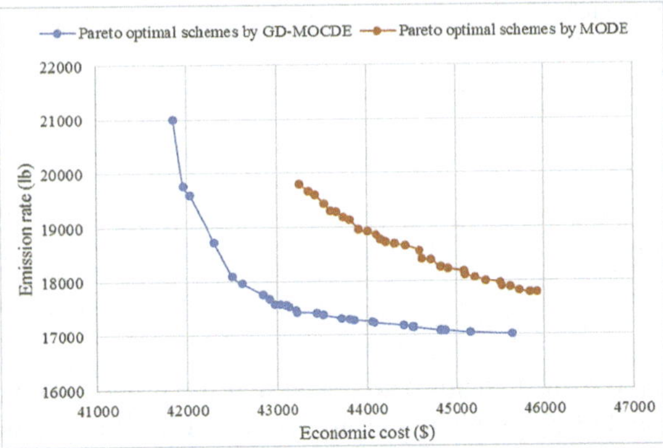

Fig. 6.10 The comparison between obtained Pareto optimal solutions by GD-MOCDE and MODE for test system 4

GD-MOCDE converges better than MODE. The obtained 30 non-dominated scheduling results are listed in Table 6.10.

Since few literature take nonlinear transmission loss into consideration, we take comparison with those obtained results without considering transmission loss, which are listed in Table 6.11. According to obtained results in Table 6.10, minimum fuel cost by GD-MOCDE is 41853$, which is smaller than ELS result of Refs. [2, 23], and minimum emission rate is 17013lb, which is also relative low in comparison to those references. Here, Scheme (15) is considered as compromise scheme with 43225$ and 17422lb, it can be obtained that fuel cost can be better than that in Ref. [2], and has lower emission rate while higher fuel cost than that in Ref. [23], 30 non-dominated schemes are generated in 1011s with each Scheme 37.3s, which is smaller than those methods in Refs. [2, 23].

For verifying the viability of obtained optimal schemes, compromise scheme is presented to take further analysis. The water discharge, hydro power output, thermal power output and transmission loss are listed in Table 6.12, the feasibility of each constraint limit can be checked, and transmission loss can't exceed 1% of system load at each time period.

The storage process of each hydro plant is presented in Fig. 6.11, it can be seen that hydro plant 3 and hydro plant 4, which are on the downstream and have larger storage ability, have obvious high storage than hydro plant 1 and hydro plant 2, then hydro resource can be made full use, more power output can be bear by hydro power to reduce the economic cost and emission volume caused by thermal power.

According to those obtained results by GD-MOCDE on SHOSEE problem, all constraint limits can be properly satisfied, GD-MOCDE can generate a set of non-dominated schemes at relative high efficiency, it can be revealed that GD-MOCDE has great convergence ability and can implement on SHOSEE problem well combing with some constraint-handling techniques.

Table 6.10 The comparison of non-dominated scheduling results between GD-MOCDE and MODE for test system 4

Scheme	GD-MOCDE		MODE		Scheme	GD-MOCDE		MODE	
	Thermalcost ($)	Emissioncost (lb)	Thermalcost ($)	Emissioncost (lb)		Thermalcost ($)	Emissioncost (lb)	Thermalcost ($)	Emissioncost (lb)
1	41853	20999	43249	19794	16	43446	17414	44589	18567
2	41960	19772	43351	19674	17	43508	17371	44616	18411
3	42035	19598	43422	19598	18	43721	17318	44715	18399
4	42302	18721	43523	19437	19	43806	17290	44827	18264
5	42497	18109	43591	19310	20	43853	17283	44904	18231
6	42604	17964	43659	19282	21	44052	17252	45092	18182
7	42847	17758	43742	19196	22	44086	17229	45108	18123
8	42913	17666	43813	19133	23	44414	17173	45217	18070
9	42972	17581	43911	18963	24	44507	17142	45337	18009
10	43039	17579	44003	18918	25	44525	17140	45498	17967
11	43096	17560	44104	18855	26	44826	17085	45523	17912
12	43105	17558	44156	18775	27	44833	17084	45619	17883
13	43135	17530	44211	18725	28	44880	17083	45717	17827
14	43210	17452	44313	18693	29	45166	17042	45836	17782
15	43225	17422	44434	18654	30	45638	17013	45922	17782

Table 6.11 Scheduling results in Refs. [2, 23]

Objective	Ref. [2] without transmission loss			Ref. [23]		
	ELS	EES	CEES	ELS	EES	SHOSEE
Fuel cost ($)	43,500	51,449	44,914	42167	45264	43165
Emission (lb)	21,092	18,257	19,615	19981	16713	17464
Comp. time (s)	72.96	72.74	74.97	–	–	1092 (30 schemes)

Table 6.12 The details of compromise schedule result by GD-MOCDE for test system 4

Hours	Water discharge($10^4 * m^3/s$)				Hydro power (MW)				Thermal power (MW)			P_L (MW)	Load (MW)
	Q_1	Q_2	Q_3	Q_4	P_{h1}	P_{h2}	P_{h3}	P_{h4}	P_{s1}	P_{s2}	P_{s3}		
1	9.754	6.068	22.141	6	84.845	49.48	34.664	131.88	175	134.706	143.363	3.938	750
2	7.066	6	23.633	6	68.958	50.127	20.884	129.027	172.105	207.008	136.54	4.649	780
3	7.41	6	23.071	6	71.866	51.258	18.289	125.744	174.54	124.888	137.143	3.728	700
4	8.14	6.01	19.323	6	76.803	52.969	34.377	121.625	103.889	209.776	53.07	2.509	650
5	7.382	6	18.232	6	71.527	54.46	37.023	115.822	128.891	124.949	140.367	3.039	670
6	6.524	6.07	18.025	6	64.995	55.971	37.041	130.943	173.253	203.412	139.048	4.663	800
7	9.454	8.025	19.341	7.125	83.612	69.462	31.881	160.994	174.959	209.686	225.802	6.396	950
8	8.828	6.837	17.96	11.844	79.793	60.788	36.124	230.947	174.914	209.817	223.976	6.359	1010
9	9.754	9.084	18.033	18.232	84.642	74.551	34.431	294.563	175	209.633	223.53	6.35	1090
10	9.775	8.79	17.175	18.025	84.833	72.342	36.721	293.098	174.935	209.706	214.526	6.161	1080
11	10.615	8.263	16.826	19.341	89.01	69.434	38.067	301.966	174.846	209.807	223.212	6.342	1100
12	7.459	6.656	18.977	17.96	72.232	59.569	30.683	292.63	175	209.74	318.799	8.653	1150
13	10.108	10.518	17.559	18.243	88.151	82.379	36.843	294.644	175	210.445	229.018	6.48	1110
14	8.905	7.445	19.699	17.127	82.198	64.181	30.826	286.236	175	126.02	271.757	6.218	1030
15	8.841	9.298	17.888	17.03	82.647	75.831	37.69	285.527	174.973	209.517	148.751	4.936	1010
16	6.238	6.069	17.831	18.611	64.955	55.471	38.626	296.821	174.474	209.699	226.355	6.401	1060
17	8.105	9.111	15.843	17.649	79.333	75.712	45.514	290.371	174.955	209.552	180.048	5.485	1050
18	8.375	10.849	15.593	19.703	81.257	83.038	46.881	304.124	175	209.823	226.286	6.409	1120
19	9.452	12.117	14.531	20	87.619	85.093	49.279	305.904	174.981	209.673	162.628	5.177	1070
20	9.51	12.314	14.925	20	87.378	82.41	48.92	303.735	175	209.642	147.838	4.923	1050
21	6.618	10.806	12.342	19.714	67.68	73.809	51.923	299.792	174.439	124.969	120.896	3.508	910
22	6.109	10.359	12.857	19.898	63.701	70.643	54.53	296.801	112.025	124.914	140.187	2.801	860
23	5.302	11.637	13.365	20	57.091	74.801	56.988	292.725	107.165	124.92	139.032	2.722	850
24	5.275	7.675	13.881	16.905	57.246	53.271	58.215	266.859	106.725	124.357	135.997	2.67	800

6.5.5 *Test System 5*

On the basis of hydrothermal system in test system 4, two wind farms and one photovoltaic field are added, and data details are shown in test system 3. In this test system, hydrothermal system represents traditional stable energy resource, wind power and photovoltaic power represent typical intermittent energy resource. The main target of this test system is to minimize the thermal cost and emission volume

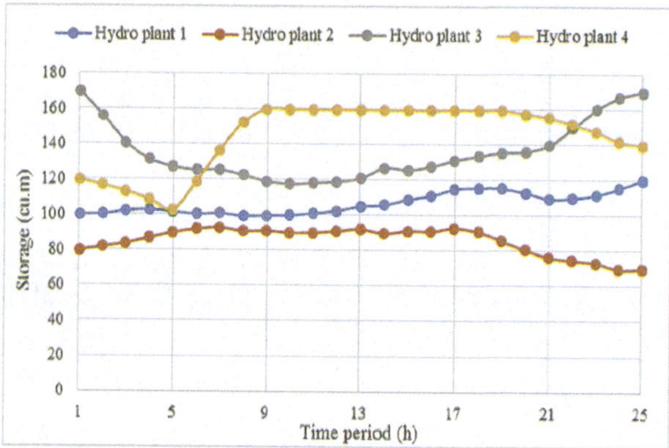

Fig. 6.11 The storage process of four hydro plants in compromise scheme

by properly assigning power output of each power generator under power generation uncertainty caused by intermittent energy resource. For properly tackling with wind power uncertainty, confidence interval is taken to describe the wind power generation process, and probability constraint is handled by the methods presented in Sects. 6.3.3.1 and 6.3.3.2. The parameter settings are shown as the same as test system 3 and test system 4.

The obtained Pareto front is shown in Fig. 6.12, 40 non-dominated schemes are produced by GD-MOCDE and MODE, which are shown in Table 6.13. It can be seen that obtained Pareto front by GD-MOCDE has wider diversity distribution and better convergence efficiency than that by MODE, and the minimum fuel cost and emission volume by GD-MOCDE is smaller than that by MODE. In comparison to hydrothermal optimal scheduling, fuel cost and emission volume of hybrid energy operation decrease sharply especially emission rate after wind farms and photovoltaic field is taken into consideration, which also presents the necessity for integration of renewable energy resources.

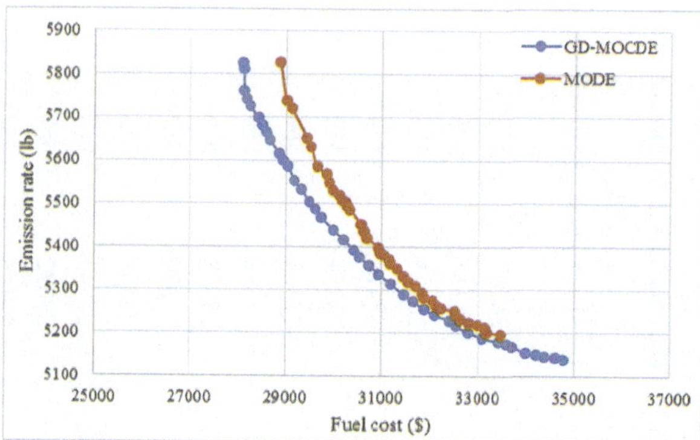

Fig. 6.12 The obtained Pareto front by GD-MOCDE and MODE for test system 5

Table 6.13 The comparison of non-dominated scheduling results between GD-MOCDE and MODE for test system 5

Scheme	GD-MOCDE		MODE		Scheme	GD-MOCDE		MODE	
	Thermalcost ($)	Emissioncost (lb)		Thermalcost ($)	Emissioncost (lb)	Thermalcost ($)	Emissioncost (lb)	Thermalcost ($)	Emissioncost (lb)
1	28107	5825	28884	5828	21	30551	5376	31136	5371
2	28117	5813	29003	5739	22	30734	5357	31174	5360
3	28123	5762	29121	5721	23	30938	5337	31312	5349
4	28178	5741	29445	5652	24	31186	5313	31446	5331
5	28254	5726	29536	5631	25	31448	5290	31549	5319
6	28423	5698	29679	5586	26	31657	5274	31705	5309
7	28498	5681	29871	5568	27	31877	5255	31833	5294
8	28578	5664	29922	5549	28	32096	5242	31882	5283
9	28670	5646	29995	5530	29	32411	5229	32057	5275
10	28853	5616	30137	5520	30	32534	5218	32114	5266
11	28943	5601	30188	5509	31	32788	5202	32159	5263
12	29032	5587	30254	5504	32	33075	5189	32234	5259
13	29165	5554	30284	5496	33	33415	5180	32512	5251
14	29329	5532	30354	5487	34	33585	5175	32542	5245
15	29488	5504	30585	5453	35	33694	5169	32640	5234
16	29611	5488	30645	5437	36	33986	5155	32805	5224
17	29740	5469	30701	5422	37	34208	5150	32981	5221
18	29992	5440	30927	5399	38	34386	5147	33141	5213
19	30209	5417	30981	5386	39	34588	5144	33177	5203
20	30421	5392	31045	5380	40	34777	5140	33471	5198

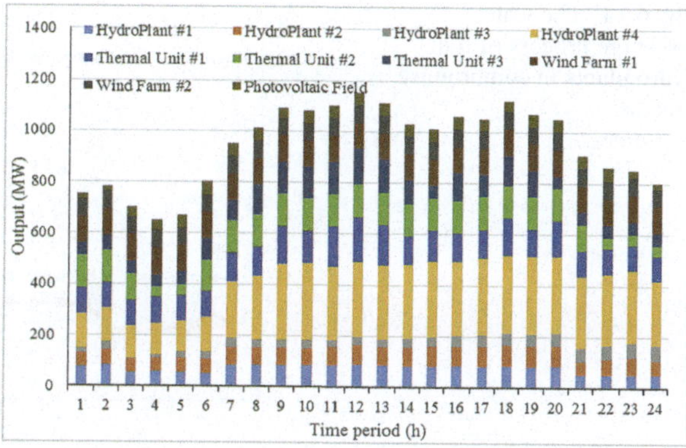

Fig. 6.13 The output process of hydro-thermal-wind-photovoltaic power in compromise scheme

Table 6.14 The transmission loss of hydro-thermal-wind-photovoltaic power system

Period (h)	1	2	3	4	5	6	7	8	9	10	11	12
Loss (MW)	1.76	1.82	1.615	1.243	1.281	2.023	2.154	2.522	3.135	2.797	3.325	3.757
Period (h)	13	14	15	16	17	18	19	20	21	22	23	24
Loss (MW)	3.349	2.24	1.979	2.403	2.218	3.039	2.343	2.23	1.586	1.248	1.247	1.224

The scheme (21) is taken as the compromise scheme for further analysis operation process, its output process of each power generator is shown in Fig. 6.13. It can be clearly seen that hydro plant #4 provides most output in the hydropower system, which can make most full use of hydropower from upstream water discharge. After checking the constraint limits of each power generator, it also can be found that output limits of each power generator and system load balance can be properly satisfied while considering transmission loss effects, which are listed in Table 6.14. The transmission loss can not exceed 4% of system load at each time interval, which also reveals that transmission loss is properly controlled.

Since hydropower generation mainly depends on the storage and water discharge of each power plant, further analysis is taken on the storage process and water discharge process, which are shown in Figs. 6.14 and 6.15. It can be seen that water discharge and storage process are similar with that without considering wind farms and photovoltaic field in shape, which reveals that wind power and photovoltaic power can merely affect the output value of hydro plant, but cannot affect the optimal operation measurement of hydro plant under system load balance constraint, those intermittent energy resources can merely bear the output burden of the whole operation system, the output measurement of each power generator doesn't relate to other energy resources, it only relates to its inner optimal scheduling measurement.

According to above results on the hydro-thermal-wind- photovoltaic power system, the proposed GD-MOCDE can solve SHOSEE-IIER problem well and can

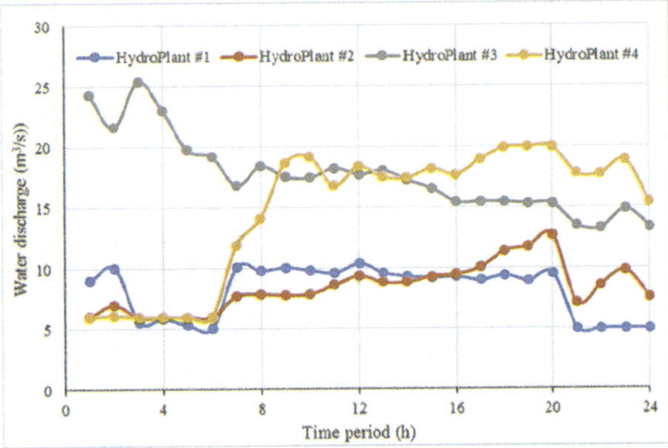

Fig. 6.14 The water discharge process of four hydro plants in compromise scheme

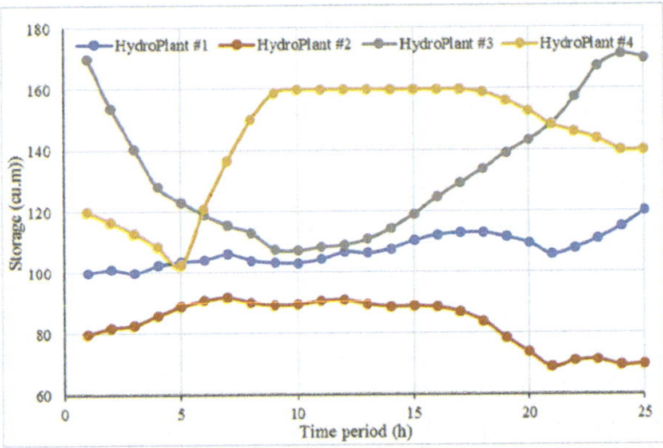

Fig. 6.15 The storage process of four hydro plants in compromise scheme

produce a set of non-dominated optimal schemes for decision-makers in the hybrid energy system. Simultaneously, it can be found that integration of wind power and photovoltaic can decrease the fuel cost and emission rate sharply, but it can't affect the operation measurement of each power generator, which also reveals that each energy resource can be considered as an independent system, and operation rule merely depends on its own inner characteristics, and we should know more about optimal scheduling rule of each energy resource.

6.6 Conclusion

This study extends hydrothermal optimal scheduling model into optimal scheduling of hybrid energy with integrating wind farms and photovoltaic fields. The hybrid energy system consists of hydro plants, thermal units, wind farms and photovoltaic fields, which makes hybrid energy system difficult to solve economic emission optimal scheduling of hybrid energy system due to its complexity, stochastic and constraint-coupled characteristics. In order to properly solve SHOSEE-IIER problem, this chapter proposes a gradient decent based multi-objective cultural differential evolution combining with several constraint-handling techniques. For improving the optimization efficiency, the proposed GD-MOCDE integrates gradient decent operator into cultural differential evolution, which can provide rapid search direction for population evolution and enhance the search ability for seeking optimal solution. Furthermore, several constraint handling techniques are utilized to tackle with probability constraint, water volume balance constraint and system load balance constraint, probability constraint is converted into deterministic constraint, heuristic constraint handling technique is used to deal with water volume water balance and system load balance constraint. Furthermore, five test systems are utilized to verify the efficiency of the proposed GD-MOCDE, it can be found that integration of wind power and photovoltaic can decrease the fuel cost and emission rate caused by thermal units sharply, and GD-MOCDE can optimize SHOSEE-IIER problem well combined with some constraint-handling techniques.

References

1. J. Cai, X. Ma, Q. Li, L. Li, H. Peng, A multi-objective chaotic particle swarm optimization for environmental/economic dispatch. Energy Convers. Manag. **50**(5), 1105–1113 (2009)
2. K. Deb, A. Pratap, S. Agarwal, T. Meyarivan, A fast and elitist multiobjective genetic algorithm: Nsga-ii. IEEE Trans. Evolut. Comput. **6**(2), 182–197 (2002)
3. M. Basu, Particle swarm optimization based goal-attainment method for dynamic economic emission dispatch. Electric Power Compon. Syst. **34**(9), 1015–1025 (2006)
4. R. Storn, K. Price, *Differential Evolution-a Simple and Efficient Adaptive Scheme for Global Optimization Over Continuous Spaces* (ICSI, Berkeley, 1995)
5. R. Reynolds, An introduction to cultural algorithms, in *Proceedings of the 3rd Annual Conference on Evolution Programming* (World Scientific Publishing, 1994), pp. 131–136
6. S.M. Saleem, *Knowledge-Based Solution to Dynamic Optimization Problems Using Cultural Algorithms*, Ph.D. thesis (Wayne State University, Detroit, 2001)
7. S. Yuenyong, A. Nishihara, A hybrid gradient-based and differential evolution algorithm for infinite impulse response adaptive filtering. Int. J. Adapt. Control Signal Process. **28**(10), 1054–1064 (2014)
8. J.Y.B. Cruz, L.R.L. Perez, J.G. Melo, Convergence of the projected gradient method for quasiconvex multiobjective optimization. Nonlinear Analy.-Theory Methods Appl. **74**(16), 5268–5273 (2011)
9. J. Desideri, Multiple-gradient descent algorithm (mgda) for multiobjective optimization. Comptes Rendus Math. **350**, 313–318 (2012)
10. G. Yu, T. Chai, X. Luo, Multiobjective production planning optimization using hybrid evolutionary algorithms for mineral processing. IEEE Trans. Evolut. Comput. **15**(4), 487–514 (2011)
11. R. Gelbard, A. Carmeli, R.M. Bittmann, S. Ronen, Cluster analysis using multi-algorithm voting in cross-cultural studies. Expert Syst. Appl. **36**(7), 10 438–10 446 (2009)
12. A.I. Selvakumar, K. Thanushkodi, A new particle swarm optimization solution to nonconvex economic dispatch problems. IEEE Trans. Power Syst. **22**(1), 42–51 (2007)
13. T. Robic, B. Filipic, Demo: Differential evolution for multiobjective optimization. Lect. Notes Comput. Sci. 520–533 (2005)

14. W.J. Zhang Haifeng, G. Feng, L. Kun, A dynamic economic dispatching model for power grid containing wind power generation system. Power Syst. Technol. **37**(5), 1298–1303 (2013)
15. S.H. Karaki, R.B. Chedid, R. Ramadan, Probabilistic performance assessment of autonomous solar-wind energy conversion systems. IEEE Trans. Energy Convers. **14**(3), 766–772 (1999). Sept
16. B. Qu, J. Liang, Y. Zhu, Z. Wang, P. Suganthan, Economic emission dispatch problems with stochastic wind power using summation based multi-objective evolutionary algorithm. Inf. Sci. **351**, 48–66 (2016)
17. Y. Lu, J. Zhou, H. Qin, Y. Wang, Y. Zhang, An adaptive chaotic differential evolution for the short-term hydrothermal generation scheduling problem. Energy Convers. Manag. **51**(7), 1481–1490 (2010)
18. J.Y. Lu, Y.H. Qin, Y. Zhang, A hybrid multi-objective cultural algorithm for short-term environmental/economic hydrothermal scheduling. Energy Convers. Manag. **52**(5), 2121–2134 (2011)
19. N. Pandit, A. Tripathi, S. Tapaswi, M. Pandit, An improved bacterial foraging algorithm for combined static/dynamic environmental economic dispatch. Appl. Soft Comput. **12**(11), 3500–3513 (2012)
20. X. Jiang, J. Zhou, H. Wang, Y. Zhang, Dynamic environmental economic dispatch using multi-objective differential evolution algorithm with expanded double selection and adaptive random restart. Int. J. Electrical Power Energy Syst. **49**, 399–407 (2013)
21. J. Aghaei, T. Niknam, R. Azizipanah-Abarghooee, J.M. Arroyo, Scenario-based dynamic economic emission dispatch considering load and wind power uncertainties. Int. J. Electrical Power Energy Syst. **47**, 351–367 (2013)
22. M. Basu, An interactive fuzzy satisfying method based on evolutionary programming technique for multiobjective short-term hydrothermal scheduling. Electric Power Syst. Res. **69**(2), 277–285 (2004)
23. H. Zhang, J. Zhou, Y. Zhang, Y. Lu, Y. Wang, Culture belief based multi-objective hybrid differential evolutionary algorithm in short term hydrothermal scheduling. Energy Convers. Manag. **65**, 173–184 (2013)

Chapter 7
Two-Layered Optimization Strategy for Hybrid Energy Systems with Price Bidding Based Demand Response

Due to uncertainty and dynamic characteristics from intermittent energy and load demand response (DR), it brings great challenge to optimal operation of hybrid energy system. This chapter proposes an event-triggered multi-agent coordinated optimization strategy with two-layered architecture, the structure of proposed strategy is presented in Fig. 7.1. Firstly, price-bidding based DR model is proposed with different stakeholders, and it also deduces optimal bidding price with Nash equilibrium theory. Then, four agents are designed to control different kind of energy resources, Agent 1 mainly analyzes the uncertainty or randomness caused by intermittent power, Agent 2 takes charge of dynamic economic dispatch (DED) within thermal units, Agent 3 manages the optimal scheduling of energy storage, and Agent 4 mainly undertakes load shifting strategy from consumers. In the upper-layer level, all agents coordinate together to ensure the stability of hybrid energy system with event-triggered mechanism, the intelligent control approach mainly depends on switching on/off power generators or curtailing system load, and consensus algorithm is utilized to optimize subsystem problem in lower-layer level. Furthermore, simulation results can further verify the efficiency of proposed method, and it also reveals that event-triggered multi-agent optimization strategy can be a promising way for solving hybrid energy system problem.

7.1 Price Bidding Strategy of Demand Response

The electricity price mainly depends on bidding among different stakeholders, which are market participants with pursuing profit for themselves. Each stakeholder can be considered as a player during price biding, generation cost and purchasing cost are considered for each player. The profit function of the $q(q = 1, 2..., Q)$th player can be described as:

D. Yue et al., *Cooperative Optimal Control of Hybrid Energy Systems*,
https://doi.org/10.1007/978-981-33-6722-7_7

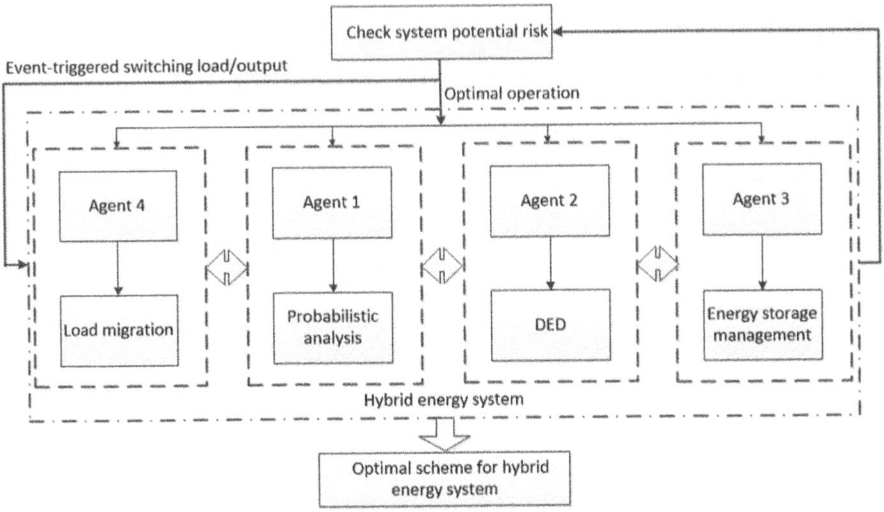

Fig. 7.1 The structure of event-triggered multi-agent optimization for hybrid energy system

$$\text{Maximize } f(x_{q,b,t}, P_{q,i}) \tag{7.1a}$$

$$\begin{cases} f_q(x_{q,b,t}, P_{q,i}(t)) = \\ \sum_{t \in T}[\gamma_{q,t}(P_{q,i}(t) - \underline{P_q}(t) - \sum_{b \in B} x_{q,b,t}) - C_{q,cost}] \end{cases} \tag{7.1b}$$

$$C_{q,cost} = \sum_{t \in T} \sum_{i \in I} [\alpha_{q,1i} + \alpha_{q,2i} P_{q,i}(t) + \alpha_{q,3i} P_{q,i}^2(t)] \tag{7.1c}$$

where $\gamma_{q,t}$ is market price, $\underline{P_q}(t)$, $\overline{P_q}(t)$ are minimum and maximum load, B is the blocking set, $x_{q,b,t} > 0$ is consumption assigned at bth block of tth time period, the size of each block is $\frac{\overline{P_q}(t) - \underline{P_q}(t)}{B}$, $C_{q,cost}$ presents the operational cost of hybrid energy system, $\alpha_{q,1i}, \alpha_{q,2i}, \alpha_{q,3i}$ are cost coefficients, $P_{q,i}(t)$ is power output, I is the set of all energy resources respectively. Some constraints should be properly satisfied:

$$0 \le B * x_{q,b,t} \le \overline{P_q}(t) - \underline{P_q}(t), \tag{7.1d}$$

$$\underline{P_{q,i}} \le P_{q,i}(t) \le \overline{P_{q,i}}, \tag{7.1e}$$

$$\underline{P_q}(t) + \sum_{b \in B} x_{q,b,t} - \underline{P_q}(t-1) - \sum_{b \in B} x_{q,b,t-1} \le R_{q,up,t}, \tag{7.1f}$$

$$\underline{P_q}(t-1)+\sum_{b\in B}x_{q,b,t-1}-\underline{P_q}(t)-\sum_{b\in B}x_{q,b,t}\le R_{q,dn,t} \tag{7.1g}$$

where $R_{q,up,t}$, $R_{q,dn,t}$ are ramp up and ramp down of total load. The Lagrangian function can be constructed with several penalty functions as follows:

$$\begin{aligned}L_q(x_{q,b,t},P_{q,i}) =& f_q(x_{q,b,t},P_{q,i})+\lambda_{q1}^{-}(Bx_{q,b,t}-\overline{P_q}(t)+\underline{P_q}(t))\\&+\lambda_{q2}^{+}(\underline{P_{q,i}}-P_{q,i}(t))+\lambda_{q2}^{-}(P_{q,i}(t)-\overline{P_{q,i}})\\&+\lambda_{q3}^{+}\left(\underline{P_q}(t)+\sum_{b\in B}x_{q,b,t}-\underline{P_q}(t-1)-\sum_{b\in B}x_{q,b,t-1}+R_{q,dn,t}\right)\\&+\lambda_{q3}^{-}\left(\underline{P_q}(t)+\sum_{b\in B}x_{q,b,t}-\underline{P_q}(t-1)-\sum_{b\in B}x_{q,b,t-1}-R_{q,up,t}\right)\end{aligned} \tag{7.2}$$

Once all players complete biding instead of corporation, Nash equilibrium optima means that there is no better scheduling scheme than current scheme $P_{q,i}^{*}$ by adjusting its own generation and bidding scheme. On the basis of Nash Equilibrium condition, it can obtain the iteration optimization algorithm as follows:

$$\frac{\partial L_q}{\partial P_{q,i}(t)}=\gamma_{q,t}-(\alpha_{q,2i}+2\alpha_{q,3i}P_{q,i}(t))-\lambda_{q2}^{+}+\lambda_{q2}^{-} \tag{7.3}$$

$$\frac{\partial L_q}{\partial x_{q,b,t}}=-\gamma_{q,t}+B\lambda_{q1}^{-}+\lambda_{q3}^{+}-\lambda_{q3}^{-} \tag{7.4}$$

The parameters λ_{q1}^{-}, λ_{q2}^{+}, λ_{q2}^{-}, λ_{q3}^{+} and λ_{q3}^{-} can be iterated with following equations:

$$\begin{cases}\lambda_{q1}^{-}=\lambda_{q1}^{-}+\beta_{q1}[Bx_{q,b,t}-\overline{P_q}(t)+\underline{P_q}(t)]^{-}\\ \lambda_{q2}^{+}=\lambda_{q2}^{+}+\beta_{q2}^{+}[\underline{P_{q,i}}-P_{q,i}(t)]^{+}\\ \lambda_{q2}^{-}=\lambda_{q2}^{-}+\beta_{q2}^{-}[P_{q,i}(t)-\overline{P_{q,i}}]^{-}\\ \lambda_{q3}^{+}=\lambda_{q3}^{+}+\beta_{q3}^{+}[\underline{P_q}(t)+\sum_{b\in B}x_{q,b,t}-\\ \underline{P_q}(t-1)-\sum_{b\in B}x_{q,b,t-1}+R_{q,dn,t}]^{+}\\ \lambda_{q3}^{-}=\lambda_{q3}^{-}+\beta_{q3}^{-}[\underline{P_q}(t)+\sum_{b\in B}x_{q,b,t}-\\ \underline{P_q}(t-1)-\sum_{b\in B}x_{q,b,t-1}-R_{q,up,t}]^{-}\end{cases} \tag{7.5}$$

where $\beta_{q1},\beta_{q2}^{+},\beta_{q2}^{-},\beta_{q3}^{+},\beta_{q3}^{-}\in\Re^{+}$. When According to Eq. (7.5), the optimal scheme can satisfy the minimum and maximum constraints after several iterations. The bidding price of the qth micro-grid can be presented as:

$$\gamma_{q,t}=\alpha_{q,2i}+2\alpha_{q,3i}P_{q,i}(t) \tag{7.6}$$

The market trading price is generally the highest bidding price among all the stakeholders, electricity price of the tth period γ_t can be obtained as:

$$\gamma_t = max\{\gamma_{q,t}, q = 1, 2, ..., Q\} \tag{7.7}$$

According to price biding, electricity market can have unified electricity price at tth period, the deduced bidding price can be taken for calculating switching cost in load shifting model.

7.2 Upper Level Problem: Event-Triggered Multi-agent Coordinated Optimization with Switching Mechanism

7.2.1 *Agent Definition*

For properly managing hybrid energy system, four agents are defined to control different energy resources. Agent 1 for intermittent energy resources, agent 2 for thermal energy resource, agent 3 for energy storage, agent 4 for system load demand.

7.2.1.1 Agent 1

Intermittent power system model. It mainly consists wind power and solar power, wind power follows Weibull distribution [1] and normalized solar power follows Beta distribution [2]. For simplicity, intermittent power output P_{Ijt} can be described as:

$$\begin{cases} P_{Ijt} = \overline{P_{Ijt}} + r_{Ijt}\widetilde{P_{Ijt}} \\ \widetilde{P_{Ijt}} \in [\widetilde{P_{Ijt,min}}, \widetilde{P_{Ijt,max}}] \end{cases} \tag{7.8}$$

where $j \in J$ is the intermittent power index, J is the number of intermittent energy resources, $\overline{P_{Ijt}}$ represents the estimated intermittent power output, $r_{Ijt} \in [0, 1]$ denotes adjustable parameter for each intermittent energy resource, $\widetilde{P_{Ijt}}$ represents power disturbance of intermittent energy, $\widetilde{P_{Ijt,min}}$ and $\widetilde{P_{Ijt,max}}$ denote lower and upper bounds of power disturbance.

7.2.1.2 Agent 2

Thermal power system model. Thermal units generate power output $P_{ck}(t)$ with consuming fuel, it is a economic issue [3, 4], economic cost can be presented as:

$$\begin{cases} minF_1 = \sum_{k\in K} f_{ck} = \sum_{k\in K}\sum_{t\in T} H_{kt}(a_k + b_k P_{ck}(t) + c_k P_{ck}^2(t)) \\ minF_2 = \sum_{k\in K}\sum_{t\in T} u_{ck} H_{kt} \end{cases} \tag{7.9}$$

where f_{ck} is the economic cost of $k \in K$th thermal unit, $H_{kt} \in 0, 1$ represents turn on/off state of thermal unit, a_k, b_k, c_k, d_k, e_k are the coefficients of economic cost of kth thermal unit, $P_{ck,min}$ is minimal output of kth thermal unit, u_{ck} is efficient of switching cost. It also subjects to several constraints as follows:

$$\begin{cases} P_{ck,min} \leq P_{ck}(t) \leq P_{ck,max} \\ DR_{ck} \leq P_{ck}(t) - P_{ck}(t-1) \leq UR_{ck} \end{cases} \tag{7.10}$$

where $P_{ck,max}$ is maximal output of $k \in K$th thermal unit, DR_{ck}, UR_{ck} are the down-ramp and down-ramp limits of kth thermal unit.

7.2.1.3 Agent 3

Energy storage model. For simplicity, battery energy storage system represents the whole energy storage of hybrid energy system, it supplements intermittent power to ensure the stability of whole power system. The charging and discharging output must satisfy some constraints [5]:

$$\begin{cases} \min F_3 = \sum_{l\in L} f_l^{store} = \sum_{l\in L}\sum_{t\in T}[\alpha_{l1} + \alpha_{l2} P_l^{store}(t) + \alpha_{l3}(P_l^{store}(t))^2] \\ V_l^{store}(t+1) = V_l^{store}(t) + \eta_l P_l^{store}(t) * \Delta T \\ V_{l,min}^{store} \leq V_l^{store}(t) \leq V_{l,max}^{store} \\ P_l^{store}(t) = P_l^{cha}(t), if\, P_l^{store}(t) \geq 0 \\ P_l^{store}(t) = -P_l^{dis}(t), if\, P_l^{store}(t) < 0 \\ 0 \leq P_l^{dis}(t) \leq P_{l,max}^{dis} \\ 0 \leq P_l^{cha}(t) \leq P_{l,max}^{cha} \\ V_l^{store}(0) = V_{l,initial}^{store} \end{cases} \tag{7.11}$$

where f_l^{store} denotes the economic cost of $l \in L$th energy storage, $P_l^{store}(t)$ represents charging/discharging output of lth battery at tth period, $V_l^{store}(t)$ is the storage of lth battery at tth time period, $\alpha_{l1}, \alpha_{l2}, \alpha_{l3}$ are coefficients of economic cost, ΔT is the time period length, $V_{l,min}^{store}$, $V_{l,max}^{store}$ are the minimum and maximum storage of the lth battery, $P_l^{dis}(t)$, $P_l^{cha}(t)$ are the output of discharging and charging state, $P_{l,max}^{dis}$, $P_{l,max}^{cha}$ are the maximum discharging and charging output at $l \in L$th battery at tth time period. $\eta_l \in (0, 1]$ is efficiency factor of charging or discharging state.

7.2.1.4 Agent 4

Load shifting model. On the demand side, DR model can be generally classified as: incentive-based model and price-based model, here it chooses price-based model to describe the state of load requirements. System load can be divided into two parts: fixed load $\overline{P_{load}}(t_i)$ and controllable load $\widetilde{P_{load}}(t_i)$ as:

$$P_{load}(t_i) = \overline{P_{load}}(t_i) + \widetilde{P_{load}}(t_i),\ t_i \in T \tag{7.12}$$

In the power system, controllable load can be adjusted to keep the system load balance when power supply cannot meet load requirement from demand side, then some load must be cut down through switching off them, it can be described as:

$$\widetilde{P_{load}}(t_i) = \sum_{s \in S} B_{s,t_i} P_s(t_i) \tag{7.13}$$

Generally, system load cannot be adjusted, it brings switching cost as:

$$\min F_4 = \sum_{t_i \in T} \gamma_{t_i} \sum_{s \in S} (1 + B_{s,t_i}) P_s(t_i) \tag{7.14}$$

$$\text{Subject to } \widetilde{P_{load}}(t_i) = \sum_{s \in S} \left(\sum_{t_j \in T, t_j \neq t_i} P_{st_jt_i} - \sum_{t_j \in T, t_j \neq t_i} P_{st_it_j} \right) \tag{7.15}$$

$$P_s(t_i) = \sum_{t_j \in T} P_{st_jt_i} - \sum_{t_j \in T, t_j \neq t_i} P_{st_it_j} \geq 0 \tag{7.16}$$

$$\sum_{t_i \in T} P_s(t_i) = M_s \geq 0 \tag{7.17}$$

$$P_{s,min} \leq P_{st_it_j} \leq P_{s,max},\ \forall t_i, t_j \in T, t_i \neq t_j \tag{7.18}$$

where B_{s,t_i} is binary number of each consumers, S represents the number of consumers, P_s is the consumption of consumer in one day, it can be assumed that it is an invariant, $P_{st_it_j}$ means that consumer s moves load consumption from t_i period to t_j period, M_s is a real number, which means that electricity consumption of each consumer is certain, $P_{s,min}$ and $P_{s,max}$ are the minimum and maximum value for moving load consumption.

7.2.2 Event-Triggered Optimization of Hybrid Energy System with Probabilistic Risk

In hybrid energy system, power supply must meet the requirement from demand side, but when power generation cannot satisfy the system load, some controllable load will be cut down, system load topology of agent 4 can be switched to a different one, which greatly increase the difficulty for optimizing hybrid energy system. Here, event-triggered method is utilized to judge switching model with probabilistic risk, which is mainly caused by imbalance between power supply and load demand. Thus, it is assumed that the expected value of total power output approximates system load as follows:

$$E(P_{total,t}) \to P_{load}(t) \tag{7.19}$$

where $P_{total,t}$ denotes the summation of power output of all energy resources. For ensuring that power generation meets load requirement, the probability of above formula needs to satisfy:

$$Prob(|P_{total,t} - P_{load}(t)| \leq \epsilon_t) \geq \delta_t \tag{7.20}$$

where $Prob()$ denotes the probability operator, $\epsilon_t \in R^+$ represents the deviation error, $\delta_t \in (0, 1)$ means the smallest permitted probability, it can also be converted into other form as follows:

$$Prob(|P_{total,t} - P_{load}(t)| \geq \epsilon_t)) \leq 1 - \delta_t \tag{7.21}$$

With considering system load balance, it can obtain:

$$\begin{aligned} Prob\Bigg(& |\sum_{j\in J} r_{Ijt}\widetilde{P_{Ijt}} - [P_{load}(t) - \sum_{j\in J} \overline{P_{Ijt}} - \sum_{k\in K} P_{ck}(t) \\ & - \sum_{l\in L} P_l^{store}(t)]| \geq \epsilon_t)\Bigg) \leq 1 - \delta_t \end{aligned} \tag{7.22}$$

Suppose that parameters r_{Ijt} are independent variables, it can obtain inequality with Chebyshev inequality as follows:

$$\frac{Var(\sum_{j\in J} r_{Ijt}\widetilde{P_{Ijt}})}{\epsilon_t^2} \leq 1 - \delta_t \tag{7.23}$$

It can deduce the permitted deviation error:

$$\epsilon_t^* = \sqrt{\frac{Var(\sum_{j\in J} r_{Ijt}\widetilde{P_{Ijt}})}{1 - \delta_t}} \tag{7.24}$$

The deviation can guide switching scheme of multi-agent system for controlling both power supply and load demand. Once generated power cannot meet load requirement (calculated deviation ϵ_t is larger than ϵ_t^*), it needs to cut off some controllable load or turn off some power generators to keep the balance. Here, an efficient switching scheme is proposed to coordinate different agent-based subsystems as follows.

Algorithm 7.1 Event-triggered based coordinated optimization

procedure S(w)itching scheme for power balance
  Check balance $|P_{total,t} - P_{load}(t)|$
  if $\epsilon_t < \epsilon_t^*$ **then**
    switch on thermal unit P_{ck}
    For $k = \xi : K$
    $P_{total,t} = P_{total,t} + P_{ck}$
    Until $P_{load}(t) \in Range(P_{total,t})$
    if $Maxtotal_t < P_{load}(t)$ **then**
      switch off controllable load P_s
      For $s = 1 : \eta$
      $P_{load}(t) = P_{load}(t) - P_s$
      Until $Maxtotal_t \in Range(P_{load}(t))$
    end if
  end if
end procedure

---

where ϵ is the number of current thermal units turned on, $Range()$ denotes the possible interval, $Maxtotal_t$ represents current maximum output of all energy resources at tth period, η is current number of consumers using the electricity. After switching scheme for power balance is made, coordinated optimization strategy is also made to optimize multi-agent system as follows:

Algorithm 7.2 Event-triggered based coordinated optimization

procedure C(o)ordination scheme of hybrid energy system
  Agent 4: Making load shifting scheme
  Agent 1: Probabilistic analysis in small intervals
  $P_{load}(t) = P_{load}(t) - P_{Intermittent,t}$
  if $P_{load}(t) < Maxtotal_t$ **then**
    Agent 2: DED on thermal units
    goto End
  end if
  $P_{load}(t) = P_{load}(t) - Maxtotal_t$
  Agent 3: Energy storage management
end procedure

---

where $P_{Intermittent,t}$ represents the total output of intermittent energy resources at tth time period.

7.3 Lower Level Problem: Convex Optimization for Multi-agent Subsystem

In upper level, event-triggered switching mechanism and coordination strategy have been made, but subsystem of each agent still needs to be properly optimized. Here, several optimization approaches are utilized for solving above problems. Agent 1 mainly analyzes probabilistic characteristics of intermittent energy resources with statistical methods, it doesn't need optimization. Actually, optimization for Agent 3 cannot be a big problem, it can arrange energy storage from big capacity to small capacity until system load is properly satisfied, which can also save switching cost. Here, it focus on the optimization of subsystems in Agent 4 and Agent 2. In Agent 4, Lagrangian relaxation approach is utilized to obtain the iterated algorithm, which can deduce the optimal load shifting scheme. Since problem formulation of Agent 2 can be a DED problem, consensus algorithm can be a better way, consensus algorithm with ADMM is utilized to assign output for each thermal unit.

7.3.1 Lagrangian Relaxation Approach for Load Shifting

The load shifting model reflects consumers' action with disturbance of electricity price, but consumers shift load from one time to another one with shifting cost, so how to make shifting scheme for consumers can be a big problem. Here, Lagrangian relaxation approach is utilized to optimize this problem. Firstly, combine with Lagrangian relaxation operator, it can obtain Lagrangian function as follows:

$$\begin{aligned} L_{load} &= F_4 + \sum_{t_i \in T}\left[\lambda^{+}_{s,t_i}\left(\sum_{t_j \in T} P_{st_jt_i} - \sum_{t_j \in T, t_j \neq t_i} P_{st_it_j}\right)\right] \\ &+ \sum_{s \in S}\lambda_s\left(\sum_{t_i \in T} P_s(t_i) - M_s\right) + \sum_{t_i \in T}\sum_{t_j \in T, t_j \neq t_i}[\lambda^{+}_{s,t_i,t_j} \\ &(P_{s,min} - P_{st_it_j}) - \lambda^{-}_{s,t_i,t_j}(P_{st_it_j} - P_{s,max})] \end{aligned} \quad (7.25)$$

where λ_{s,t_i^+}, λ_s, λ^{+}_{s,t_i,t_j} and λ^{-}_{s,t_i,t_j} are Lagrangian parameters. It can obtain following equations:

$$
\begin{cases}
\frac{\partial L_{load}}{\partial P_{st_i t_j}} = -\sum_{t_i \in T} \gamma_{t_i} \sum_{s \in S} (1 + B_{s,t_i}) + \lambda^{+}_{s,t_i} + \\
\lambda_s + \sum_{t_i \in T} \sum_{t_j \in T, t_j \neq t_i} (\lambda^{-}_{s,t_i,t_j} - \lambda^{+}_{s,t_i,t_j}) \\
\frac{\partial L_{load}}{\partial \lambda^{+}_{s,t_i}} = \sum_{t_j \in T} P_{st_j t_i} - \sum_{t_j \in T, t_j \neq t_i} P_{st_i t_j} \geq 0 \\
\frac{\partial L_{load}}{\partial \lambda_s} = \sum_{t_i \in T} P_s(t_i) - M_s \\
\frac{\partial L_{load}}{\partial \lambda^{+}_{t_i,t_j}} = P_{st_i t_j} - P_{s,min} \geq 0 \\
\frac{\partial L_{load}}{\partial \lambda^{-}_{t_i,t_j}} = P_{s,max} - P_{st_i t_j} \geq 0
\end{cases}
\tag{7.26}
$$

With above equations, the best $P_{st_i t_j}$, λ_{s,t_i^+}, λ_s, λ^{+}_{s,t_i,t_j} and λ^{-}_{s,t_i,t_j} can be deduced after several iterations, so optimal load shifting scheme can be properly made.

7.3.2 Consensus with Regularization Algorithm for DED

Combined with Lagrangian operator, since switching strategy has been made from upper-level mechanism, it merely needs to take fuel cost into consideration, it can be converted into following mathematical model:

$$
\begin{aligned}
L_{ck} = & \sum_{k \in K} f_{ck} + \lambda_{c1}(P_{ck}(t) - P_{ck,min} - d_1) \\
& + \lambda_{c2}(P_{ck,max} - P_{ck}(t) - d_2) + \lambda_{c3}(P_{ck}(t) \\
& - P_{ck}(t-1) - DR_{ck} - d_3) + \lambda_{c4}(UR_{ck} \\
& + P_{ck}(t-1) - P_{ck}(t))
\end{aligned}
\tag{7.27}
$$

where λ_{c1}, λ_{c2}, λ_{c3}, λ_{c4} represent the Lagrangian parameters. For accelerating search ability, it is converted into a distributed way with consensus theory. With equal increment criterion, it can obtain:

$$
\frac{\partial f_{ck}}{\partial P_{ck}(t)} = -\lambda_{c1} + \lambda_{c2} - \lambda_{c3} + \lambda_{c4} \tag{7.28}
$$

For output of each thermal unit, it can be deduced:

$$
P^{*}_{ck}(t) = (\lambda_{c2} - \lambda_{c1} - \lambda_{c3} + \lambda_{c4} - b_k)/(2c_k) = z \tag{7.29}
$$

where $P^{*}_{ck}(t)$ is the utopia optima of $P_{ck}(t)$, z is common global variable. Combined with alternating direction method of multipliers (ADMM) algorithm [6], equal increment criterion can be treated as constraint limit, regularization operator is taken in iterations to achieve synchronization, it can obtain following iterative procures:

$$\begin{cases} P_{ck}^{n+1} := \underset{P_{ck}}{\arg\min}[L_{ck} + (\rho/2)||P_{ck} - z^n + \mu_{ck}^n||_2^2] \\ z^{n+1} := \underset{z}{\arg\min}[g(z) + (K\rho/2)||z - \bar{P_c}^{n+1} - \bar{\mu_c}^n||_2^2] \\ \mu_{ck}^{n+1} := \mu_{ck}^n + P_{ck}^{n+1} - z^{n+1} \end{cases} \tag{7.30}$$

where n is iteration number, μ_{ck}^n represents scaled dual variable, $\rho > 0$ is augmented Lagrangian parameter, $\bar{P_c}$ is average value of thermal units, $\bar{\mu_c}$ is average value of μ_{ck}. Since L_{ck} and $g(z)$ are differentiable and KKT conditions can be properly satisfied, the $argmin[\cdot]$ operator here is mainly implemented with derivation to deduce the extreme value of P_{ck} and z, which are taken as P_{ck}^{n+1} and z^{n+1} for iterations. With consideration of feasibility of iterations, the procedure is implemented as follows:

$$P_{ck}^{n+1} = \begin{cases} \overline{P_{ck}} & P_{ck}^n > \overline{P_{ck}} \\ \underline{P_{ck}} & P_{ck}^n < \underline{P_{ck}} \end{cases} \tag{7.31}$$

where $\overline{P_{ck}}$ and $\underline{P_{ck}}$ represent upper bound and lower bound of feasible domain. The convergence can be ensured with satisfying two assumptions in literature [7]: (1) f_{ck} and $g(z)$ (actually $g(z) = 0$ when it is implemented) are both closed, proper and convex; (2) The function L_{ck} has at least one saddle point, because f_{ck} is monotonically increasing function. Once the above iteration converges, it means $P_{ck}^{n+1} \to P_{ck}^*$, optimal scheme can be made for dispatch problem of thermal units.

7.4 Case Study

For verifying feasibility and efficiency of proposed algorithm, it is implemented in two test systems: hybrid energy system without switching mode and hybrid energy system with switching model. Test system 1 can be considered as traditional optimal operation without considering DR, and event-triggered multi-agent optimization is not involved. While in test system 2, all factors are taken into consideration, and comparison with test system 1 can reflect the priority of event-triggered multi-agent optimization for hybrid energy system.

7.4.1 *Test System 1: Hybrid Energy System Without Switching Mode*

This test system includes 4 wind farms, 3 photovoltaic fields, 10 thermal units and 4 energy storage, all details can be found in literature [8, 9]. The wind power can calculated with wind speed, and photovoltaic power is closely related to illumination intensity. The predicted wind and PV power output at least 85% confidence interval are presented in Tables 7.1 and 7.2, which list upcoming output interval for 24 h. The system load includes five different kinds of load: Load #1, Load #2, Load #3, Load #4 and Load #5, which can be found in Fig. 7.2. For minimizing the expected value of total economic cost, it needs to find optimal scheme of ten thermal units and 4 energy storage, here it is presented in Figs. 7.3 and 7.4. Since state of thermal units

Table 7.1 85% confidence interval of wind power generation

Period	Wind 1	Wind 2	Wind 3	Wind 4	Period	Wind 1	Wind 2	Wind 3	Wind 4
00:00–00:59	[32, 45]	[30, 42]	[30, 40]	[25, 34]	12:00–12:59	[16, 22]	[15, 21]	[12, 16]	[13, 19]
01:00–01:59	[35, 45]	[35, 41]	[32, 38]	[29, 35]	13:00–13:59	[20, 26]	[20, 26]	[17, 23]	[17, 23]
02:00–02:59	[35, 44]	[34, 40]	[30, 36]	[25, 43]	14:00–14:59	[25, 31]	[22, 30]	[22, 28]	[21, 27]
03:00–03:59	[29, 35]	[27, 35]	[23, 29]	[18, 24]	15:00–15:59	[30, 38]	[28, 36]	[27, 35]	[25, 33]
04:00–04:59	[20, 28]	[20, 26]	[16, 24]	[12, 18]	16:00–16:59	[26, 34]	[24, 32]	[24, 30]	[22, 28]
05:00–05:59	[15, 21]	[13, 19]	[12, 18]	[10, 18]	17:00–17:59	[24, 30]	[22, 26]	[20, 26]	[19, 25]
06:00–06:59	[18, 26]	[15, 23]	[13, 20]	[13, 20]	18:00–18:59	[22, 28]	[19, 25]	[18, 24]	[17, 23]
07:00–07:59	[22, 28]	[19, 25]	[17, 23]	[14, 22]	19:00–19:59	[15, 20]	[17, 23]	[15, 21]	[15, 21]
08:00–08:59	[22, 30]	[22, 28]	[20, 24]	[17, 23]	20:00–20:59	[22, 28]	[23, 29]	[20, 26]	[19, 25]
09:00–09:59	[20, 26]	[18, 24]	[15, 23]	[15, 21]	21:00–21:59	[25, 32]	[28, 34]	[25, 33]	[23, 29]
10:00–10:59	[17, 23]	[15, 19]	[12, 18]	[12, 18]	22:00–22:59	[31, 39]	[27, 35]	[25, 33]	[23, 30]
11:00–11:59	[17, 23]	[15, 21]	[12, 18]	[12, 16]	23:00–23:59	[33, 43]	[32, 40]	[30, 38]	[27, 35]

Table 7.2 85% confidence interval of PV output

Period	PV 1	PV 2	PV 3	Period	PV 1	PV 2	PV 3
00:00-00:59	[0, 0]	[0, 0]	[0, 0]	12:00–12:59	[28, 36]	[24, 32]	[26, 34]
01:00–01:59	[0, 0]	[0, 0]	[0, 0]	13:00–13:59	[25, 35]	[23, 29]	[27, 33]
02:00–02:59	[0, 0]	[0, 0]	[0, 0]	14:00–14:59	[23, 29]	[20, 24]	[23, 29]
03:00–03:59	[2, 4]	[0, 0]	[1, 3]	15:00–15:59	[20, 24]	[16, 20]	[20, 24]
04:00–04:59	[4, 6]	[1, 3]	[2, 6]	16:00–16:59	[15, 19]	[14, 18]	[15, 21]
05:00–05:59	[8, 12]	[6, 10]	[7, 11]	17:00–17:59	[10, 14]	[11, 15]	[10, 14]
06:00–06:59	[11, 15]	[10, 14]	[8, 12]	18:00–18:59	[6, 8]	[8, 12]	[6, 10]
07:00–07:59	[15, 21]	[13, 19]	[12, 16]	19:00–19:59	[1, 3]	[3, 5]	[4, 6]
08:00–08:59	[16, 22]	[17, 23]	[17, 23]	20:00–20:59	[0, 0]	[0, 0]	[0, 2]
09:00–09:59	[20, 26]	[20, 26]	[17, 23]	21:00–21:59	[0, 0]	[0, 0]	[0, 0]
10:00–10:59	[23, 29]	[22, 28]	[20, 24]	22:00–22:59	[0, 0]	[0, 0]	[0, 0]
11:00–11:59	[23, 29]	[25, 31]	[24, 30]	23:00–23:59	[0, 0]	[0, 0]	[0, 0]

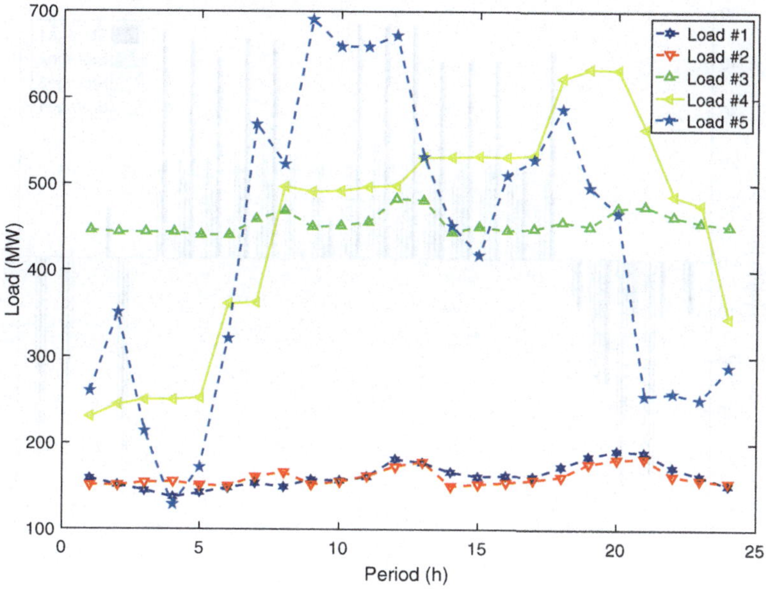

Fig. 7.2 System load before load shifting strategy

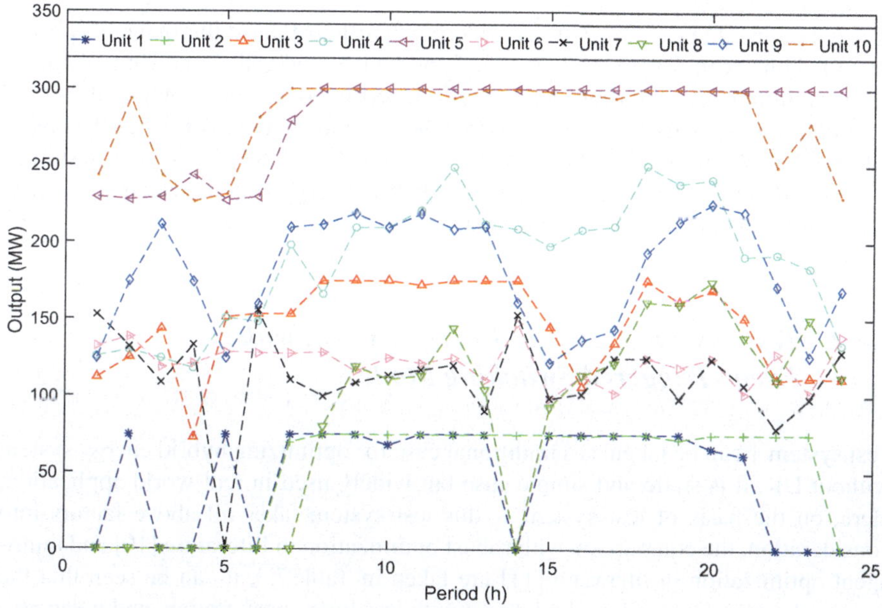

Fig. 7.3 The Output of ten thermal units for minimizing economic cost

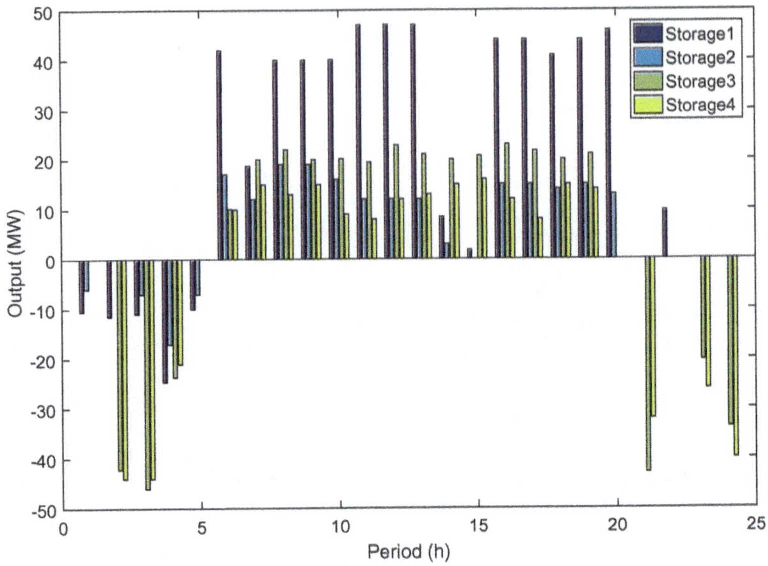

Fig. 7.4 The charging and discharging process of energy storage

is closely related to capacity with considering on/off cost, most thermal units always be turned on except Unit 1, Unit 2, Unit 7 and Unit 8, and Unit 5 and Unit 10 almost keep maximum output during 24 h. In Fig. 7.3, permitted minimum output of thermal units does not equal to 0 MW, it means that thermal unit is turned off when the output achieve 0 MW, and it is also the same for energy storage. In Fig. 7.4, energy storage is frequently utilized to keep the stability of hybrid energy system, which also means that it will generate more economic cost.

7.4.2 *Test System 2: Hybrid Energy System with Event-Triggered Switching Mode*

Test system 1 can be taken as a traditional case for optimizing hybrid energy system without DR, it is static and simple case but widely used in real-world application. Here, on the basis of test system 1, this test system takes all above factors into consideration, the comparison with robust optimization in Literature [10] and multi-agent optimization in literature [11] are taken in Table 7.3, it can be seen that the proposed method has minimal total cost with less time consumption, and it can also ensure the safety of hybrid energy system with high average confidence degree. Since load migration is taken into consideration, consumers can arrange proper timing for electricity consumption, it can be seen in Fig. 7.5, in which system load can be more stable, electric peak has been curtailed and electricity trough has been supplemented

Table 7.3 The comparison with other optimization methods

Methods	Literature [2]	Literature [18]	The proposed method
Fuel cost ($)	33463	31766	30071
On/off cost ($)	7382	7124	7233
Charging/discharging cost ($)	10531	6254	5677
Load shifting cost ($)	0	1011	2135
Total cost ($)	51376	46155	45116
Time (s)	68	59	56
Average confidence degree (%)	87	82	91

in comparison to original load, five kinds of system load after shifting have been presented in Fig. 7.6. The obtained output of thermal units has been presented in Fig. 7.7, it can be found that Unit 1 and Unit 2 are almost turned off during the whole period, which can save total economic cost better than that in test system 1. With considering charging and discharging cost, energy storage can be used merely when other power generation cannot meet load requirement in Fig. 7.8, obviously energy storage is seldom used in comparison that in test system 1. It can be noted that there are three key periods, where energy storage has been used. Actually, it adjusts potential risk to minimum extent, which can be found in Fig. 7.9. Here, δ_t can be set as 0.85, once confidence degree is smaller than it, event-triggered switching mechanism can be utilized to decrease the potential risk or improve the confidence degree. As it is shown in Fig. 7.9, there are five dangerous periods with low confidence level in original load, but they are improved after utilizing event-triggered switching mechanism, which also proves the feasibility of proposed method. In subsystem level, consensus with ADMM is employed to optimize economic dispatch of ten thermal units, those obtained results are listed in Table 7.4. It can be found that total cost has been greatly reduced especially charging and discharging cost in comparison to test system 1. In addition, convergence process has also been analyzed in comparison to DP and QP methods in Fig. 7.10, though DP and QP have good performance before 50 iterations, QP fall into premature problem after 100 iterations and search ability is still not good enough. With above comparison and analysis, four designed agents can work properly in hybrid energy system, the proposed event-triggered switching mechanism based multi-agent optimization can improve optimal efficiency for reducing the total economic cost as well as decreasing potential risk, which can ensure the reliability of hybrid energy system.

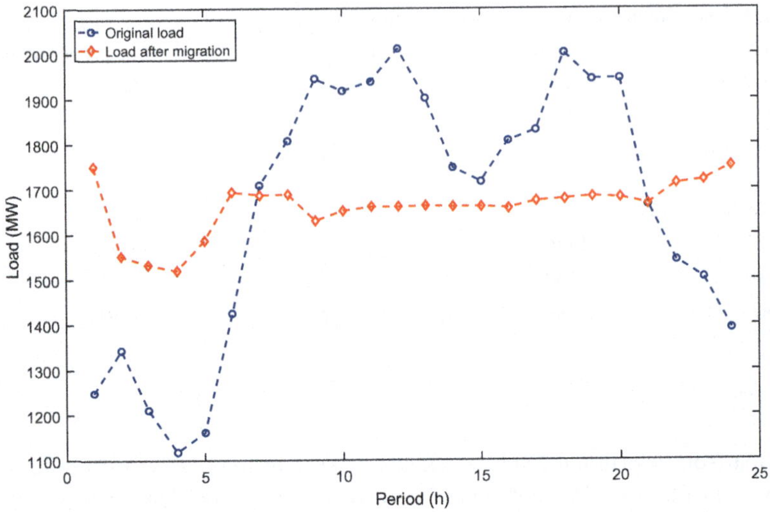

Fig. 7.5 The original load and load after load migration

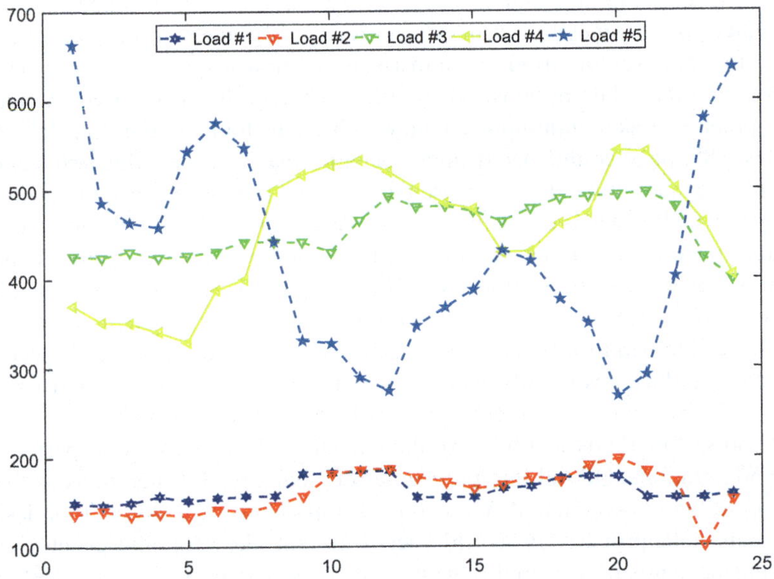

Fig. 7.6 System load after load shifting strategy

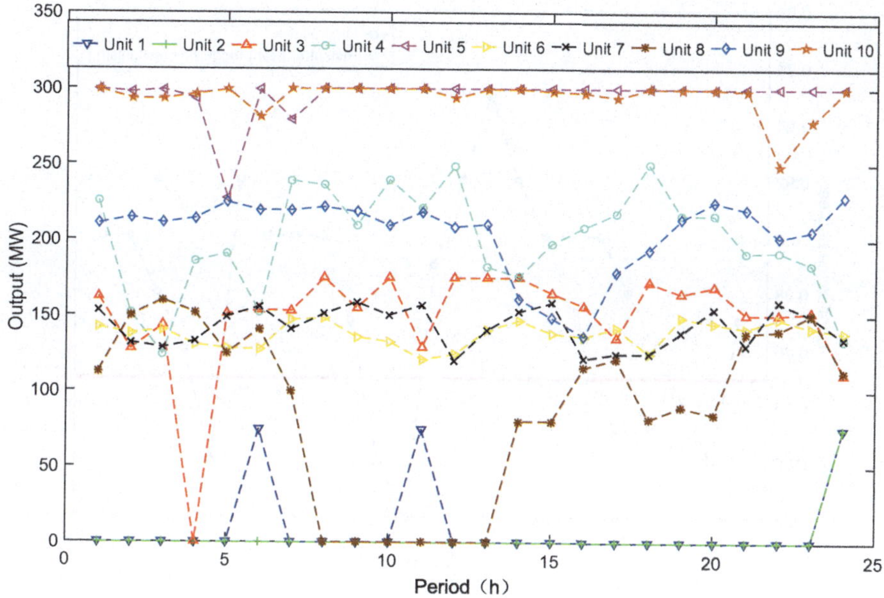

Fig. 7.7 The output of thermal units with event-triggered mechanism

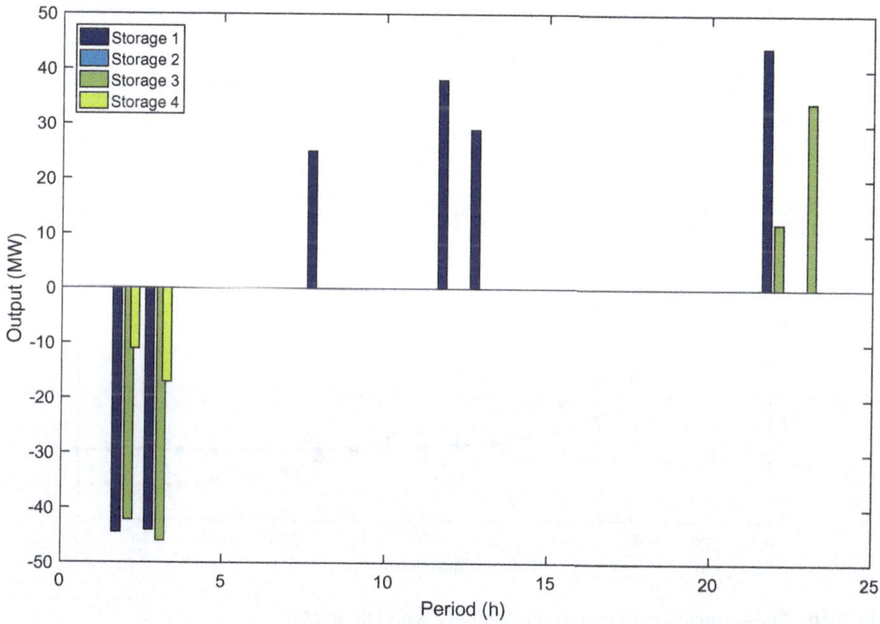

Fig. 7.8 The charging and discharging of energy storage with event-triggered mechanism

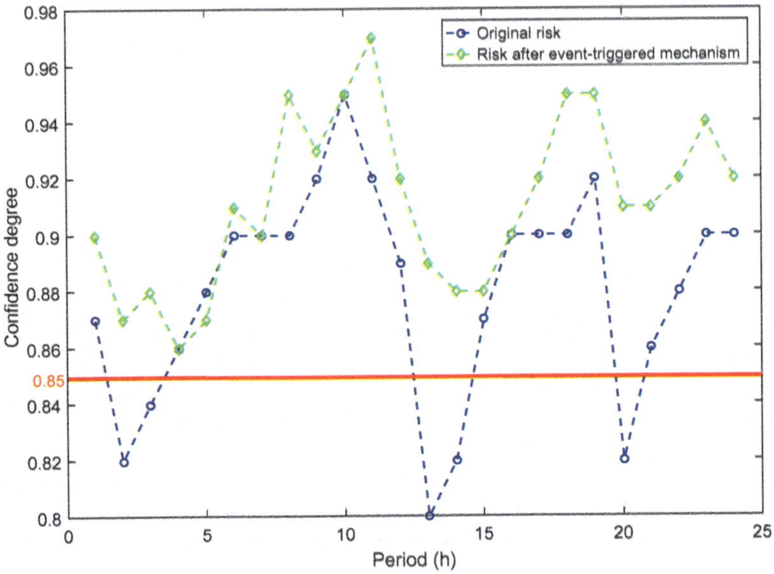

Fig. 7.9 The confidence degree for exempting potential risk

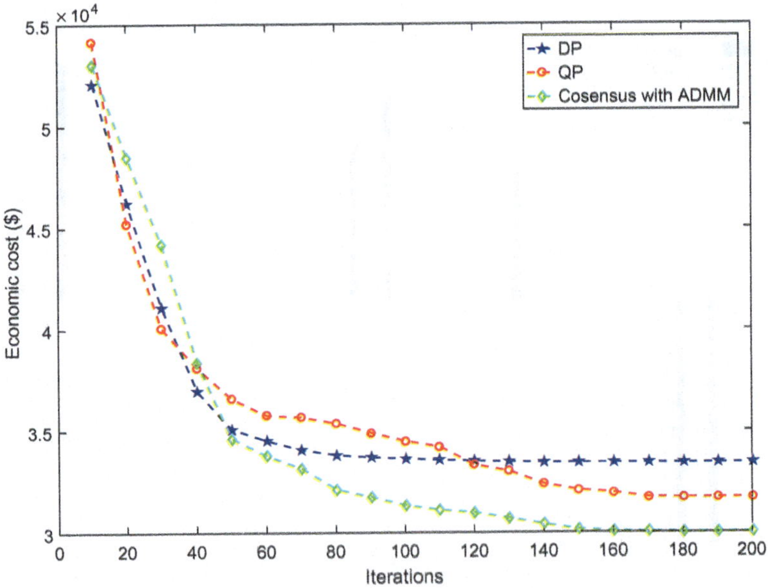

Fig. 7.10 The comparison of convergence ability with DP and QP

7.5 Conclusion

Due to the strong uncertainty and coupled complexity of hybrid energy system, optimal operation has become a great challenge for both system modeling and optimization methodology. This chapter proposes a two-layered multi-agent optimization with event-triggered switching mechanism. After simulation on two test systems, some merits can be concluded as follows:

(1) Since different energy resources have different characteristics, four agents are designed for different purposes. Agent 1 analyzes probabilistic characteristics and provides the probability interval for power dispatch. Agent 2 assigns output of thermal units to minimize fuel cost and on/off cost. Agent 3 manages energy storage with minimizing the charging and discharging cost. Agent 4 provides load shifting/migration model for consumers' consumption in DR.

(2) In upper level, event-triggered switching mechanism is proposed to decrease potential risk caused by intermittent energy, the switching of power supply's on/off state and load curtailment can be controlled to properly arrange power generator state or load curtailment. For proper coordination among different energy resources, some norms are designed for properly optimizing of four subsystems.

(3) In lower level, ADMM is developed with regularized consensus algorithm to optimize DED model in subsystem. Combined with equal increment criterion, different power generators can achieve a utopia optima after several iterations.

Finally, those obtained simulation results can support above view points, and it also reveals that the proposed event-triggered multi-agent optimization can be a viable and promising approach for optimal operation of hybrid energy system.

References

1. B. Qu, J. Liang, Y. Zhu, Z. Wang, P. Suganthan, Economic emission dispatch problems with stochastic wind power using summation based multi-objective evolutionary algorithm. Inf. Sci. **351**, 48–66 (2016)
2. S.H. Karaki, R.B. Chedid, R. Ramadan, Probabilistic performance assessment of autonomous solar-wind energy conversion systems. IEEE Trans. Energy Convers. **14**(3), 766–772 (1999). Sept
3. A. Rabiee, B. Mohammadi-Ivatloo, M. Moradi-Dalvand, Fast dynamic economic power dispatch problems solution via optimality condition decomposition. IEEE Trans. Power Syst. **29**(2), 982–983 (2014). March
4. V. Loia, A. Vaccaro, Decentralized economic dispatch in smart grids by self-organizing dynamic agents. IEEE Trans. Syst. Cybern.: Syst. **44**(4), 397–408 (2014). April
5. M. Mahmoodi, P. Shamsi, B. Fahimi, Economic dispatch of a hybrid microgrid with distributed energy storage. IEEE Trans. Smart Grid **6**(6), 2607–2614 (2015). Nov
6. Y. Zheng, Y. Song, D.J. Hill, Y. Zhang, Multiagent system based microgrid energy management via asynchronous consensus admm. IEEE Trans. Energy Convers. **33**(2), 886–888 (2018). June
7. S. Boyd, N. Parikh, E. Chu, B. Peleato, J. Eckstein, *Distributed Optimization and Statistical Learning via the Alternating Direction Method of Multipliers*. Now Foundations and Trends (2011)

8. H. Zhang, D. Yue, X. Xie, Distributed model predictive control for hybrid energy resource system with large-scale decomposition coordination approach. IEEE Access **4**, 9332–9344 (2016)
9. J. Aghaei, T. Niknam, R. Azizipanah-Abarghooee, J.M. Arroyo, Scenario-based dynamic economic emission dispatch considering load and wind power uncertainties. Int. J. Electr. Power Energy Systems **47**, 351–367 (2013)
10. Y. Xiang, J. Liu, Y. Liu, Robust energy management of microgrid with uncertain renewable generation and load. IEEE Trans. Smart Grid **7**(2), 1034–1043 (2016). March
11. H. Nunna, S. Doolla, Multiagent-based distributed-energy-resource management for intelligent microgrids. IEEE Trans. Ind. Electron. **60**(4), 1678–1687 (2013). April

Part III
Distributed Optimization for Energy Management of Microgrid

Chapter 8
Consensus-Based Economic Hierarchical Control Strategy for Islanded MG Considering Communication Path Reconstruction

In the islanded microgrid (MG), to improve the control effect of the output voltages of DERs and keep the economical operation of MG, a consensus-based economic hierarchical control strategy is proposed in this chapter. The corresponding control structure is divided into physical and cyber layers, where the cyber layer is mainly composed by communicators. In these two layers, there are two main designs: the consensus-based primary control (CBPC) method and the secondary control (CBSC) method. In CBPC: Firstly, the undirected communication path among communicators is designed by path planning. Then, based on the constructed path, an economic P–U droop control method considering with line loss and flexible load is realized by using consensus control. In CBSC: Firstly, a cyber-physical vulnerability assessment method is designed. Based on this method, the DER with the lowest vulnerability is selected to be connected with the "virtual leader". And when the data transmission behavior of the undirected communication path is selected, the undirected network can be transferred as a directed network. Thus, it can make the virtual leader-following consensus control (VLFCC) and the secondary control on voltage accomplished. Finally, the simulation results show the effectiveness of the proposed strategy.

8.1 The Control Structure and Control Process

In this section, the proposed consensus-based economic hierarchical control structure is divided into physical and cyber layers. The purpose of this chapter is to use the data of the cyber layer to solve the problem in the physical layer.

D. Yue et al., *Cooperative Optimal Control of Hybrid Energy Systems*,
https://doi.org/10.1007/978-981-33-6722-7_8

8.1.1 The Control Structure

The specific control structure is as Fig. 8.1. And in this structure, the parts with deep background are the designed parts in this study.

The cyber layer is composed by the following parts: (1) the communication network (including sensors and communicators) where the sensor is used to receive data from the local DER and send it to the corresponding communicator; each communicator is a node of the communication network. And each one is used to exchange the communication data with its neighbor communicators. (2) the path planning and reconstruction part, which is used to (re-)design the appropriate communication path among communicators. (3) the consensus control part, which is used to adjust the incremental costs (ICs) of DERs and the incremental benefits (IB) of flexible load; (4) the VLFCC part, which is used to accomplish the secondary control on voltage. (5) the vulnerability assessment part, which is used to design the directed communication path reasonably.

The physical layer is consisted by the following parts: DER, inverter, LC filter, the Park transformation part, the power calculation part, the CBPC part, the virtual impedance, the secondary control part, the voltage synthesis part, the double closed loop part and the PWM signal generator etc. The control loops which is consisted by PQ calculation part and P–U droop control part are also called as "outer loops". The double closed loops are also called as "inner loops". There are two parts needed to design in this layer: (1) The CBPC part, which is used to optimize the operation benefit of MG. (2) The secondary control part, which is used to adjust the output voltages of DERs to the reference value.

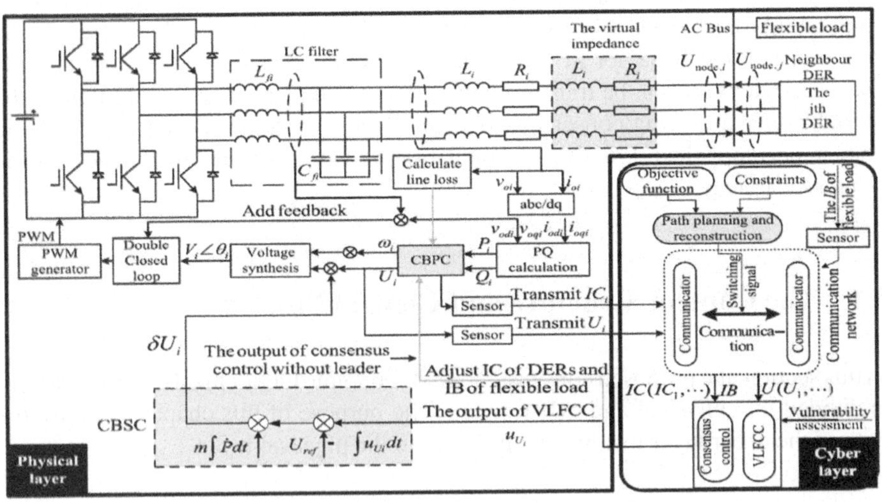

Fig. 8.1 Hierarchical control structure of the ith DER

Remark 8.1 In traditional high-voltage power grid, the line is inductive, so P–f droop control is used more extensively. However, for low-voltage MG, the line is resistive. Therefore, if the P–f control is still used in MG, the line impedance needs to be adjusted from resistive to inductive. But in the actual MG, this will bring harmonics, and it is difficult to be applied. Therefore, the P–U droop control suitable for low-voltage MG is adopted in this chapter.

8.1.2 Explain the Overall Control Process

For example, the control process of the ith DER is as below: (1) At first, the output data (voltage and frequency) of outer loops will be transmitted from the local sensor to the corresponding communicator as the communication data. (2) Then, the communication data will be used for VLFCC. (3) And then, the obtained data by VLFCC will be used to generate the corresponding feedback. And the CBSC will be accomplished by adding feedback to the output of CBPC. (4) Afterwards, the obtained voltage and frequency will be transferred through the voltage synthesis part to generate a real-time voltage. (5) Next, the real-time voltage will be transmitted to the inner loops to generate a trigger voltage. Thus, the trigger voltage will be used to generate a PWM signal to control the operation of inverter and the ith DER. (6) Finally, through the LC filter, the output voltage and current of DER will be transmitted to the outer loops again. At this moment, a control process is done. The definitions of symbols in Fig. 8.1 will be given in the following text.

8.2 The CBPC

In this section, the CBPC is designed. And our main purpose is to use the CBPC to maximize the benefit of MG with considering line loss and flexible load. The specific design process is as follows.

8.2.1 Theory Basic

To better explain the designs of controllers, there are following theories given at first.

Graph theory [1]: In a multi-agent system, its topology can be usually represented by a directed graph. And it is composed by a set of nodes and lines. Each node represents an agent. If any node to any other node is connected by a path, the directed graph is called "connected strongly". If there is a directed path connecting all nodes, the path is called "directed tree". If there are n nodes, the adjacency matrix of the directed graph is defined as $\mathbf{A} = a_{ij} \in R^{N\times N}$. If the jth node connects to the ith node, a_{ij} is positive and $a_{ij} = 0$ otherwise. The degree matrix of a graph is

$\mathbf{D} = \text{diag}\{d_1, \ldots, d_N\} \in R^{N\times N}$, where $d_i = \sum_{j\in N_i} a_{ij}$; N_i is the collection of the neighbor nodes of the ith node. The corresponding Laplace matrix is defined as $\mathbf{L} = \mathbf{D} - \mathbf{A}$.

Consensus control theory [1, 2]: In a directed graph, the state of the ith node is marked as x_i. Meanwhile, if the consensus controller (8.1) is applied to control the ith node, the $\lim_{t\to\infty} \left|x_i(t) - x_j(t)\right| = 0$ can be realized.

$$\dot{x}_i = k_i \sum_{j\in N_i} a_{ij}\left(x_j - x_i\right) \tag{8.1}$$

Moreover, if there is only one node can transfer its state to any other nodes along a directed path (i.e., directed spanning tree exists in the network), the only one node is called as the "leader" or "root node". The corresponding state of the leader is defined as x_L, and the other nodes are called as "followers". And if the controller (8.2) is used for the ith node, the $\lim_{t\to\infty} |x_i(t) - x_L(t)| = 0$ will be established.

$$\dot{x}_i = k_{i1} \sum_{j\in N_i} a_{ij}\left(x_j - x_i\right) - k_{i2} b_i \left(x_L - x_i\right) \tag{8.2}$$

where k_i, k_{i1} and k_{i2} are gain coefficients; b_i represents the relationship between the leader and the ith node ("1" is connected, "0" otherwise).

Remark 8.2 Each DER in MG can be regarded as an agent [1]. But in actually, it is not convenient to select a real "leader" in MG. Therefore, a "virtual leader" is taken in this chapter to solve the problem. Its implementation needs two conditions: 1. The x_L in consensus controller is designed as a constant (e.g., voltage or frequency reference values); 2. Design a_{ij} to guarantee the existence of the directed spanning tree; 3. Design the b_i of the DER connected to leader as "1".

8.2.2 *Economic Optimization Strategy*

At present, the stability of MG operation is gradually improved. On this basis, the economy of MG should be considered more. Economic control is based on the stable operation of MG system. It is achieved by real-time adjustment of economic parameters. In this chapter, the economic control considering with line loss and flexible load is designed. And it is accomplished by using consensus control. Meanwhile, in the control process, the incremental cost (*ICs*) of DERs considering line loss and the incremental benefit (*IB*) of flexible load are taken as controlled variables. In addition, the cost function and income function can be all regarded as quadratic function. The specific functions are as below [2–4]:

$$C_i\left(P_{Gi}\right) = \alpha_i P_{Gi}^2 + \beta_i P_{Gi} + \gamma_i \tag{8.3}$$

$$B_i\left(P_{Dj}\right)=a_jP_{Dj}^2+b_jP_{Dj}+c_j \tag{8.4}$$

where C_i is the generation cost; α_i, β_i and γ_i are cost coefficients; P_{Gi} is the output active power; i refers to the ith DER. B_i is the running benefit; a_j, b_j and c_j are benefit coefficients; P_{Dj} is the required power for load; j refers to the jth flexible load.

In actually, the economic control problem refers to the optimization problem of DERs and flexible loads. The objective function can be designed as

$$\max\sum_{j\in S_D}B_j\left(P_{Dj}\right)-\sum_{i\in S_G}C_j\left(P_{Gi}\right) \tag{8.5}$$

If the fixed load is defined as $P_{D.fix}$, the constraints are

$$\begin{aligned}&P_{D.\text{fix}}+\sum_{j\in S_D}P_{Dj}+P_L-\sum_{i\in S_G}P_{Gi}=0\\&P_{Gi,\min}\le P_{Gi}\le P_{Gi,\max},\ i\in S_G\\&P_{Dj,\min}\le P_{Dj}\le P_{Dj,\max},\ j\in S_D\end{aligned} \tag{8.6}$$

where P_L is the line loss; S_G is the number of DERs; S_D is the number of flexible loads. Afterwards, based on the Lagrange multiplier method, the above optimization problem can be transformed into Eq. (8.7). The specific process is as follows: 1. Construct the constrained single-objective optimization function based on Eqs. (8.5) and (8.6); 2. Add the Lagrange multiplier λ in the optimization process. Thus, these constraints can be transformed into the parts of the objective function. And then, a new objective function can be generated; 3. Based on the newly obtained objective function, the corresponding partial derivative equations of variables requiring solution are calculated; 4. For these partial derivative equations, the optimal value of variables can be obtained by joint solution.

$$\begin{aligned}\min G\left(P_{Gi},\lambda\right)=&-\sum_{j\in S_D}B_j\left(P_{Dj}\right)+\sum_{i\in S_G}C_i\left(P_{Gi}\right)\\&+\lambda\left(P_{D.\text{fix}}+\sum_{j\in S_D}P_{Dj}+P_L-\sum_{i\in S_G}P_{Gi}\right)\end{aligned} \tag{8.7}$$

The corresponding partial derivative equations are

$$\begin{aligned}\frac{\partial G}{\partial P_{Gi}}&=\frac{\partial C_i\left(P_{Gi}\right)}{\partial P_{Gi}}-\lambda+\lambda\frac{\partial P_L}{\partial P_{Gi}}\\\frac{\partial G}{\partial P_{Dj}}&=-\frac{\partial B_j\left(P_{Dj}\right)}{\partial P_{Dj}}+\lambda\end{aligned} \tag{8.8}$$

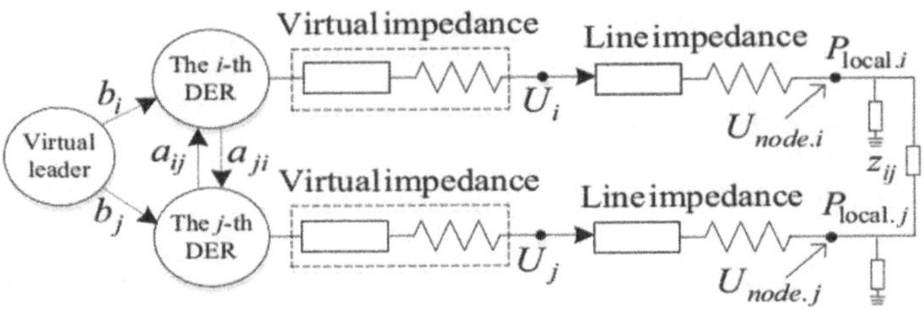

Fig. 8.2 The typical electrical structure of DERs

where $\frac{\partial P_L}{\partial P_{Gi}}$ is line loss correction factor. It can be found that when Eq. (8.9) is satisfied, the economy of MG is optimal.

$$\lambda = \frac{\partial C_i\left(P_{Gi}\right)}{\partial P_{Gi}\left(1 - \frac{\partial P_L}{\partial P_{Gi}}\right)} = \frac{\partial B_j\left(P_{Dj}\right)}{\partial P_{Dj}} \tag{8.9}$$

where $\frac{\partial C_i(P_{Gi})}{\partial P_{Gi}\left(1-\frac{\partial P_L}{\partial P_{Gi}}\right)}$ is IC_i; $\frac{\partial B_j(P_{Dj})}{\partial P_{Dj}}$ is IB_i. Meanwhile, to obtain the line loss correction factor, a typical electrical topology of DERs is given as Fig. 8.2 at first.

There are two DERs in the Fig. 8.2, i.e., the ith and the jth DERs. The line loss correction factor can be obtained by the following method.

For the ith DER, its output active power is consisted by two parts: 1. The injection power between the ith node and its connecting nodes; 2. The line loss on corresponding line impedance (including the virtual impedance); 3. The required power of local load. Therefore, the line loss on the electrical line corresponding to the ith DER can be represented as $P_{L.i} = P_{Gi} - \sum_{j\in S_G} P_{ij} - P_{\text{Local}.i}$, where $P_{L.i}$ is the line loss of the corresponding line of the ith DER; P_{ij} is the injection power between the ith and the jth nodes; $P_{\text{Local}.i}$ is the local load of the ith DER. And generally speaking, for the branch i-j, there is $P_{ij} = U_iU_j\left(g_{ij}\cos\theta_{ij} + b_{ij}\sin\theta_{ij}\right)$, where $U_ie^{j\theta_i}$ and $U_je^{j\theta_j}$ are the exponential form of $U_{\text{node.i}}$ and $U_{\text{node.j}}$, respectively; θ_i and θ_j are the angles of line impedance of the ith and jth DRRs, respectively; z_{ij} is the connection impedance between these two nodes; θ_{ij} is the angle of z_{ij}; g_{ij} and b_{ij} are conductance and susceptance of this branch. Furthermore, in MG, θ_{ij} is usually very small. Therefore, there are usually $\cos\theta_{ij} \approx 1$ and $\sin\theta_{ij} \approx \theta_{ij}$. Therefore, the upper formula can be simplified as

$$P_{ij} = U_iU_j\left(g_{ij} + b_{ij}\theta_{ij}\right) \tag{8.10}$$

Therefore, the line loss on the electrical line corresponding to the ith DER is $P_{L.i} = U_i^2 g_{ij} - \sum_{j\in S_G} U_iU_j(g_{ij} + b_{ij}\theta_{ij}) - P_{\text{local}.i}$. If there is no power injection between the ith node and the jth node, the corresponding line impedance is equal to infinity. Based on analysis above, the total line loss of MG is the sum of the line loss of each node, i.e., $P_L = \sum_{i\in S_G} P_{L.i}$. And finally, based on the analysis above, the line loss correction factor of the ith DER is

$$\frac{\partial P_L}{\partial P_{Gi}} = \frac{\partial P_L}{\partial U_i}\frac{\partial U_i}{\partial P_{Gi}} = \frac{\sum_{j\in S_G} 2U_i g_{ij} - U_j\left(g_{ij} + b_i\theta_{ij}\right)}{2U_i g_{ij}} \tag{8.11}$$

As everyone knows, under the action of droop control, the output voltages of DERs are time-varying. However, when the system is stable, the output voltage of each DER will also be stable. Moreover, due to the effect of secondary control, the voltage of each DER will be eventually adjusted to the reference value. In addition, since the use of droop control is an internal loop control on inverter, so it does not affect the electrical topology of MG. In other words, the power calculation formula presented in this chapter is still valid. Therefore, the line loss correction factor proposed in this chapter can still be obtained by Eq. (8.11). Based on the above analysis, the obtained optimal power is

$$\begin{cases} P_{Gi}^{\text{Optimal}} & = \left[\frac{\partial G}{\partial P_{Gi}}'\left(1 - \frac{\partial P_L}{\partial P_{Gi}}\right) - \beta_i\right]/2\alpha_i \\ P_{Dj}^{\text{Optimal}} & = \left[\frac{\partial G'}{\partial P_{Dj}} - b_i\right]/2a_i \end{cases} \tag{8.12}$$

where P_{Gi}^{Optimal} and P_{Dj}^{Optimal} are obtained by the consensus control without leader. In this situation, each DER can output its optimal active power by adjusting the virtual impedance.

$$\begin{aligned} S_{Gi} &= \frac{U_{\text{DER}i} - U_{\text{node}i}}{R_i + R_i' + j\left(X_i + X_i'\right)} U_{\text{DER}i} \\ &= \frac{\left(U_{\text{DER}i} - U_{\text{node}i}\right)\left(R_i + R_i'\right)}{\left(R_i + R_i'\right)^2 + \left(X_i + X_i'\right)^2} U_{\text{DER}i} \\ &\quad + j\frac{\left(U_{\text{DER}i} - U_{\text{node}i}\right)\left(X_i + X_i'\right)}{\left(R_i + R_i'\right)^2 + \left(X_i + X_i'\right)^2} U_{\text{DER}i} \\ \Rightarrow P_{Gi} &= \frac{\left(U_{\text{DER}i} - U_{\text{node}i}\right)\left(R_i + R_i'\right)}{\left(R_i + R_i'\right)^2 + \left(X_i + X_i'\right)^2} U_{\text{DER}i} \end{aligned} \tag{8.13}$$

where S_{Gi} is the apparent power; $U_{\text{DER}i}$ is the voltage of the ith DER; $U_{\text{node}i}$ is the node voltage; $R' + jX'$ is the virtual impedance. After combining Eqs. (8.12) and (8.13), there is:

$$\frac{\left(U_{\text{DER}i} - U_{\text{node}i}\right)\left(R_i + R_i'\right)}{\left(R_i + R_i'\right)^2 + \left(X_i + X_i'\right)^2} U_{\text{DER}i} = \left[\frac{\partial L'}{\partial P}\left(1 - \frac{\partial P_L}{\partial P_{Gi}}\right) - \beta_i\right]/2\alpha_i \tag{8.14}$$

However, each DER and flexible load has its power limitations. Therefore, in a actual MG, there should be

$$\begin{cases} P_{Gi}^{\text{Optimal}} = P_{Gi.\min} \left[\frac{\partial L'}{\partial P} \left(1 - \frac{\partial P_L}{\partial P_{Gi}}\right) - \beta_i \right] / 2\alpha_i \leq P_{Gi.\min} \\ P_{Gi}^{\text{Optimal}} = \left[\frac{\partial L'}{\partial P} \left(1 - \frac{\partial P_L}{\partial P_{Gi}}\right) - \beta_i \right] / 2\alpha_i \, P_{Gi.\min} \leq \left[\frac{\partial L'}{\partial P} \left(1 - \frac{\partial P_L}{\partial P_{Gi}}\right) - \beta_i \right] / 2\alpha_i \leq P_{Gi.\max} \\ P_{Gi}^{\text{Optimal}} = P_{Gi.\max} \, P_{Gi.\max} \leq \left[\frac{\partial L'}{\partial P} \left(1 - \frac{\partial P_L}{\partial P_{Gi}}\right) - \beta_i \right] / 2\alpha_i \end{cases} \tag{8.15}$$

$$\begin{cases} P_{Dj}^{\text{Optimal}} = P_{Dj.\min} \left[\frac{\partial G}{\partial P_{Dj}} - b_j \right] / 2a_j \leq P_{Dj.\min} \\ P_{Dj}^{\text{Optimal}} = \left[\frac{\partial G}{\partial P_{Dj}} - b_j \right] / 2a_{kj} \quad P_{Dj.\min} \leq \left[\frac{\partial G}{\partial P_{Dj}} - b_j \right] / 2a_j \leq P_{Dij\max} \\ P_{Dj}^{\text{Optimal}} = P_{Dj.\max} \quad P_{Di.\max} \leq \left[\frac{\partial L}{\partial P_{Dj}} - b_j \right] / 2a_i \end{cases} \tag{8.16}$$

In summary, the equation of the economic *P–U* droop control is $U_i = U_{ref} - m_i P_{Gi}^{\text{Optimal}}$. In addition, to make Eq. (8.9) realized, the communication path among communicators should be designed at first. Thus, the consensus controller without leader can be used to control the *IC* (or *IB*) of each DER (or flexible load). The design process of the communication path is shown in the following subsection.

8.2.3 Path Planning of the Undirected Communication Path

The meaning of communication path planning is as follows: 1. The designed undirected path can meet the requirements to accomplish the consensus control; 2. According to the designed undirected communication path, as long as the data transmission direction is selected, the directed path required to complete the VLFCC can be selected (i.e., directed spanning tree). So far, the existing consensus-based approaches usually presuppose the existence of a communication path and do not give the specific design method. Moreover, when the path scheme is reconstructed, there is few people have analyzed the effect on control process. In actually, the meaning of undirected path is to make the communication network of the whole MG "strongly connected", i.e., the data from each controlled DER can be transmitted to the any other DERs. Therefore, based on graph theory, the construction process can be regarded as an optimization problem. In this study, the objective function is designed to select the shortest and cheapest path among communicators. If there are n DERs and n communicators in a MG, the objective functions are

$$\begin{cases} Z_1 = \min \sum_{i=1}^{n} \left[\sum_{j \in N_i} L_{ij} sgn\left(a_{ij}\right) \right] \\ Z_2 = \min \sum_{i=1}^{n} \left[\sum_{j \in N_i} C_{\text{line}.ij} L_{ij} sgn\left(a_{ij}\right) \right] \end{cases} \tag{8.17}$$

where L_{ij} is the distance from the jth communicator to the ith communicator; sgn is the sign function; L_{i0} is the distance from the communicator of the virtual leader to the ith communicator; $C_{\text{line}.ij}$ is the cost of constructing the line from the jth communicator to the ith communicator.

There are also some constraints in optimal progress:

(1) The capacity constraints of the ith communicator

$$\sum_{j \in N_i} sgn\left(a_{ij}\right) \leq C_{\max i}, \sum_{j \in N_i} sgn\left(a_{ji}\right) \leq C'_{\max i} \tag{8.18}$$

where $C_{\max i}$ is the maximum allowed number of lines that other communicators access to the ith communicator; $C'_{\max i}$ is the maximum allowed number of lines that the ith communicator access to the others.

(2) Ensure that the minimum number of lines for communication is met.

$$\begin{cases} \sum_{i=1}^{n}\left[\sum_{j \in N_i} sgn\left(a_{ij}\right)\right] = 2(n-1) \\ a_{ij} \odot a_{ji} = 1 \end{cases} \tag{8.19}$$

(3) Ensure that each DER can be connected to the path.

$$\sum_{j \in N_i} sgn\left(a_{ij}\right) = 1 \tag{8.20}$$

(4) Ensure that each communicator is connected to at least one undirected line.

$$2 \leq \sum_{j \in N_i}\left[sgn\left(a_{ij}\right) + sgn\left(a_{ji}\right)\right] \leq 2(n-1) \tag{8.21}$$

Based on the designed optimization problem above, the optimal solution can be solved by iterative calculation with intelligent algorithms.

8.2.4 Path Reconstruction Method

Based on Eqs. (8.15) and (8.16), it has found that some DERs and flexible load may be removed from the communication process. To deal with this situation, we design a path reconstruction method. The contributed method can reselect the best communication scheme based on the existing communication path. Thus, according to the reconstructed communication path, the consensus control can be re-implemented. This can ensure that each DER accomplishes the CBPC normally. The specific design process of the reconstruction method is as follows: At first, we retain the normal lines in the original communication path and abandon the interrupted lines. Meanwhile, according to the abandoned lines, the corresponding elements in $\mathbf{A}$ and b_i are eliminated. Then, based on the reserved lines, the corresponding elements in $\mathbf{A}$ and b_i are set as "1". Afterwards, the number of normal lines and the communicators needed to be connected are set as o and η, respectively. Thus, Eq. (8.19) can be adjusted as $\sum_{i=1}^{n}\left[\sum_{j\in N_i} sgn\left(a_{ij}\right)\right] = o + \eta$ in the solution process.

In summary, through the methods of path planning and reconstruction, the designer can design the appropriate communication path scheme in advance. And the above process can be regarded as a constrained multi-objective optimization problem. Thus, the existing intelligent algorithms are needed to solve the problem at first. Then, based on the results of solution, the CBPC and CBSC can be implemented in MG. It can be also found that the path planning method is just a solution method of optimization problem. Moreover, the planning method is not a real hardware part associated with physical MG and therefore it does not make the structure of MG more complex.

8.2.5 Consensus Controller

Based on the designed communication path, the communication among DERs can be realized. Thus, the application of consensus control on DERs can be also ensured. And the specific design process of the consensus controller is as follows.

To optimize the efficiency of MG operation, it is necessary to make Eq. (8.9) set up. This coincides with the effect of consensus control [5, 6]. Based on the theory in Sect. 8.3.1, the consensus controller of the ith DER is designed in discrete form as

$$\mathbf{X}(k+1) = \mathbf{X}(k) + \mathbf{KLX}(k) \tag{8.22}$$

where $\mathbf{X}(k)$ is the matrix form of controlled variables at t = KT; $\mathbf{K}$ is gain matrix. The selection of $\mathbf{K}$ will directly affect the effect of consensus control, so there is a theorem proposed as below.

Theorem 8.1 *If there exists* $\mathbf{K}$ *such that* $l_{ii}(k) + \frac{\mathbf{K}}{2}\left\langle l_i(k), l_i(k)\right\rangle \leq 0$, *the whole controlled system can be stable.*

See the proof as follows.

Take the following function as the Lyapunov function at first.

$$V(\mathbf{X}(k)) = \mathbf{X}^T(k)\mathbf{X}(k) \tag{8.23}$$

It's found that Eq. (8.18) is a positive definite function. After combining with Eq. (8.22), there is

$$
\begin{aligned}
\dot{V}(\mathbf{X}(k)) &= V(\mathbf{X}(k+1)) - V(\mathbf{X}(k)) \\
&= \mathbf{X}^T(k+1)\mathbf{X}(k+1) - \mathbf{X}^T(k)\mathbf{X}(k) \\
&= ((\mathbf{I}+\mathbf{KL}(k))\mathbf{X}(k))^T(\mathbf{I}+\mathbf{KL}(x))\mathbf{X}(k) - \mathbf{X}^T(k)\mathbf{X}(k) \\
&= \mathbf{X}^T(k)(\mathbf{I}+\mathbf{KL}(k))^T(\mathbf{I}+\mathbf{KL}(k))\mathbf{X}(k) - \mathbf{X}^T(k)\mathbf{X}(k) \\
&= \mathbf{X}^T(k)\left(\mathbf{I}+2\mathbf{K}L(k)+\mathbf{K}^2\mathbf{L}^2(k)-\mathbf{I}\right)\mathbf{X}(k) \\
&= 2\mathbf{K}\mathbf{X}^T(k)\left(\mathbf{L}(k)+\frac{\mathbf{K}}{2}\mathbf{L}^2(k)\right)\mathbf{X}(k)
\end{aligned}
\tag{8.24}
$$

Make $\mathbf{J}(k) = \mathbf{L}(k) + \frac{\mathbf{K}}{2}\mathbf{L}^2(k)$ at first. According to Lyapunov stability theory, as long as the matrix $J(k)$ is negative semidefinite, then Eq. (8.22) is stable. $\mathbf{L}(k)$ can be also represented as row vectors as

$$
\begin{cases}
\mathbf{L}(k) = [l_1(k), l_2(k), \ldots, l_n(k)]^T \\
\mathbf{L}^2(k) = [l_1(k), l_2(k), \ldots, l_n(k)]^T\,[l_1(k), l_2(k), \ldots, l_n(k)]
\end{cases}
\tag{8.25}
$$

Moreover, $L^2(k)$ can be described as

$$
\begin{bmatrix}
\langle l_1(k), l_1(k)\rangle & \langle l_1(k), l_2(k)\rangle & \cdots & \langle l_1(k), l_n(k)\rangle \\
\langle l_2(k), l_1(k)\rangle & \langle l_2(k), l_2(k)\rangle & \cdots & \langle l_2(k), l_n(k)\rangle \\
\vdots & \vdots & \ddots & \vdots \\
\langle l_n(k), l_1(k)\rangle & \langle l_n(k), l_2(k)\rangle & \cdots & \langle l_n(k), l_n(k)\rangle
\end{bmatrix}
\tag{8.26}
$$

Because $\mathbf{L}(k)$ is a symmetric matrix, $\mathbf{L}^2(k)$ and $\mathbf{J}(\mathbf{k})$ are also symmetric matrix. Meanwhile, the elements of each row in $\mathbf{L}^2(k)$ will meet $\sum_{j=1,j\neq i}^{n}\left\langle l_i(k), l_j(k)\right\rangle = -\langle l_i(k), l_i(k)\rangle$. Therefore, the elements of each row in $\mathbf{J}(\mathbf{k})$ will meet

$$
\mathbf{J}_{ii}(k) = -\sum_{j=1,j\neq i}^{n}\mathbf{J}_{ij}(k) = -\left[\sum_{j=1,j\neq i}^{n}\left(l_{ij}(k) + \frac{\mathbf{K}}{2}\left\langle l_i(k), l_j(k)\right\rangle\right)\right]
\tag{8.27}
$$

According to Gersgorin theorem, the eigenvalues of $\mathbf{J}(\mathbf{k})$ are in the following union:

$$
|\lambda - \mathbf{J}_{ii}(k)| \le \sum_{j\neq i}\left|\mathbf{J}_{ij}(k)\right|
\tag{8.28}
$$

To make $\mathbf{J}(\mathbf{k})$ as a negative definite matrix, all eigenvalues of $\mathbf{J}(\mathbf{k})$ must be negative. Therefore, as long as $\mathbf{J}_{ii}(k) \le 0$, the $\mathbf{J}(\mathbf{k})$ can be ensured as a negative definite matrix. The analysis of $\mathbf{J}_{ii}(k)$ is carried out as follows:

$$\begin{aligned}\mathbf{J}_{ii}(k) &= -\left[\sum_{j=1,j\neq i}^{n}\left(l_{ij}(k) + \frac{\mathbf{K}}{2}\left\langle l_i(k), l_j(k)\right\rangle\right)\right] \\ &= l_{ii}(k) - \frac{\mathbf{K}}{2}\sum_{j=1,j\neq i}^{n}\left[-\left\langle l_i(k), l_j(k)\right\rangle\right] \\ &= l_{ii}(k) + \frac{\mathbf{K}}{2}\left\langle l_i(k), l_i(k)\right\rangle\end{aligned} \tag{8.29}$$

Therefore, as long as Eq. (8.30) is satisfied, the whole controlled system can be stable.

$$l_{ii}(k) + \frac{\mathrm{K}}{2}\left\langle l_i(k), l_i(k)\right\rangle \leq 0 \tag{8.30}$$

The proof is completed.

Moreover, based on Eqs. (8.15) and (8.16), it has found that some DERs or flexible loads may exist from the consensus control process. To maintain the control effect, it needs to use the method of path reconstruction. Therefore, it is necessary to ensure the control effect of consensus controller when the path is reconstructed. The specific analysis is as below.

When the communication path is reconstructed, the communication topology will be also changed. This will cause the change of $\mathbf{A}$ in consensus controller. Therefore, the communication network can be regarded as a switched system. And in this study, there is a feedback controller $\mathbf{K}_{\sigma(t)}\mathbf{x}(t)$ designed to maintain the stability of the MG system. The specific design process is as below. At first, the equation of the switched system is constructed as

$$\begin{cases}\mathbf{x}(t) = (\mathbf{A}_\sigma + \mathbf{K}_\sigma)\,\mathbf{x}(t) + \mathbf{D}\mathbf{w}(t) \\ \mathbf{z}(t) = \mathbf{C}\mathbf{x}(t)\end{cases} \tag{8.31}$$

where $\mathbf{x}$ refers to the controlled variables in consensus controller, e.g., *IC* and *IB* etc.; $\mathbf{D}\mathbf{w}(t)$ is the disturbance data; $\sigma = 1, 2$. $\sigma = 1$ represents the state before switching, $\sigma = 2$ represents the state after switching. After setting $\hat{\mathbf{A}}_\sigma = \mathbf{A}_\sigma + \mathbf{K}_\sigma$, there is

$$\begin{cases}\dot{\mathbf{x}}(t) = \hat{\mathbf{A}}_\sigma\mathbf{x}(t) + \mathbf{D}\mathbf{w}(t) \\ \mathbf{z}(t) = \mathbf{C}\mathbf{x}(t)\end{cases} \tag{8.32}$$

Therefore, there are two tasks in this design: 1. Select the gain matrix (i.e., $\mathbf{K}_\sigma$) and the appropriate Lyapunov function to confirm the system (8.32) is asymptotically stable; 2. Make the controlled system satisfy the H_∞ robust characteristic: $\|\mathbf{z}\|_2 < \gamma\|\mathbf{w}\|_2$, where γ is constant. In summary, the following theorem is given out.

Theorem 8.2 *Given $\gamma > 0$ at first, and assume there are two constant values: β_1, β_2 (both positive or negative) and two positive real numbers: μ, ε. Then, the closed-loop system (8.32) is asymptotically stable and satisfies the H_∞ robust control perfor-*

mance when the communication path is reconstructed, if and only if there are two positive definite confrontation matrices ($\mathbf{P}_1$ *and* $\mathbf{P}_2$) *satisfying Eq. (8.33).*

$$\begin{gathered}\begin{bmatrix}\boldsymbol{\Phi}_1 & \mathbf{P}_1\mathbf{M}\\ \mathbf{M}^{\mathrm{T}}\mathbf{P}_1 & -J_2\end{bmatrix}<0,\ \begin{bmatrix}\boldsymbol{\Phi}_2 & \mathbf{P}_2\mathbf{M}\\ \mathbf{M}^{\mathrm{T}}\mathbf{P}_2 & -J_2\end{bmatrix}<0\\ \Delta\mathbf{A}\mathbf{P}_\sigma+\mathbf{P}_\sigma\Delta\mathbf{A}<\varepsilon\mathbf{F}^{\mathrm{T}}\mathbf{F}+\varepsilon^{-1}\mathbf{P}_\sigma\mathbf{E}\mathbf{E}^{\mathrm{T}}\mathbf{P}_\sigma+\gamma^{-2}\mathbf{P}_\sigma\mathbf{D}\mathbf{D}^{\mathrm{T}}\mathbf{P}_\sigma+\mathbf{C}^{\mathrm{T}}\mathbf{C}\end{gathered} \tag{8.33}$$

where $\boldsymbol{\Phi}_1=\mathbf{P}_1\mathbf{A}_1+\mathbf{A}_1^{\mathrm{T}}\mathbf{P}_1+\beta_1(\mathbf{P}_1-\mathbf{P}_2)+\mathbf{N}^T\mathbf{J}_1\mathbf{N}-\mu^{-2}\mathbf{P}_1^{\mathrm{T}}\mathbf{P}_1$; $\Phi_2=\mathbf{P}_2\mathbf{A}_2+\mathbf{A}_2^{\mathrm{T}}\mathbf{P}_2+\beta_2(\mathbf{P}_2-\mathbf{P}_1)+\mathbf{N}^{\mathrm{T}}\mathbf{J}_1\mathbf{N}-\mu^{-2}\mathbf{P}_1^{\mathrm{T}}\mathbf{P}_2$; $\mathbf{M}=[\mathbf{D}\quad\mathbf{E}]$; $\mathbf{N}=[\mathbf{C}^{\mathrm{T}}\quad\mathbf{F}^{\mathrm{T}}]$; $\mathbf{J}_1=diag[\mathbf{I}\quad\varepsilon\mathbf{I}]$; $\mathbf{J}_2=diag\left[\gamma^2\mathbf{I}\quad\varepsilon\mathbf{I}\right]$. In addition, the gain matrices of state feedback are $\mathbf{K}_1=-\frac{1}{2\mu^2}\mathbf{P}_1$ and $\mathbf{K}_2=-\frac{1}{2\mu^2}\mathbf{P}_2$.

See the proof of Theorem 8.2 as follows.

In order not to lose generality, it's assumed that $\beta_1,\beta_2\geq 0$. After taking $\hat{\mathbf{A}}_1$ and $\hat{\mathbf{A}}_2$ into Eq. (8.33), there are:

$$\begin{bmatrix}\hat{\mathbf{A}}_1^{\mathrm{T}}\mathbf{P}_1+\mathbf{P}_1\hat{\mathbf{A}}_1+\mathbf{N}^{\mathrm{T}}J_1\mathbf{N}+\beta_1(\mathbf{P}_1-\mathbf{P}_2) & \mathbf{P}_1\mathbf{M}<0\\ \mathbf{M}^{\mathrm{T}}\mathbf{P}_1 & -J_2\end{bmatrix}<0 \tag{8.34}$$

$$\begin{bmatrix}\hat{\mathbf{A}}_2^{\mathrm{T}}\mathbf{P}_2+\mathbf{P}_2\hat{\mathbf{A}}_2+\mathbf{N}^{\mathrm{T}}J_1\mathbf{N}+\beta_2(\mathbf{P}_2-\mathbf{P}_1) & \mathbf{P}_2\mathbf{M}<0\\ \mathbf{M}^{\mathrm{T}}\mathbf{P}_2 & -J_2\end{bmatrix}<0 \tag{8.35}$$

By using Schur complementary Lemma, there is:

$$\begin{cases}\mathbf{x}^{\mathrm{T}}\left[\hat{\mathbf{A}}_1^{\mathrm{T}}\mathbf{P}_1+\mathbf{P}_1\hat{\mathbf{A}}_1+\varepsilon\mathbf{F}^{\mathrm{T}}\mathbf{F}+\varepsilon^{-1}\mathbf{P}_1\mathbf{E}\mathbf{E}^{\mathrm{T}}\mathbf{P}_1+\gamma^{-2}\mathbf{P}_1\mathbf{D}\mathbf{D}^{\mathrm{T}}\mathbf{P}_1+\mathbf{C}^{\mathrm{T}}\mathbf{C}\right]\mathbf{x}<0\\ \mathbf{x}^{\mathrm{T}}\left[\hat{\mathbf{A}}_2^{\mathrm{T}}\mathbf{P}_2+\mathbf{P}_2\hat{\mathbf{A}}_2+\varepsilon\mathbf{F}^{\mathrm{T}}\mathbf{F}+\varepsilon^{-1}\mathbf{P}_2\mathbf{E}\mathbf{E}^{\mathrm{T}}\mathbf{P}_2+\gamma^{-2}\mathbf{P}_2\mathbf{D}\mathbf{D}^{\mathrm{T}}\mathbf{P}_2+\mathbf{C}^{\mathrm{T}}\mathbf{C}\right]\mathbf{x}<0\end{cases} \tag{8.36}$$

Verify asymptotically stability: Construct Lyapunov function as $V_\sigma(\mathbf{x}(t))=\mathbf{x}^{\mathrm{T}}\mathbf{P}_\sigma\mathbf{x}$ at first. When $\mathbf{w}\equiv 0$, there is

$$\begin{aligned}\dot{V}_\sigma(\mathbf{x}(t))&=\mathbf{x}^{\mathrm{T}}\left[\left(\hat{\mathbf{A}}_\sigma+\Delta\hat{\mathbf{A}}\right)^{\mathrm{T}}\mathbf{P}_\sigma+\mathbf{P}_\sigma\left(\hat{\mathbf{A}}_\sigma+\Delta\hat{\mathbf{A}}\right)\right]\mathbf{x}\\ &=\mathbf{x}^{\mathrm{T}}\left(\hat{\mathbf{A}}_\sigma^{\mathrm{T}}\mathbf{P}_\sigma+\mathbf{P}_\sigma\hat{\mathbf{A}}_\sigma+\Delta\hat{\mathbf{A}}\mathbf{P}_\sigma+\mathbf{P}_\sigma\Delta\hat{\mathbf{A}}\right)\mathbf{x}\\ &\leq\mathbf{x}^{\mathrm{T}}\left(\hat{\mathbf{A}}_\sigma^{\mathrm{T}}\mathbf{P}_\sigma+\mathbf{P}_\sigma\hat{\mathbf{A}}_\sigma+\varepsilon\mathbf{F}^{\mathrm{T}}\mathbf{F}+\varepsilon^{-1}\mathbf{P}_\sigma\mathbf{E}\mathbf{E}^{\mathrm{T}}\mathbf{P}_\sigma\right)\mathbf{x}\end{aligned}$$

Combine with Eqs. (8.34), (8.35) and (8.36), there is $\dot{V}_\sigma(\mathbf{x}(t))<0$. The switched system satisfies the Lyapunov stability.

Verify H_∞ robustness: When the initial state of the system is zero (i.e. $\mathbf{x}(0)=0$), For all allowable uncertainties and for all non-zero $\mathbf{w}$(t), $\|\mathbf{z}\|<\gamma\|\mathbf{w}\|$ are established. The specific proof process is as follows.

Given the initial time t_0 and initial state $\mathbf{x}_0$, a switching sequence can be defined as below:

$$\Sigma = \{\mathbf{x}_0; (i_1, t_1), \ldots, (i_n, t_n), \ldots\} \tag{8.37}$$

where the first coordinate sequence of Σ (i.e. $\{i_1, i_2, \ldots i_n, \ldots\}$) represents the switching sequence of the subsystem; The second coordinate sequence of Σ (i.e. $\{t_1, t_2, \ldots t_n, \ldots\}$) represents the starting time of the corresponding subsystem. For example, (i_k, t_k) indicates that the kth subsystem starts at $t = t_k$ and ends at $t = t_k + 1$. When $t = t_j$, there is:

$$V_{\sigma(t_j)}\left(\mathbf{x}\left(t_j\right)\right) \le \lim_{t \to t_j^-} V_{\sigma(t)}(\mathbf{x}(t)) \tag{8.38}$$

To make the controlled object Eq. (8.32) have H-inf robustness, the following function is designed:

$$J = \int_0^\infty \left[\mathbf{z}(t)^T \mathbf{z}(t) - \gamma^2 \mathbf{w}(t)^T \mathbf{w}(t)\right] dt \tag{8.39}$$

Meanwhile, there is

$$\begin{aligned}
J = & \int_{t_{k1}}^{t_{k1}+1} \left[\mathbf{z}^{\mathrm{T}}\mathbf{z} - \gamma^2 \mathbf{w}^{\mathrm{T}}\mathbf{w} + \frac{d}{dt}\left(\mathbf{x}^{\mathrm{T}}\mathbf{P}_1\mathbf{x}\right)\right] dt - \mathbf{x}^{\mathrm{T}}\mathbf{P}_1\mathbf{x}\Big|_{t_{k1}}^{t_{k1}+1} \\
& + \int_{t_{k1}+1}^{t_{k2}} \left[\mathbf{z}^{\mathrm{T}}\mathbf{z} - \gamma^2 \mathbf{w}^{\mathrm{T}}\mathbf{w} + \frac{d}{dt}\left(\mathbf{x}^{\mathrm{T}}\mathbf{P}_2\mathbf{x}\right)\right] dt - \mathbf{x}^{\mathrm{T}}\mathbf{P}_2\mathbf{x}\Big|_{t_{k1}+1}^{t_{k2}} \\
= & \int_{t_{k1}}^{t_{k1}+1} \mathbf{x}^T \left(\hat{\mathbf{A}}_1^{\mathrm{T}}\mathbf{P}_1 + \mathbf{P}_1\hat{\mathbf{A}}_1 + \Delta\hat{\mathbf{A}}_1\mathbf{P}_1 + \mathbf{P}_1\Delta\hat{\mathbf{A}}_1 + \mathbf{C}^{\mathrm{T}}\mathbf{C}\right) \mathbf{x} dt \\
& - \int_{t_{k1}}^{t_{k1}+1} \left(\gamma^{-1}\mathbf{D}^{\mathrm{T}}\mathbf{P}_1\mathbf{x} - \gamma\mathbf{w}\right)^T \left(\gamma^{-1}\mathbf{D}^{\mathrm{T}}\mathbf{P}_1\mathbf{x} - \gamma\mathbf{w}\right) dt \\
& + \int_{t_{k1}+1}^{t_{k2}} \mathbf{x}^T \left(\hat{\mathbf{A}}_2^{\mathrm{T}}\mathbf{P}_2 + \mathbf{P}_2\hat{\mathbf{A}}_2 + \Delta\hat{\mathbf{A}}_2\mathbf{P}_2 + \mathbf{P}_2\Delta\hat{\mathbf{A}}_2 + \mathbf{C}^{\mathrm{T}}\mathbf{C}\right) \mathbf{x} dt \\
& - \int_{t_{k1}+1}^{t_{k2}} \left(\gamma^{-1}\mathbf{D}^{\mathrm{T}}\mathbf{P}_2\mathbf{x} - \gamma\mathbf{w}\right)^{\mathrm{T}} \left(\gamma^{-1}\mathbf{D}^{\mathrm{T}}\mathbf{P}_2\mathbf{x} - \gamma\mathbf{w}\right) dt \\
& - \mathbf{x}^{\mathrm{T}}(t_{k1}+1)(\mathbf{P}_1 - \mathbf{P}_2)\mathbf{x}(t_{k1}+1) - \mathbf{x}^{\mathrm{T}}(t_{k1})(\mathbf{P}_1 - \mathbf{P}_2)\mathbf{x}(t_{k1}) \\
\le & \int_{t_{k1}}^{t_{k1}+1} \mathbf{x}^T \left(\hat{\mathbf{A}}_1^{\mathrm{T}}\mathbf{P}_1 + \mathbf{P}_1\hat{\mathbf{A}}_1 + \mathbf{F}^{\mathrm{T}}\Sigma^{\mathrm{T}}(t)\mathbf{E}^{\mathrm{T}}\mathbf{P}_1 + \mathbf{P}_1\mathbf{E}\Sigma(t)\mathbf{F} + \gamma^{-2}\mathbf{P}_1\mathbf{D}\mathbf{D}^{\mathrm{T}}\mathbf{P}_1 + \mathbf{C}^{\mathrm{T}}\mathbf{C}\right) \mathbf{x} dt \\
& + \int_{t_{k1}+1}^{t_{k2}} \mathbf{x}^T \left(\hat{\mathbf{A}}_2^{\mathrm{T}}\mathbf{P}_2 + \mathbf{P}_2\hat{\mathbf{A}}_2 + \mathbf{F}^{\mathrm{T}}\Sigma^{\mathrm{T}}(t)\mathbf{E}^{\mathrm{T}}\mathbf{P}_2 + \mathbf{P}_2\mathbf{E}\Sigma(t)\mathbf{F} + \gamma^{-2}\mathbf{P}_2\mathbf{D}\mathbf{D}^{\mathrm{T}}\mathbf{P}_2 + \mathbf{C}^{\mathrm{T}}\mathbf{C}\right) \mathbf{x} dt
\end{aligned} \tag{8.40}$$

Based on Eq. (8.40), there is $J < 0$ and $\|\mathbf{z}\|_2 < \gamma\|\mathbf{w}\|_2$.

So far, the proof is done.

Moreover, please note that: In fact, these two Eqs. (8.22) and (8.32) are designed for consensus control. In other words, the controlled objective is consensus controller. For Eq. (8.22), it is used to confirm that the controlled MG is asymptotically stable under the action of consensus control. And the form of consensus controller is as Eq. (8.1). Therefore, after determining the control parameters, the consensus controllers of all DERs can be described as the matrix form of Eq. (8.22). Moreover, because of some DER or flexible load will exit the consensus control process, it is necessary to design appropriate controller $\mathbf{k}_\sigma$ to maintain stability. And the exit of DER or flexible load is also a switch process. Therefore, when the external disturbance is considered, the switched consensus controller can be written as the form of Eq. (8.32).

8.3 The CBSC

It can be found that the output voltage by CBPC still has errors with the reference value. To address this problem, the VLFCC theory is used to design the secondary controller in this section. The mainly related contents are the designs of vulnerability assessment and secondary controller on voltage. The specific designs are as below.

8.3.1 Cyber-Physical Vulnerability Assessment

In this subsection, the main purpose is to find a DER with the lowest prospective vulnerability in the constructed undirected path in Sect. 8.3.3. And then, the selected DER will be connected with the virtual leader directly. Thus, the completion of VLFCC can be ensured. The specific design is as follows.

In CPS, the physical devices such as DER can be classified as the physical nodes, and the communication devices such as communicators can be classified as the cyber nodes. And in actually, all nodes have certain vulnerabilities [7, 8]. Furthermore, since each DER corresponds to a communicator, the vulnerability of physical nodes and cyber nodes is related to each other. Thus, each DER and its corresponding communicator can constitute a cyberphysical node (CPN). For a CPN, the cyber and physical vulnerabilities should be combined. And if the vulnerability is not considered in path planning, the constructed path is probably not very stable. Therefore, the specific steps of assessment method on the jth CPN are as follows.

(1) Select the vulnerability index of physical node. For physical node, the source-flow path partition method is a simple and effective method to analyze the vulnerability. Through this method, the set of all power paths can be obtained as $L\{Lq, q = 1 \sim N_L\}$. The vulnerability index of the lth branch of the path Lq is defined as $V_{Lq(l)}$, and there is

$$V_{Lq(l)} = d_{Lq(l)} I_{1,Lq(l)} I_{2,Lq(l)} \lambda_{Lq(l)} \tag{8.41}$$

where N_L is the total number of electrical paths; $d_{Lq(l)}$ is the electrical length of the lth branch; $I_{1,Lq(l)}$ and $I_{2,Lq(l)}$ are the importance degree of the source and flow node in the lth branch respectively; $\lambda_{Lq(l)}$ is the utilization ratio of the lth branch on Lq. Based on Eq. (8.41), the cumulative vulnerability value of the lth branch in all connected electrical paths can be defined as the vulnerability of the branch. Meanwhile, the maximum vulnerability of the branch connected to the jth node can be defined as the vulnerability of that node. Therefore, the specific equation is

$$V_p[branch(l)] = \sum_{q=1}^{N_L} V_{Lq(l)} \tag{8.42}$$

$$V_p[N(j)] = \max_l \{C_p[branch(l)]\}, l \text{ includes the jth node} \tag{8.43}$$

where $V_p[branch(l)]$ is the vulnerability of the lth branch; $V_p[N(j)]$ is the vulnerability of the jth node.

(2) Select the vulnerability index on cyber node. In this case, the main purpose is to assessment the vulnerability of communicators. The indexes include three aspects: the importance degree of this node, the number of failures of this node during the year and the protective measures (e.g., firewall) of this node. The specific values can be obtained by the expert scoring method, the specific equation is

$$V_C[N(j)] = V_{C1j} + V_{C2j} + V_{C3j} \tag{8.44}$$

where V_{C1j} is the importance degree of the jth cyber node; V_{C2j} is the number of failures at this node during the year; V_{C3j} is the protective measures of the jth cyber node. The specific score is obtained by the expert according to the corresponding scoring criteria as Table 8.1

(3) Pretreatment on vulnerability indexes (dimensionless): Among the many methods, the min–max standardization method is easy to be utilized. Therefore, it is used in this study. For the jth CPN, the calculation equation on the ith index is $\xi_{ij} = \left(V_{cij} - V_{i\,\min}\right) / (V_{ci\,\max} - V_{ci\,\min})$, where ξ_{ij} is the value after pretreatment; V_{cij} is the value of the ith index; $V_{i\,\min}$ and $V_{i\,\max}$ are the minimum and maximum values of the ith index, respectively.

(4) Use entropy weight method to assign the weights of indexes: For each CPN, to avoid the further affecting by subjective factors, the objective method is used to assign weights of indexes. Among the many methods, the entropy weight method is selected in this study because of its simplicity and practicability. Through the step 4), the weights of vulnerability indexes are as Table 8.2.

(5) By applying the prospective theory [9], the prospective vulnerability value of each CPN is obtained. The prospective theory is a decision theory, which can consider the uncertainty of the object to be evaluated and the subjective preference of the decision-maker. In this theory, the core is the prospective value of

Table 8.1 Scoring criteria on cyber node

Importance degree of cyber node	The number of failures	Protective measures	Score
Node is not important. It doesn't damage the cyber layer after it's broken	Basically no failures	Protective measures are sufficient	1
Node isn't too important. Node damage can cause a lower level of damage	The number of failures during the past year is less than 5	Protective measures are basically perfect	3
Node is important. Node damage can cause moderate damage	The number of failures during the past year is between 5 and 10	Protective measures are general	5
Node is important. Node damage can cause serious damage	The number of failures during the past year is between 10 and 20	Protective measures are poor	7
Node is very important. Node damage can cause very serious damage	The number of failures during the past year is more than 20	Basically no protective measures	9

Table 8.2 Weights of vulnerability indexes

Vulnerability indexes	Physical vulnerability index(ξ_{1j})	Cyber node importance (ξ_{2j})	The number of failures (ξ_{3j})	Protective measures (ξ_{4j})
Weight	ψ_1	ψ_2	ψ_3	ψ_4

node, which is determined by the value function $v(\xi_{ij})$ and the weight function $\pi(\psi_i)$. The specific equation is

$$V_j = \sum_{i=1}^{n} \pi_i^+ v(\xi_i) + \sum_{i=1}^{n} \pi_i^- v(\xi_i) \tag{8.45}$$

where V_j is the prospective value; π_i^+ is the decision-making weights for decision-makers facing incomes; π_i^- is the decision-making weights for decision makers facing losses. The corresponding equations are

$$\begin{cases} v(\xi_{ij}) = \xi_{ij}^{\alpha_v}, \xi \geq 0 \\ v(\xi) = -\lambda(-\xi)_{ij}^{\beta_v}, \xi < 0 \end{cases} \tag{8.46}$$

$$\pi(\psi_i) = \begin{cases} \pi_i^+ = \dfrac{\psi_i}{\left(\psi_i^{\gamma^+} + (1-\psi_i)^{\gamma^+}\right)^{1/\gamma^+}}, & \Delta\xi \geq 0 \\ \pi_i^- = \dfrac{\psi_i^{\gamma^-}}{\left(\psi_i^{\gamma^-} + (1-\psi_i)^{\gamma^-}\right)^{1/\gamma^-}}, & \Delta\xi < 0 \end{cases} \tag{8.47}$$

where $\Delta\xi$ is the difference between ξ_{ij} and the selected reference value; α_v and β_v are sensitivity coefficients; λ is the loss avoidance coefficient; γ^+ is the income attitude coefficient; γ^- is the loss attitude coefficient. In general, α_v is 0.88; β_v is 0.9; γ^+ and γ^- are all selected as 0.61.

(6) Obtain the prospective vulnerability value of each CPN. Among these CPNs in MG, the smaller the prospective value, the smaller the vulnerability of the CPN.

Based on the above theories, the prospective vulnerability of each CPN can be obtained. And then, the CPN with the lowest vulnerability can be selected to receive the data from the virtual leader. In summary, the main innovations of this subsection are as follows. 1. In vulnerability assessment, the physical node (e.g., DER) and its corresponding node (e.g., communicator) are regarded together as a cyber-physical node (CPN). Then, the physical vulnerability and cyber vulnerability are considered comprehensively. In this way, the incomplete assessment from only one vulnerability alone have been addressed; 2. The cumulative prospective theory is applied to vulnerability assessment. Through this theory, the future vulnerability of CPNs can be considered to make the selected scheme forward-looking. Thus, it can make the constructed directed communication path is strong. So far, there are few related studies proposed.

8.3.2 Path Planning of the Directed Communication Path

For directed path, it needs to build a directed spanning tree in actually, where the leader is the "root node". This is because only the data of leader can be transmitted to other controlled DERs via a directed path, the VLFCC can be accomplished. Thus, the secondary control can be accomplished.

In actually, based on the previous constructed undirected path in the Sect. 8.3.3, the directed path can be constructed by only determining the data transmission direction and setting the leader as "root node". Moreover, because the constructed undirected path has achieved the goal of shortest path and minimum cost, so long as the node connected to virtual leader is determined, the directed path is also the optimal. And in this study, the CPN with the lowest vulnerability is selected to connect with the virtual leader. Because of this, the communication vulnerability between path and virtual leader is the lowest.

Through the above process, the directed communication path can be constructed. And the contributions of our method are: 1. The directed communication path we built is based on the previously built undirected path. Since the constructed undirected path is already optimal (e.g. the total length of the path is the shortest or the economy is the best etc.), so the directed path we construct is also optimal. 2. Through selecting the appropriate communication direction in the constructed undirected path, so that the undirected graph can be transformed into a directed graph. This reduces the process of re-optimizing to select the directed path. 3. Based on the prospective vulnerability assessment in Sect. 8.3.1, it is possible to quickly select the CPN connected to the leader. In this way, our method can construct the directed path directly and effectively.

8.3.3 The Secondary Controller on Voltage

In islanded MG, droop control is often used to realize the plug and play of DER. However, the output voltage of this control is often deviated from the voltage reference value for two reasons: 1. The droop control is obtained by many approximate process, and there is a deviation in itself; 2, As can be seen from Eq. (8.48), there are always deviations (e. g., $m_i p_i$) between the output voltages and the reference value. If these deviations are not compensated, the output voltages of DERs will be unbalanced. And it will cause the collapse of the whole MG. Therefore, it is necessary to design the secondary controller on voltage to address the above problem [10]. In the following, we first give the P–U droop [11] control of the ith DER as

$$\begin{cases} U_i = U_{ref} - m_i P_i \\ \omega_i = \omega_{ref} + n_i Q_i \end{cases} \tag{8.48}$$

where m_i is the droop coefficients; U_{ref} is the voltage reference value; P_i is the active power. Afterwards, if an auxiliary variable is designed as u_{Ui}, there is

$$\dot{U}_i = \dot{U}_{ref} - m_i \dot{P}_i \equiv u_{Ui} \tag{8.49}$$

Based on Eq. (8.2) and the proposed methods above, the corresponding secondary controller can be designed as

$$\begin{cases} \text{1. In normal situation:} \\ u_{Ui} = K_{Ui1} \sum_{j \in N_i} a_{ij} \left(\eta_{ji} U_j(t) - U_i \right) + K_{Ui2} b_i \left[\eta_{Li} U_L - U_i \right] \\ \text{2. Path reconstruction occurs in MG} \\ u_{Ui} = \breve{K}_{\text{Uil}} \sum_{j \in N_i} \breve{a}_{ij} \left(\eta_{ji} U_j(t) - U_i(t) \right) + \breve{K}_{Ui2} \breve{b}_i \left[\eta_{Li} U_L - U_i(t) \right] \end{cases} \tag{8.50}$$

where U_i and U_j represent the voltage of the ith DER and the jth DER respectively; U_L is the voltage of the virtual leader. It's constant and known; K_{Ui1} and K_{Ui2} are the gains in the controller; $\breve{a}_{ij}$ and $\breve{b}_i$ are the new consensus coefficients after path reconstruction; $\breve{K}_{\text{Uil}}$ and $\breve{K}_{Ui2}$ are the new gains in the controller after path reconstruction.

8.3.4 Stability Analysis

In fact, the influence of secondary control on MG is mainly the effect of consensus control on MG. It includes three aspects: 1. The effect of adding secondary control on the stability of overall control system; 2. The influence on the whole MG when the communication topology is changed; 3. In the secondary control process, the influence of disturbance data on the stability of the overall MG.

For the first aspect, our analysis is as follows. For a single DER, the overall control process is as below: 1. The obtained data by consensus controller will be used to generate the voltage regulators (i.e., feedback values); 2. These feedback values will be added to the output voltages of droop control to accomplish the secondary control. This makes the output voltages of droop control adjusted to their reference value; 3. The output voltage and frequency will be used as the input of the double closed loop. And next, a trigger signal will be output by the double closed loop; 4. After using the trigger signal, a PWM signal will be generated to control the DER output the voltage reference value. Moreover, it can control the DER output the corresponding current; 5. Through the Park transformation part and power calculation part, the output voltage and current by DER will be input into the droop control part as power form; 6. The output voltage and frequency of droop control part will be re-used as the input of consensus controller. And in the next control process, the output of consensus controller will be used to obtain the regulators again.

In summary, the addition of secondary control can ensure that the output voltage from droop control part adjusted to U_{ref}, and then maintain the output voltage of each DER as U_{ref} at all times. This ensures that the output voltages are always stable. Therefore, the use of secondary control can make up for the defects of droop control and strengthen the stability of the whole control process.

For the second aspect, the method of adding state feedback has been designed in the Sect. 8.2.5 to maintain the stability. For the third aspect, the small signal model (SSM) is used to accomplish the stability analysis. The specific discuss is as below: In this study, the MG model in Fig. 8.2 is taken as the analyzed object. If there is no connection between the ith and the jth nodes, the output active and reactive power of the ith DER are

$$\begin{cases} P_{oi} = \frac{U_{\text{node}.i}}{Z_i}\left[U_i \cos(\theta_i - \varphi_i) - U_{\text{node}.i} \cos\theta_i\right] + P_{\text{Local}.i} \\ Q_{oi} = \frac{U_{\text{node}.j}}{Z_i}\left[U_i \sin(\theta_i - \varphi_i) - U_{\text{node}.j} \sin\theta_i\right] + Q_{\text{Local}.j} \end{cases} \tag{8.51}$$

where Z_i and θ_i are value and impedance angle of line impedance, respectively; φ_i is the angle of output voltage U_i; $P_{\text{Local}.i}$(or $P_{\text{Local}.j}$) and $Q_{\text{Local}.i}$(or $Q_{\text{Local}.j}$) are active and reactive power of local loads; $U_{\text{node}.i}$ and $U_{\text{node}.j}$ are node voltages. After filtering (filter time is T_{filter}), there are

$$\begin{cases} P'_{oi} = \frac{1}{T_{filter}s+1} P_{oi} \\ Q'_{oi} = \frac{1}{T_{filter}s+1} Q_{oi} \end{cases} \tag{8.52}$$

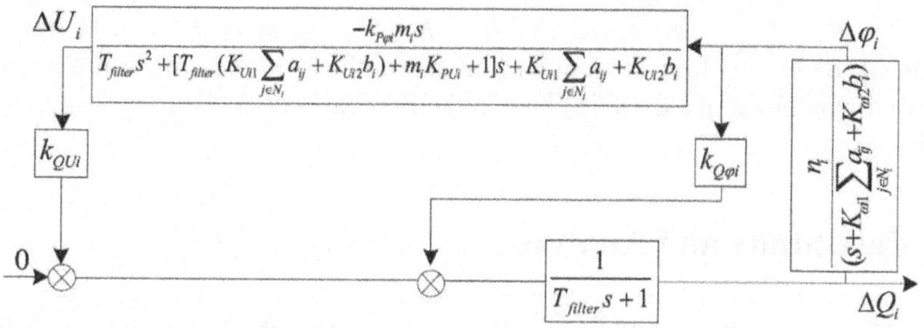

Fig. 8.3 The SSM of the ith DER

If there are small disturbances ΔU_i and $\Delta\varphi_i$ assumed in the MG, the influences on power can be described as

$$\begin{cases} \Delta P_i = \frac{1}{T_{filter}s+1}\left(k_{P\varphi i}\Delta\varphi_i + k_{PUi}\Delta U_i\right) \\ \Delta Q_i = \frac{1}{T_{filter}s+1}\left(k_{Q\varphi i}\Delta\varphi_i + k_{QUi}\Delta U_i\right) \end{cases} \tag{8.53}$$

where $k_{P\varphi i} = \partial P_i/\partial\varphi_i$; $k_{PUi} = \partial P_i/\partial U_i$; $k_{Q\varphi i} = \partial Q_i/\partial\varphi_i$; $k_{QUi} = \partial Q_i/\partial U_i$. After combining with the droop controller and secondary controllers, there are

$$\begin{cases} \Delta\omega_i = n_i\Delta Q_i - \int\left(K_{\omega i1}\sum_{j\in N_i}a_{ij} + K_{\omega i2}b_i\right)\Delta\omega_i dt \\ \Rightarrow s\Delta\varphi_i = n_i\Delta Q_i - \left(K_{\omega i1}\sum_{j\in N_i}a_{ij} + K_{\omega i2}b_i\right)\Delta\varphi_i \\ \Rightarrow \Delta\varphi_i = n_i\Delta Q_i/\left(s + K_{\omega i1}\sum_{j\in N_i}a_{ij} + K_{\omega i2}b_i\right) \\ \Delta U_i = -m_i\Delta P_i - \int\left(K_{Ui1}\sum_{j\in N_i}a_{ij} + K_{Ui2}b_i\right)\Delta U_i dt \\ \Rightarrow \frac{-m_i}{T_{filter}s+1}\left(k_{P\varphi i}\Delta\varphi_i + k_{PUi}\Delta U_i\right) - \left(K_{Ui1}\sum_{j\in N_i}a_{ij} + K_{Ui2}b_i\right)\Delta U_i/s \\ = -k_{P\varphi i}m_i\Delta\varphi_i s/\{T_{filter}s^2 + \left[T_{filter}\left(K_{Uil}\sum_{j\in N_i}a_{ij} + K_{Ui2}b_i\right) + m_iK_{PUi} + 1\right]s \\ +K_{Ui1}\sum_{j\in N_i}a_{ij} + K_{Ui2}b_i\} \end{cases} \tag{8.54}$$

According to Eqs. (8.51)–(8.54), the SSM of the ith DER with the newly proposed secondary controllers is as Fig. 8.3.

Based on Fig. 8.3, it has found that the SSM is a closed-loop system. And its corresponding closed-loop characteristic equation is

$$\begin{aligned} &Ts^4 + \left[(A+B)T + m_iK_{PUi} + 2\right]Ts^3 + \left[(A+B)T + (AT + m_iK_{PUi} + 1)(BT+1)\right. \\ &\left.-n_iK_{Q\varphi i}T\right]s^2 + \left[(1 - n_iK_{Q\varphi i})(AT + m_iK_{PUi} + 1) - K_{P\varphi i}K_{QUi}m_i + BT + 1\right]s \\ &+AB + m_iK_{P\varphi i}A = 0 \end{aligned} \tag{8.55}$$

where $A = K_{Ui1}\sum_{j\in N_i} a_{ij} + K_{Ui2}b_i$; $B = K_{\omega i1}\sum_{j\in N_i} a_{ij} + K_{\omega i2}b_i$. Therefore, the stability of the closed-loop system can be guaranteed by selecting suitable gains to make all eigenvalues of Eq. (8.55) have negative real roots.

8.4 Case Study and Analysis

In this study, to confirm the effect of the designed methods through carrying out the simulation experiments, the MG system in Fig. 8.4 is taken as the experimental model. And the related parameters are as Table 8.3.

In Fig. 8.4, there are four kinds of DERs in this experiment model, including battery, photovoltaic power generation, wind turbine and fuel cell. These four DERs are connected to the AC bus via converters (including rectifiers and inverters). Moreover, their related data (i.e., controlled variations) will be interacted through the communication network to accomplish the proposed hierarchical control in this chapter. Under the control of the contributed strategy, the MG model provides the power required by flexible load.

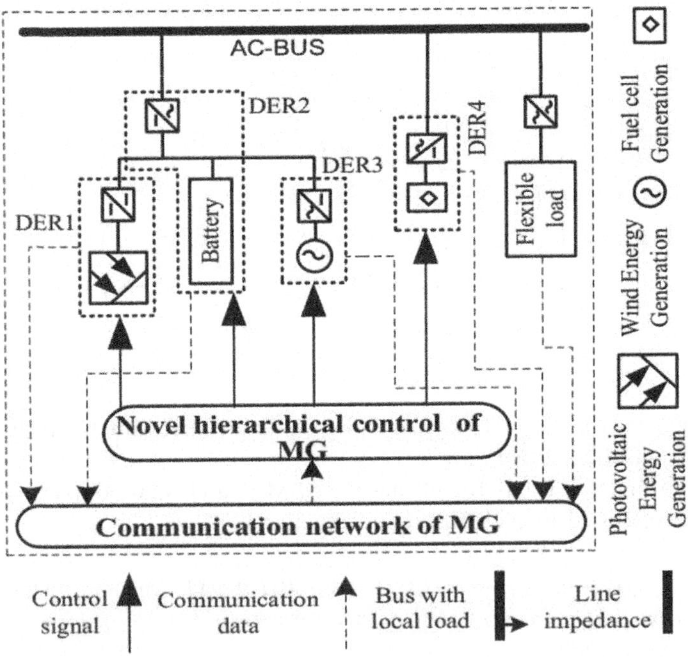

Fig. 8.4 The MG structure used for experiments

Table 8.3 Parameters in simulation

Parameters	Values
Sample time (i.e., step size in simulation)	5e−6 s
Power	$V_1 = V_2 = V_3 = V_4 = 700$ V, $f_1 = f_2 = f_3 = f_4 = 50$ Hz
Load (i.e., the flexible load)	P = 300 kW Q = 30 kVar
a_{ij}, b_i	$a_{14} = 1, a_{21} = 1, a_{32} = 1, a_{43} = 1,$ $b_1 = 1$, other $a_{ij} = 0$
Droop coefficient	$\mathrm{m} = 1.6 \times 10^{-5}, \mathrm{n} = 2 \times 10^{-4}$
Line impedance	$Z_{\text{Line.DER1-AC-BUS}} = 0.612 + j3.18e - 3$
	$Z_{\text{Line.DER2-AC-BUS}} = 0.613 + j2.99e - 3$
	$Z_{\text{Line.DER3-AC-BUS}} = 0.614 + j3.18e - 3$
	$Z_{\text{Line.DER4-AC-BUS}} = 0.615 + j3.18e - 3$
$U_L(t), \omega_L(t)$	311 V, 314 Rad/s
Gains in consensus protocol	$k_{Ui1} = 4, k_{Ui2} = 400$
	$k_{\omega i1} = 0.4, k_{\omega i2} = 40$
$V_{max}(path)$	2.5
γ^+, γ^-	0.61

Table 8.4 The physical vulnerability of CPNs

Vulnerability indexes	MG			
	Line 1	Line 2	Line 3	Line 4
$d_{Lq(l)}$	0.612	0.613	0.614	0.615
$I_{1,Lq(l)}$	0.250	0.250	0.250	0.250
$I_{2,Lq(l)}$	0.250	0.250	0.250	0.250
$\lambda_{Lq(l)}$	1.000	1.000	1.000	1.000
Vulnerability value	0.0383	0.0383	0.0383	0.0384

8.4.1 *Case 1 Verify the Effectiveness of Cyber-Physical Vulnerability Assessments*

In this subsection, the four CPNs composed by DERs and their respective communicators in Fig. 8.4 are taken as the experiment objection. And the relevant data of CPNs are as Tables 8.4, 8.5 and 8.6.

Based on the above Tables 8.4, 8.5 and 8.6, the corresponding weight parameters can be obtained as follows:$\Psi_1 = 0.503$; $\Psi_2 = 0.120$; $\Psi_3 = 0.126$; $\Psi_4 = 0.251$. The prospective vulnerability of each node is shown in Table 8.7.

To verify the effectiveness of vulnerability assessment more intuitively, the vulnerability of each CPN is shown as Fig. 8.5. In this figure, the horizontal and vertical

Table 8.5 The cyber vulnerability of CPNs

Vulnerability indexes	MG			
	CPN 1	CPN 2	CPN 3	CPN 4
Node importance degree	1	3	5	5
The number of failures	1	3	5	3
Protective measures	1	3	3	3

Table 8.6 The vulnerability indexes of CPNs after dimensionless

Vulnerability indexes	MG			
	CPN 1	CPN 2	CPN 3	CPN 4
Physical vulnerability index(ξ_1)	0.000	0.000	0.000	1.000
Node importance degree (ξ_2)	0.000	0.500	1.000	1.000
The number of failures (ξ_3)	0.000	0.500	1.000	0.500
Protective measures (ξ_4)	0.000	0.000	1.000	1.000

Table 8.7 Obtained prospective vulnerability of CPNs

Vulnerability indexes	MG			
	CPN 1	CPN 2	CPN 3	CPN 4
$v(\xi_1)$	0.000	0.000	0.000	1.000
$v(\xi_2)$	0.000	0.543	1.000	1.000
$v(\xi_3)$	0.000	0.543	1.000	0.543
$v(\xi_4)$	0.000	0.000	1.000	1.000
Prospective value	0.000	0.134	0.497	0.942

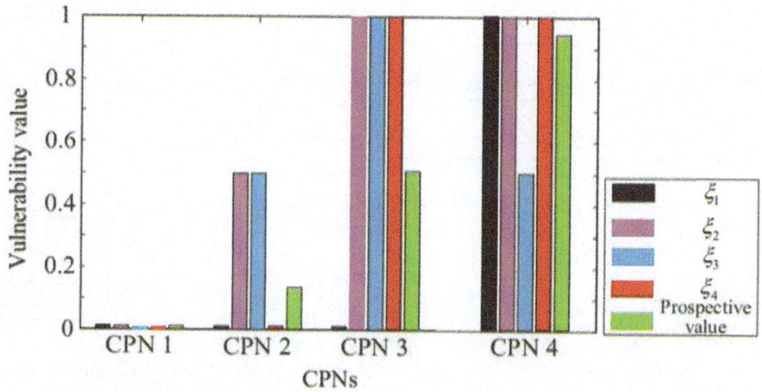

Fig. 8.5 The prospective vulnerability of CPNs

axes represent CPNs and vulnerability, respectively. It can be seen that: the physical vulnerability of CPN 4 is the highest which closes to 1; the physical vulnerability of the other three CPNs the lowest which is zero. After considering both cyber vulnerability and physical vulnerability, the vulnerability of CPN 1 and CPN 4 are still the highest and the lowest, respectively. However, the prospective vulnerabilities of CPN 2 and CPN 3 are changed. The prospective vulnerabilities of these two CPNs are 0.134 and 0.497, respectively. Therefore, Fig. 8.5 has shown that it is not comprehensive to consider only physical or cyber vulnerabilities. There should be a combination of considerations.

8.4.2 *Case 2 Verify the Effectiveness of Path Planning and Reconstruction Method*

In this case, the communication network of Fig. 8.4 is taken as the example, and the corresponding length of communication lines is as Table 8.8. And because the communication line cost of MG is generally calculated in terms of length, the shorter

Table 8.8 Length of communication lines

Number	Number			
	1	2	3	4
Length (dm)				
1	0	14	18	23
2	14	0	20	22
3	18	20	0	31
4	23	22	31	0

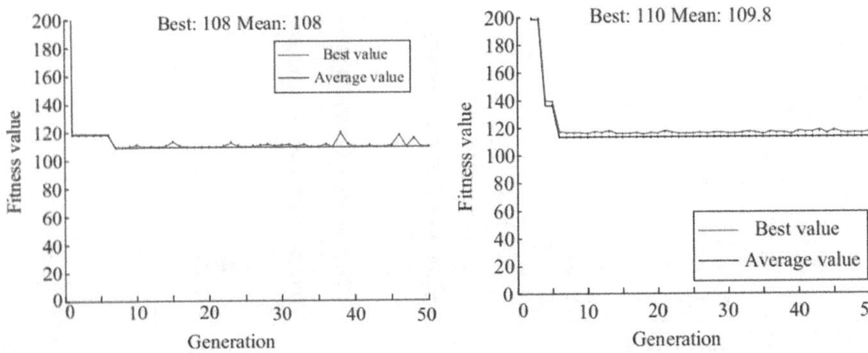

(a) Path planning under 1)

(b) Path reconstruction under 2)

Fig. 8.6 The effectiveness of path planning and reconstruction method

the total line length, the better the economy of line. To facilitate the experiment, only the goal of making line length shortest is taken as the objective function in this case.

Furthermore, in this subsection, there are four situations are assumed to carry out the experiments: (1) Construct the undirected communication path; (2) Based on (1), the communicator 2 is set to be unable to output data, and then the undirected communication path is reconstructed; (3) Construct the directed communication path based on (1); (4) On the basis of (2), the directed communication path is constructed. The simulation results are as Fig. 8.6.

Through the above two subfigures, it has found that the appropriate path schemes can be found effectively. The lengths of two communication paths are 108 dm and 110 dm, respectively. The specific schemes are

In (1): Communicator 1 $\begin{cases} \leftrightarrow \text{Communicator 2} \leftrightarrow \text{communicator 4} \\ \leftrightarrow \text{Communicator 3} \end{cases}$

In (2): Communicator 1 $\leftrightarrow \begin{cases} \text{Communicator 2} \\ \text{Communicator 3} \\ \text{Communicator 4} \end{cases}$

Based on the constructed undirected path and the obtained prospective vulnerability in Case 1, the communicator 1 is selected as the node to be connected with the virtual leader. For (3) and (4), the obtained directed path schemes are:

In (3): Virtual leader $\rightarrow$

Communicator 1 $\begin{cases} \rightarrow \text{Communicator 2} \rightarrow \text{Communicator 4} \\ \rightarrow \text{Communicator 3} \end{cases}$

In (4): Virtual leader $\rightarrow$ Communicator 1 $\begin{cases} \rightarrow \text{Communicator 2} \\ \rightarrow \text{Communicator 3} \\ \rightarrow \text{Communicator 4} \end{cases}$

The length of these two schemes is 54 and 55, respectively. In summary, through the proposed path planning and reconstruction method, the appropriate undirected and directed communication paths can be constructed.

8.4.3 *Case 3 Verify the Effectiveness of CBPC*

In this case, the related cost coefficients of DERs are as Table 8.9. The simulation results are as Fig. 8.7. The physical devices are represented by different colors in simulation results as below: __ DER 1, __ DER 2, __ DER 3, __ DER 4, __ Flexible load. The experiments are composed by three kinds of situations as the follows.

(1) The variations of DERs and flexible load without using consensus control

To illustrate the effectiveness of the proposed CBPC by comparison, the experiments without using the consensus control are also carried out in this subsection. The simulation results are as Fig. 8.7.

The analysis is as below: As shown in Fig. 8.7, the *IC* of each DER is not consistent when the consensus control is not adopted, which are almost 0.48, 0.32, 0.29 and 0.22, respectively. The *IC* of DER with large cost coefficient is also large. The *IB* of the flexible load is –0.35. Meanwhile, under the action of *P–U* droop control, the power of flexible load is always 300 kW. The output power of each DER is about 75 kW. Because the output power of each DER is same, the output voltages of DERs by *P–U* droop control are all almost 309.85 V.

(2) The variations of DERs and flexible load without considering power limitation.

In this subsection, the consensus control without leader is used to adjust the *ICs* of DERs and the *IB* of flexible load. Furthermore, the power limitations of devices are not taken into account. After using consensus control at $t = 0.1$ s, the corresponding simulation results are as Fig. 8.8.

The analysis is as follows: Based on Fig. 8.8a, it has found that the *IC* of each DER will tend to be consistent after using the consensus control at $t = 0.1$ s. When $t = 0.3$ s, the *IC* of each DER will be adjusted to about 0.191. It can be seen from Fig. 8.8b that the power of each DER will be output according to its own optimal power. Moreover, the optimal power of each DER is inversely proportional to its own cost coefficient. The output power of these DERs is almost 2.75 kW, 28.3 kW, 33.75 kW and 60.5 kW, respectively. The flexible load is about 125 kW. In Fig. 8.8c, it has found that the output voltages of DERs will be different. In addition, the output voltage of each DER is inversely proportional to its optimal power. The voltages of DERs are about 310.96 V, 310.57 V, 310.46 V and 310.03 V, respectively.

(3) The variations of DERs and flexible load with considering power limitations.

Table 8.9 Cost coefficients of devices

Device	$\alpha/[\$; \cdot(kW \cdot h)^{-2}]$	$\beta/[\$; \cdot(kW \cdot h)^{-1}]$	$\gamma/(\$; \cdot h^{-1})$
DER 1	0.0020	0.18	0.0400
DER 2	0.0013	0.12	0.0140
DER 3	0.0012	0.11	0.0150
DER 4	0.0010	0.07	0.0015
Flexible load	−0.0015	0.56	0.0060

Fig. 8.7 Variations of DERs and flexible load in (1)

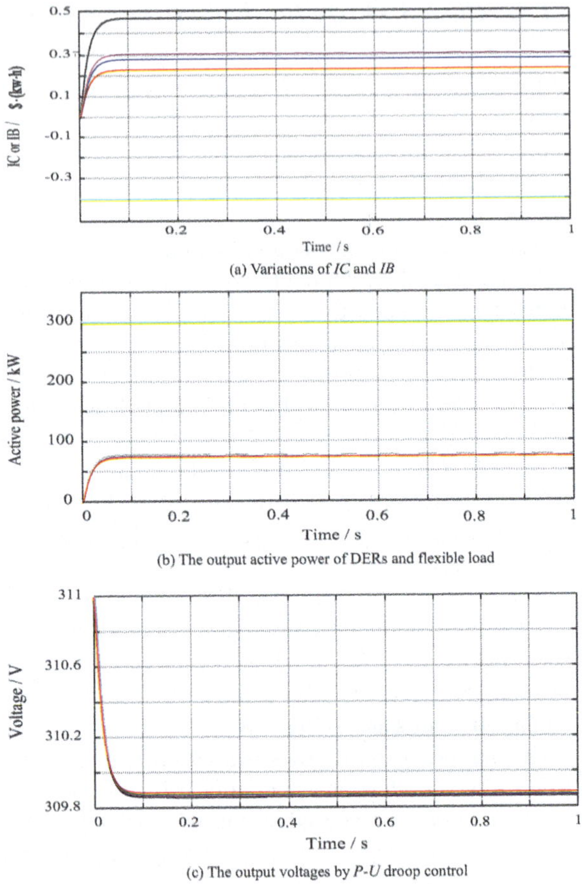

Situation 1: There is power limitation on DER.

In this situation, the minimum output power of DER 1 is designed as 30 kW. And the corresponding simulation results are as Fig. 8.9.

The analysis is follows: From Fig. 8.9a, it can be seen that the *IC* of each DER will begin to be consistent by using the consensus control at $t = 0.1$ s. However, due to there is power limitation on DER1, the path reconstruction will occur in the consensus control process. Moreover, the *IC* of DER1 will withdraw from the consensus control, and the ultimate value is the *IC* corresponding to the limited power. This makes the *ICs* of the other three DERs and the *IB* of flexible load need to converge again (the new consensus value is about 0.215). It can be seen from Fig. 8.9b that the output power of DER1 will eventually be its maximum power. The power of the remaining three DERs and flexible load will be output according to their new optimal power. They are about 36.76 kW, 43.75 kW, 72.51 kW and 182.67 kW, respectively. It can be seen from Fig. 8.9c that the output voltages of DERs will be re-output according

Fig. 8.8 Variations of DERs and flexible load in (2)

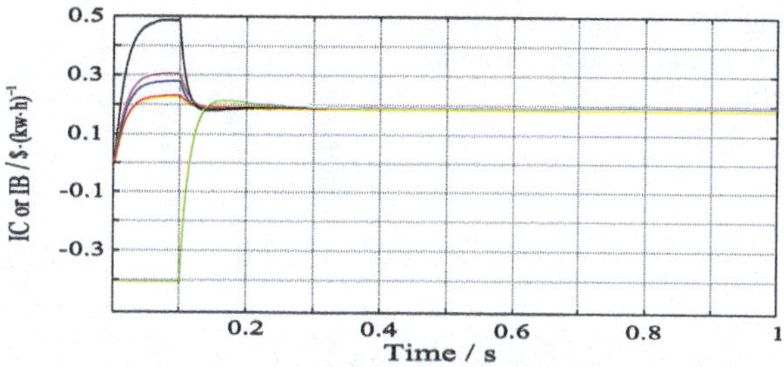

(a) Variations of *ICs* and *IB*

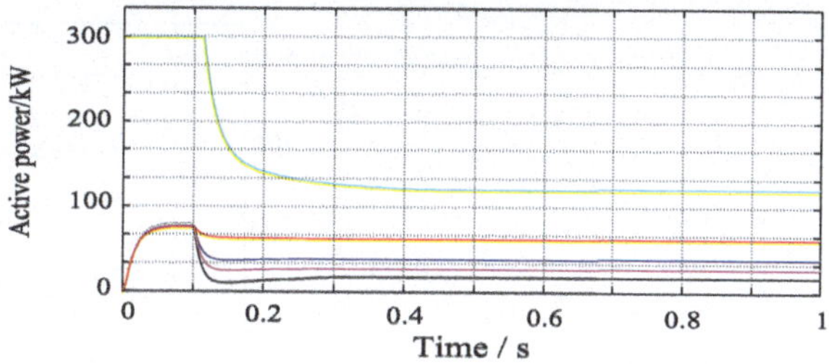

(b) The output active power of DERs and flexible load

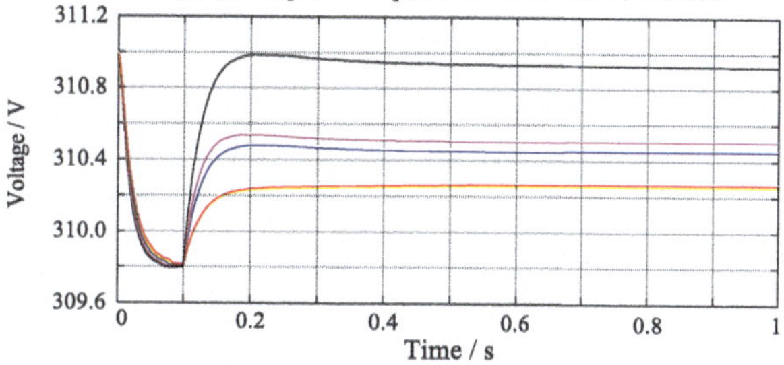

(c) The output voltages by *P-U* droop control

to their respective new active power. They are almost 310.88 V, 310.45 V, 310.33 V and 309.88 V.

Situation 2: There is power limitation on flexible load

In this situation, the maximum allowable removal power of flexible load is designed as 150 kW. And the corresponding simulation results are as Fig. 8.10.

The analysis is as follows: From Fig. 8.10a, it can be seen that the IB and ICs will begin to be consistent due to the use of consensus control at $t = 0.1$ s. However, due to the power limitation of flexible load, the IB of flexible load will withdraw from the consensus control process at $t = 0.125$ s. Meanwhile, the path reconstruction will occur in MG, which makes the *IC* of each DER tend to a new consistency value again. And the new consensus value is about 0.215. In Fig. 8.10b, it has found that the output power of flexible load will ultimately be about 150 kW. Meanwhile, the output power of each DER will be output according to its new optimal power. They are almost 7.5 kW, 34.61 kW, 41.66 kW and 72.67 kW. The power is It can be seen from Fig. 8.10c that the output voltages of DERs will be re-output according

Fig. 8.9 Variations of DERs and flexible load in situation of (3)

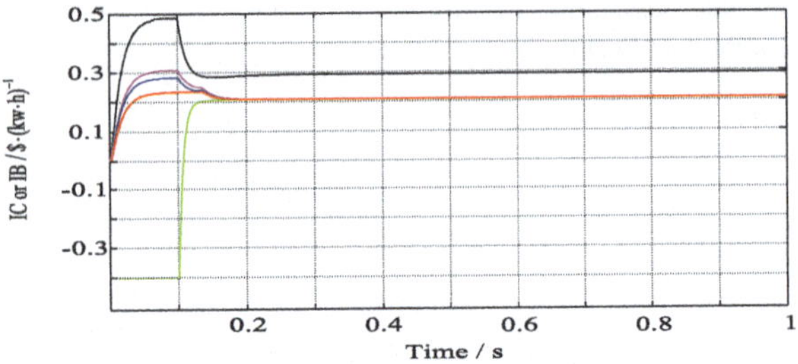

(a) Variations of *IC*s and *IB*

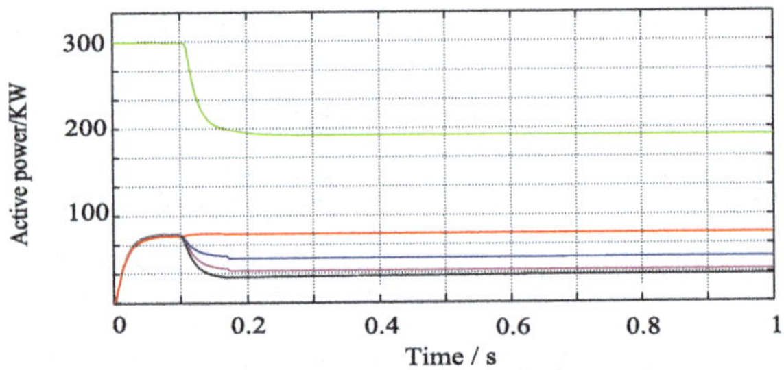

(b) The output active power of DERs and flexible load

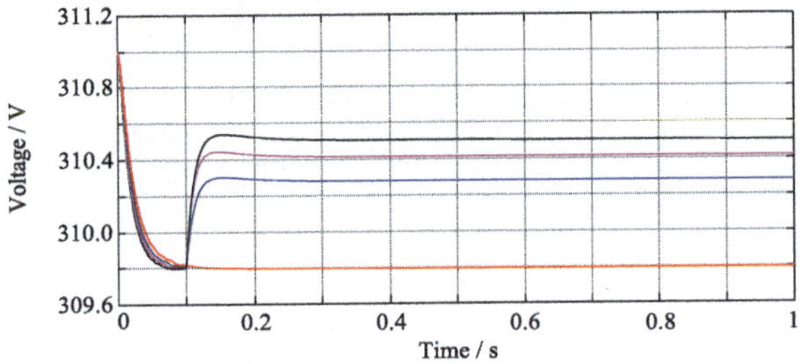

(c) The output voltages by *P-U* droop control

to their new optimal power. They are 310.88 V, 310.45 V, 310.33 V and 309.88 V, respectively.

8.4.4 Case 4 Verify the Effectiveness of CBSC

In this case, the effect of secondary control is verified for all kinds of situations in Case 3. In addition, the second control is added in all situations at $t = 0.5$ s. The simulation results are as Fig. 8.11.

The analysis is as below: For all situations in Fig. 8.11, it can be seen that the P–U droop control will cause deviations between the output voltages of DERs and U_{ref}. The voltages are almost 309.8 V. After $t = 0.5$ s, the deviation will be gradually eliminated by using secondary control method. Moreover, it has found that the output

Fig. 8.10 Variations of DERs and flexible load in situation 2 of (3)

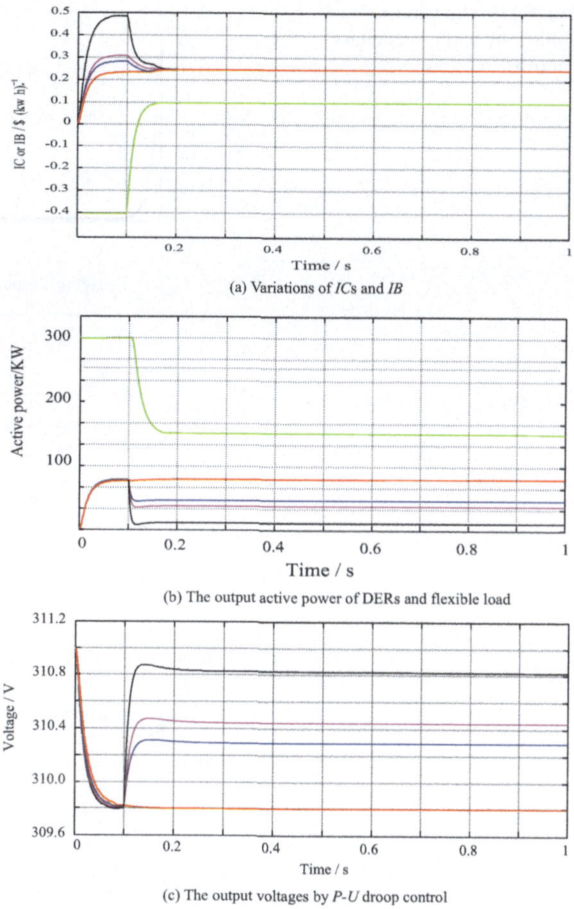

(a) Variations of *ICs* and *IB*

(b) The output active power of DERs and flexible load

(c) The output voltages by *P-U* droop control

voltages of DERs will be adjusted to be consistent after $t = 0.7$ s. And finally, the output voltage of each DER can still be maintained at 311 V. This also shows the necessity of studying the secondary control.

8.5 Conclusion

In this study, a consensus-based economic hierarchical control strategy is proposed for islanded MG. The purpose of this chapter is to maintain the stability of MG while maximizing the benefits. In this control strategy, we mainly design the CBPC and CBSC methods as follows: 1. In the CBPC, based on consensus control theory and virtual impedance, an economic P–U droop control method is proposed; 2. In the

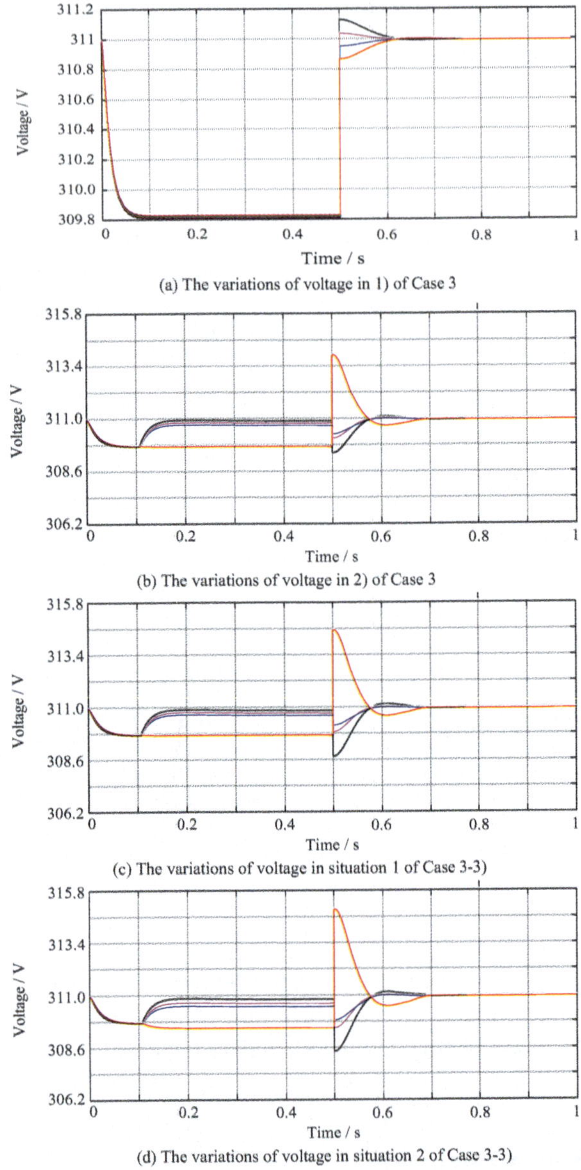

Fig. 8.11 The variations of voltage by using secondary control

CBSC, based on the cyber-physical vulnerability assessment and VLFCC theory, a novel secondary control method on voltage is designed.

And finally, the Case 1 has confirmed that the effectiveness of vulnerability assessment. The vulnerability of CPNs can be fully evaluated by this method; the Case 2 has confirmed that the appropriate path scheme can be obtained by path planning and reconstruction method; the Case 3 has confirmed that the effectiveness of CBPC. Through this method, each DER can output its own optimal active power. In addition, the flexible load is also controlled reasonably; the Case 4 has confirmed that the output voltages of all DERs can be adjusted to U_{ref} by using CBSC. And in the future, we will study more solutions to other communication problems.

References

1. B. Zhang, C. Dou, D. Yue, Z. Zhang, Response hierarchical control strategy of communication data disturbance in micro-grid under the concept of cyber physical system. IET Gener. Transm. Distrib. **12**(21), 5867–5878 (2018)
2. G. Hug, S. Kar, C. Wu, Consensus + innovations approach for distributed multiagent coordination in a microgrid. IEEE Trans. Smart Grid **6**(4), 1893–1903 (2017)
3. Z. Tang, D.J. Hill, T. Liu, A novel consensus-based economic dispatch for microgrids. IEEE Trans. Smart Grid **9**(4), 3920–3922 (2018)
4. Z. Zhang, M. Chow, Convergence analysis of the incremental cost consensus algorithm under different communication network topologies in a smart grid. IEEE Trans. Power Syst. **27**(4), 1761–1768 (2012)
5. C. Dou, B. Zhang, D. Yue, Z. Zhang, Cyber-physical cooperative control strategy for islanded micro-grid considering communication interruption. Int. Trans. Electr. Energy Syst. **29**(1) (2019)
6. X. Yin, D. Yue, S. Hu, Distributed event-triggered control of discrete-time heterogeneous multi-agent systems. J. Frankl. Inst.-Eng. Appl. Math. **350**(3), 651–669 (2013)
7. Z. Li, M. Shahidehpour, F. Aminifar, Cybersecurity in distributed power systems. Proc. IEEE **105**(7), 1367–1388 (2017)
8. J.C. Balda, A. Mantooth, R.S. Blum, P. Tenti, Cybersecurity and power electronics: addressing the security vulnerabilities of the internet of things. IEEE Power Electron. Mag. **4**(4), 37–43 (2017)
9. L. Xiao, N.B. Mandayam, H.V. Poor, Prospect theoretic analysis of energy exchange among microgrids. IEEE Trans. Smart Grid **6**(1), 63–72 (2015)
10. X. Liu, H. He, Y. Wang, Q. Xu, F. Guo, Distributed hybrid secondary control for a dc microgrid via discrete-time interaction. IEEE Trans. Energy Convers. **33**(4), 1865–1875 (2018)
11. R. Wai, Q. Zhang, Y. Wang, A novel voltage stabilization and power sharing control method based on virtual complex impedance for an off-grid microgrid. IEEE Trans. Power Electron. **34**(2), 1863–1880 (2019)

Chapter 9
Multi-agent-system-based Bi-level Bidding Strategy of Microgrid with Game Eheory in the Electricity Market

This chapter focuses on multi-agent-system-based bilevel bidding strategy of microgrid to make all power producers participate in electricity allocation as rational agents under electricity market transaction. Firstly, the multi-agent-system-based bidding system is constructed as bi-level MAS. Then, the bidding model is formulated as a bi-level optimization problem, which makes the best of renewable energy source, and subjects it to the object of obtaining maximum profits for all power producers. According to this model, electricity market is divided into two subsidiary markets, which involve the upper and the lower electricity market. The lower electricity market is composed of microgrids, which ally together in the upper electricity market to offset their shortages compared to traditional power producers, such as efficient market influence, the reliability of the energy supplying. Additionally, game theory is introduced for modeling and Nash equilibrium as the solution of bidding optimization problem is found through a series of iterative process based on the bi-level MAS. Finally, both the upper market and the lower market are illustrated in the case studies and corresponding proper conclusions are received.

9.1 The Framework of MAS-Based Bi-level Bidding

9.1.1 Introduction

An agent is an entity (software or hardware) that is situated in some environment and is able to autonomously react to change in the environment [1]. In the upper market, due to the diversity of power producers, it is difficult for each producer to know the characteristics of every opponent (the cost function, strategic space, bidding model, and so on). Power producer agent could change its own behavior only based on the

D. Yue et al., *Cooperative Optimal Control of Hybrid Energy Systems*,
https://doi.org/10.1007/978-981-33-6722-7_9

change of the market and its own status, without knowing the other power producer's relevant information. The interaction among the different agents enables the system to carry out its main objectives, which is also beneficial to the system safety.

In this chapter three types of agents have been defined:

1. Load agents: They represent loads that expend electricity energy in the whole market and they are not able to change their own behavior according to the change of environment (assuming the demand side submits inelastic demand).
2. Generation agents: they represent power producers, involving the MGs, to bid for the quantity and price according to corresponding strategy.
3. Market agents: the upper market agents are responsible for electricity transaction between all market members under the System security constraints, while the lower market agents are in charge of transaction between MGs in the lower market and bid as a power alliance in the upper market simultaneously.

Each agent solves a particular optimization problem to realize the local target. For different agents, they will decide how react to the environment according to the specific knowledge, which is encapsulated into the agent, such as its own status, optimize arithmetic. This concrete agent function will be given below.

9.1.2 The Framework of Bi-level Bidding MAS-Based

In China, the electricity market has just started, where the supplying of electricity for the market is closed to be monopolized by several large power producers. Microgrid that takes renewable energy as the main source, generation capacity is relatively small, and the output stability of single microgrid is relatively poorer than most traditional power producers. All of these features discussed above lead to a lack of sufficient market influence for a single microgrid, which is adverse when it comes to obtaining more benefit according to strategy bidding in the market.

This chapter formulates a bi-level bidding system based on MAS, in which the MGs will ally together to obtain more capacity in the market competition, as shown in Fig. 9.1.

EMAG: its main responsibility is collecting information from all agents, and then schedule transaction plan under the constraints of system security with a specific optimization method. The MSMAG is now considered as a power producer agent to send bidding information to EMAG.

MSMAG: its responsibility contains two parts. Firstly, in the upper-market, it will collect the relevant information about the capacity, price, costing, and so on, from all MAG. Then it will apply a suitable bidding to EMAG according to its own situation and learn to adjust strategy according to the data offered by the EMAG. Secondly, it will schedule transaction plan in the lower-market under some limited conditions, which is similar to the function of EMAG. Obviously, their optimization method and algorithms are different.

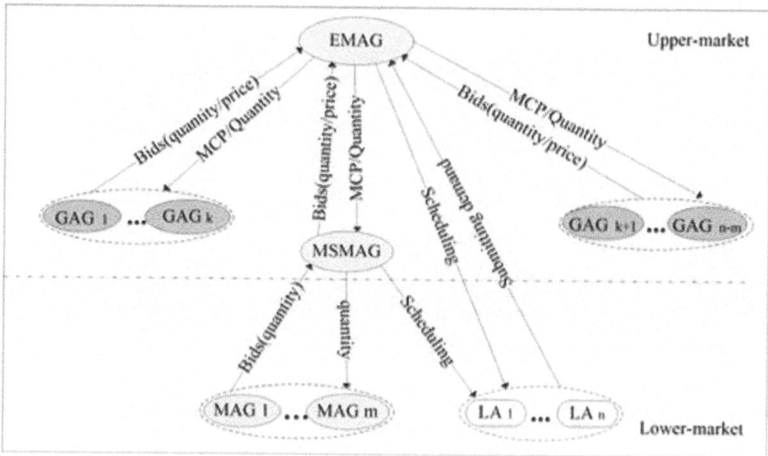

Fig. 9.1 Market bidding structure of bi-level MAS

GAG: the main responsibility of this agent is submitting bids according to its own status and other data such as minimum and maximum power limits. This agent also need learn to adjust the strategy dynamically according to the feedback data from EMAG and respond to it for generating and obtaining profit.

MAG: the main responsibility of this agent also consists of two parts. In the upper-market, it will send information about its own generating capacity, corresponding cost, several constraint conditions, and so on, whereas in the lower market, it will submit bids to MSMAG according to its own status and other data, and learn to adjust the strategy simultaneously. Finally, it also needs to respond to MSMAG for generating and obtaining profit.

LA: it is responsible for forecasting the load demand in the market and then sending it to EMAG. Then it will accept the scheduling of EMAG in the upper-market and of MSMAG in the lower-market.

There are some more explanations about EMAG and MSMAG here. This chapter studies the electricity market of the power generation side, so there is no consideration for the response of the demand side and large user bilateral trade issues. The character of the market model is powertrading center on behalf of all power users to buy electricity from the power generators. In general, there is only one power trading center in the market, no matter how many power generators there are. MSMAG and EMAG are respectively the market agent (the agent of power trading center in the market) of the upper-market and the lower-market. The number would not increase with the increase of the number of power generators. In addition, though the MSMAG is considered as the alliance agent in the upper-market, the number of MSMAG would not also increase with the increase of the number of MAG. So this chapter doesn't consider the condition that there are more than one MSMAG and EMAG.

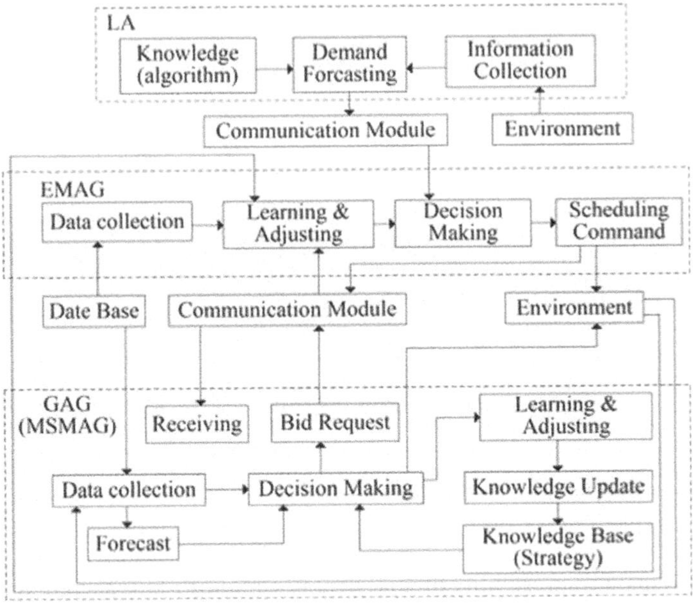

Fig. 9.2 Interaction process of agents in the upper-market

The proposed multi-agent system was developed according to IEEE FIPA standards [2]. JADE framework that conforms to FIPA standards for intelligent agent is applied in this chapter. After operating JADE, MAS will automatically simulate according to commands and interactions [3]. Figure 9.2 shows the interaction process of agents in the upper-market, in which MSMAG is considered as a power producer to participate in bidding. For simplicity, Fig. 9.2 only shows the process of a single GAG and LA.

The process can be described simply as follows: firstly, LA forecasts the load demand on the basis of some knowledge, data, and informs demand to EMAG; at the same time, GAG and MSMAG make suitable bidding decision and submit bids to EMAG according to various information which include several local knowledge; secondly, EMAG collects various information including bids, load demand, its own local knowledge, and so on; simultaneously, EMAG calculates the dispatched results and market clearing prices and transfers the scheduling command toMAG and MSMAG; thirdly, after MAG and MSMAG receive the results, they will check the results and compare them with the last results. If the two results are not equal, they will make bids again with learning and adjusting the strategy until the last two bidding results are consistent, which is the symbol of equilibrium.

The interaction process of agents in the lower-market is similar to the process in the upper-market, except that the load demand and market clearing price is fixed in the lower- market, which is given by the result of upper-market biding. In the upper-market, MAG will participate in market transaction in the form of MSMAG.

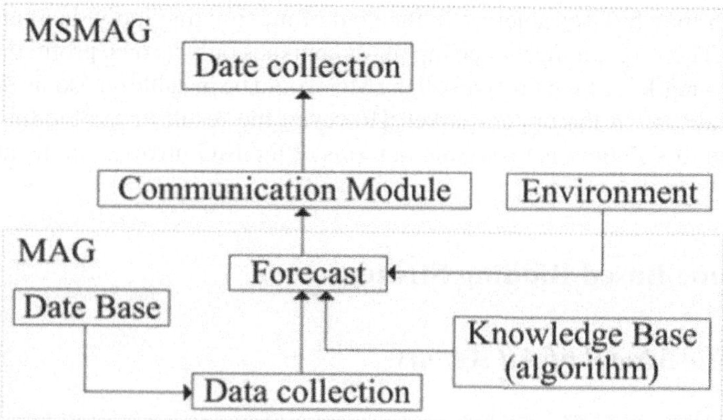

Fig. 9.3 Communication between MAG and MSMAG

Therefore the MAG needs to communicate with MSMAG, the process of which is described in Fig. 9.3.

In the alliance of MAG, MAG needs to accurately forecast the generating capacity according to various information and send it to MSMAG. The strategy proposed in this chapter also solves the uncertainties of microgrid to a certain extent. The uncertainties of MG output is mainly caused by the wind, solar, and other distributed generators, of which the output is unstable. The economic optimization operation of MG first refers the optimal dispatch of the distributed generators. MG takes into account all the characters of distributed generators (involving wind, solar, and other generators whose output is stable but cost may be greater) and other factors, and then elicits an optimal scheduling result which includes the type (selling electricity or buying) and the quantity of market transaction. This chapter studies the bidding strategy when the MG needs to sell electricity to market, and considers the quantity which is determined by the optimal scheduling results of MG as the upper limit of bidding capacity And making microgrids gain more market share, in other words, realizing the optimal scheduling results of MG as mentioned above to the greatest extent is one of the goals of the bi-level bidding strategy proposed in this chapter. This chapter assumes the motivation of cooperation for MAG is increasing the market power and overtaking the shortage discussed above to enhance competitiveness, so the other factors which will influence the stability of ally are not considered.

However, as a self-financing market entity, pursuing the maximum profit should be the main goal of participating in market competing for every microgrid. That this chapter proposes increasing the market power as the main object is based on two main factors. 1) The upper limit of bidding capacity for every MAG is determined by the result of optimal economic dispatch, so the goal of bidding should maximize the quantity as possible as it can; 2) MAG takes renewable energy as its primary energy source, so the goal can improve the proportion of renewable energy and increase the holistic benefit at the same time, which is conform to the national energy development strategy but may be contradictory to the goal of maximizing individual interests for some MG. Therefore, if the ownership of MG pursues only its own profit, the alliance in the upper-market which involves all the MG would be instable and some MG would bidindependently in the upper-market. However, the result of the bidding strategy proposed in this chapter is invariable in terms of the MG involved in the alliance.

9.2 Game Based Bidding Strategy

9.2.1 *The Model of All Agents*

In the electricity market transaction of supply side, the objective of all power producers is to obtain the maximum profit through rational bidding strategy. Game theory is a decisionmaking problem of multi-person or multi-firms who pursue their own maximum profits based on a set of interdependent and interactional strategies [3]. This section provides the mathematical model of the optimization problems based on game theory and that are solved by each agent in the market transaction. For different types of agents, the models can be described as follows:

(1) MSMAG: as the ally of MAG, it will integrate the local information of all MAG to make a suitable bidding strategy. The relevant function can be formulated as follows:

$$q_{MS}^{\min} = \min\left\{q_{Mj}^{\min},\quad j = 1, 2, \ldots, m\right\} \tag{9.1}$$

$$q_{MS}^{\max} = \sum_{j=1}^{m} q_{Mj}^{\max} \tag{9.2}$$

$$q_{MS}^{\min} \le q_{MSM} \le q_{MS}^{\max} \tag{9.3}$$

where$q_{MSM}$is the bidding quantities of MSM;$\left\{q_{Mj}^{\max}, q_{Mj}^{\min}\right\}$are the upper and lower limits of the ith MAG respectively;$\{q_{MS}^{\max}, q_{MS}^{\min}\}$are the upper and lower limits ofq_{MSM}.

After collecting the local information about the MAG, the MSMAG will make a rational bid according to a specific strategy, which can be considered as the intelligence of an agent. The mathematical model can be described as follows:

$$U_{MSM}^{pf} = q_{MSM} \cdot P_{MCP} - CLMSM \tag{9.4}$$

$$P_{MCP} = P_{\max} - \alpha \cdot q_{all} \tag{9.5}$$

$$C_{MSM} = a_{MSM} + b_{MSM} \cdot q_{MSM} + c_{MSM} \cdot q_{MSM}^2 \tag{9.6}$$

$$q_{all} = \sum_{i=1}^{n-m} qGi + q_{MSM} \tag{9.7}$$

where$\{P_{MCP}, C_{MSM}\}$are the market clearing price and the cost function of MSMAG; a_{MSM}, b_{MSM} and c_{MSM} }are the cost coefficients, which are different from every other power producer;$q_{all}$is the total quantity supplied by the power producers, including all GAG and MAG; p_{max}s the upper limit of market clearing price; U_{MSM}^{pf} is the profit of MSMAG that is obtained from the electricity market; q_{Gi} and q_{MSM} are the bidding quantities of the ith GAG and of the MSMAG, respectively (this chapter assumes there are $n - m$ GAG). The objective function of optimization problem for MSMAG can be described as follows:

$$\text{Max}\, U_{MSM}^{pf} \tag{9.8}$$

Associated with Eqs. (9.4)–(9.7), the revenue function of MSMAG which is a function of the variable q_{Gi} and q_{MSM} can be calculated as follows:

$$\begin{aligned} U_{MSM}^{pf} &= f\,(q_{MSM}, q_{Gi}) \\ &= \left(P_{\max} - \alpha \cdot \sum_{i=1}^{n-m} q_{Gi} - b_{MSM}\right) \cdot q_{MSM} \\ &\quad - (\alpha + C_{MSM}) \cdot q_{MSM}^2 - a_{MSM} \\ &\quad (i = 1, 2 \ldots, n - m) \end{aligned} \tag{9.9}$$

In the upper-market, the number of power producers is large, and there are many factors that will influence the strategy bidding, which makes it difficult for every bidder to obtain competitor's information. So, the variables for the revenue of MSMAG are $\{a_{MSM}, b_{MSM}, c_{MSM}$ and $q_{MSM}\}$, and q_{MSM} is the variable of biding simultaneously. As the variable related to the revenue of MSMAG, P_{MCP} is decided by all power producers in the market, the maximum revenue can be obtainedby solving the equation as follows:

$$\frac{\partial U_{MSM}^{pf}}{\partial q_{MSM}} = 0 \tag{9.10}$$

$$\begin{aligned} i \cdot e., q_{MSM} &= \frac{P_{\max} - \alpha \cdot \left(\sum_{i=1}^{n-m} q_{Gi} + q_{MSM}\right) - b_{MSM}}{2c_{MSM} + \alpha} \\ &= \frac{P_{\max} - b_{MSM} - \alpha \cdot q_{all}}{2c_{MSM}} + \alpha \end{aligned} \tag{9.11}$$

From the expression above, it can be known that the optimization bid for MSMAG is not only relevant to its own quantity, but also decided by the quantity of opponents. In the incomplete market, we can establish a perfect information market through multiple games dynamically base on the communication of MAS, in which the equilibrium could be found by repeatedly iterative process. The process of solving will be given in next section.

In addition, formula (9.11) shows the bidding quantity of MSMAG is related tightly to$\{b_{MSM}, c_{MSM}\}$. As long as the coefficients decrease, the quantity will usually increase. In upper-market, the main objective of bidding is to obtain more quantity in competing, and the allocation of quantity will be completed in lower- market. Therefore, the coefficients of MSMAG should be as small as possible to obtain more quantity. This chapter adopts the method of undetermined coefficients based on the minimum total cost at the maximum quantity. Marking $q_{Mj}^{max} < q_{M(j+1)}^{max}$, $j = 1, 2, ..., j + 1$, then calculate the cost of every MAG at the point of $q_{Mj}^{max} \forall j = 1 : m$.The C_j can be formulated as follows:

$$C_j = \min\left\{C_{Mj}\left(q_{Mj}^{\max}\right), \forall j = 1 : m\right\}, \quad \forall j = 1 : m \tag{9.12}$$

where C_j is the minimum cost of the q_{Mj}^{max}; $C_{Mj}^{max}(q_{Mj}^{max})$ is the cost of ith MAG at the quantity of the q_{Mj}^{max}. For simplicity, this chapter assumes the MSMAG always preferentially allocates power to the lowest coat MAG. If the cost of MSMAG is less than or equal to q_{M1}^{max}, the MAG corresponding to the cost marked c_1 is preferentially allocated the power; if the cost of MSMAG is less than or equal to q_{M2}^{max} and greater than q_{M1}^{max}, the MAG corresponding to the cost marked C_2 is preferentially allocated the power, and the other sequential analogies. In the process of calculating C_j, the microgrid corresponding to the cost marked C_j must meet its capacity limit, otherwise the second low-cost MAG and its cost will be considered as the C_J. Assuming the cost of MSMAG can be formulated in the form of formula (9.6), the cost function curve of the MSMAG should pass through the point of (q_{Mj}^{max}, C_j), through which it can obtain the cost function of the MSMAG.

(2) GAG: this chapter assumes there are (n-m) GAG in the upper-market. Similar to MSMAG, GAG will also make a rational biding strategy to EMAG to achieve the maximum profit in the market. The mathematical model can be formulated as follows:

$$U_{Gi}^{pf} = q_{Gi} \cdot P_{MCP} - C_{Gi}(i = 1, 2, \ldots, n - m) \tag{9.13}$$

$$\begin{gathered} C_{Gi} = a_{Gi} + b_{Gi} \cdot q_{Gi} + c_{Gi} \cdot q_{Gi}^2 \\ (i - 1, 2, \ldots, n - m) \end{gathered} \tag{9.14}$$

Subject to

$$q_{Gi}^{\min} \le q_{Gi} \le q_{Gi}^{\max} \tag{9.15}$$

where $\{q_{Gi}, U_{Gi}^{pf}\}$ are the quantity and the revenue of the ith GAG; C_{Gi} is the cost function of the ith GAG, and$\{a_{Gi}, b_{Gi}$ and $c_{Gi}\}$are the corresponding coefficient; $\{q_{Gi}^{min}, q_{Gi}^{max}\}$ are the upper and lower limits of the ith GAG. The objective function of optimization problem for the ith GAG can be described as follows:

$$Max U_{Gi}^{pf} \tag{9.16}$$

Associated with the above Eqs. (9.5) and (9.12)–(9.13), the revenue of the ith GAG can be described as follows:

$$U_{Gi}^{pf} = \left(P_{\max} - \alpha q_{MSM} - \sum_{\substack{l=1 \\ l \neq i}}^{n-m} q_{Gl} - b_{Gi} \right) \cdot q_{Gi} - (\alpha + c_{Gi}) \cdot q_{Gi}^2 - a_{Gi} \quad (i = 1, 2, \ldots, n-m) \tag{9.17}$$

Similar to the MSMAG, the maximum revenue can be obtained by solving the equation

$$U_{Gi}^{pf} / \partial q_{Gi} = 0 \tag{9.18}$$

$$\text{i.e., } q_{Gi} = \frac{P_{\max} - \alpha \cdot q_{MSM} - \alpha \cdot \sum_{\substack{i=1 \\ l \neq i}}^{n-m} q_{Gl} - b_{Gi}}{2(\alpha + c_{Gi})} = \frac{P_{\max} - b_{Gi} - \alpha \cdot q_{all}}{\alpha + 2c_{Gi}} \tag{9.19}$$

(3) EMAG and LA: After receiving the bids of all bidders, EMAG will make a decision, which would be regarded as the result and the information of feedback to the bidders. This is a dynamic interaction process, which will continue until the formation of equilibrium. Because of the fact that power cannot be stored and the generated energy must match the load always, the final generating output of participant should meet the power balance. The formulation can be described as follows:

$$\sum_{i=1}^{n-m} q_{Gi}^* + q_{MSM}^* = q_{\text{load}} \tag{9.20}$$

$$q_{\text{load}} = \sum_{I=1}^{n} \lambda_I \cdot q_{LI} \tag{9.21}$$

$$q_{MSM}^* = \sum_{j=1}^{m} q_{Mj}^* \tag{9.22}$$

where$\{q_{Gi}^*, q_{MSM}^*\}$are the final power output of the ith GAG and the MSMAG ;$\{q_{load}, q_{LI}, and \lambda_I\}$are the total load demand, load of the Ith LA, and corresponding sensitive coefficient in the upper-market;$q_{Mj}^*$is the final power output of the jth MAG. Formula (9.20) shows the final total quantities of all bidders are equal to the load demand because of the power balance. However, formula (9.5), (9.11) and (9.19) show the p_{MCP}, q_{MSM}, and q_{Gi} are determined by the total quantities of all bidders. It seems that the p_{MCP}, q_{MSM}, and q_{Gi} can be formulated directly by load demand. In fact, thc load is an unknown variable for all bidders, and formula (9.5) reflects the relationship between$p_{MCP}$and the total power supplied in the market. At the beginning of bid, the quantities of all bidders which represent the total power supply in market are not equal to the load demand. So formula (9.20) is only considered as the object constrain. And even though the market dynamic game reaches Nash equilibrium, the constraints of Eq. (9.20) may not be satisfied because of the contradiction between (9.19) and (9.20). Therefore, this chapter adopts the method of multiobjective planning to meliorate the results, which will be presented in Sect. 9.4.

(4) MAG: this chapter assumes the number of MSMAG in the market is m. The responsibilities of MSMAG in different level of market are unlike. In upper-market, as the agent of the microgrid alliance, MSMAG needs to predict its own status information and report it to EMAG, which involves generating capacity, cost function, the upper, and lower limits of bid and so on. In lower-market, the load demand is the final quantity of MSMAG obtained from the upper-market, and the market clear price is also equal to the market clear price in upper-market due to the uniform price in electricity market. So, the responsibility of MAG is maximizing its own profit by a rational strategy bidding of quantity, which is a variable related to the market clear price, and all MAG must satisfy the constraint of power balance. The concrete process of bidding will be presented detailed in Sect. 9.4.

9.2.2 *Game Theory*

In recent years, game theory is used to solve the strategic behavior of generating units. This chapter solves the optimization problem of bidding based on the game theory. Therefore it is necessary to know the basic concepts of game theory and the relevant definition.

Game theory is the study of multi-person or multi-firm decision-making problems, in which their own maximum profits based on several interdependent strategies are pursued, and usually used extensively to study auction behavior, bargaining, product differentiation, and strategic behavior by firms [4]. According to different criteria, game theory can be divided into many types, while some elements are necessary, such as player, strategy, and the payoff function. Generally, the number of players in the game is assumed to be finite. The key to realize the object of player is making a suitable strategy, which depends on not only its own behavior but also the action of the other participants in the game. A payoff function (or revenue function) can represent the utility that a player obtains in view of a strategy profile for all players [3].

In game theory, Nash equilibrium is a widely used concept. A Nash equilibrium point is a strategy combination, in which every player will be unwilling to change their strategies, otherwise its revenue will reduce. Definition 9.1 [4] gives a formal definition of Nash equilibrium.

Definition 9.1 In the n-player strategic form game, the profile strategies $s_1^*, ...s_n^*$ re a Nash equilibrium if, for each player i, $s_1^* \in S_i (S_i$ is strategy space) is the best response of player i to the strategies specified for the other (n - 1) players $s_{-i}^* =$

$(s_1^*, \ldots, s_{i-1}^*, s_{i+1}^*, \ldots s_n^*)$ such that utility function $u_i(s_i^*, s_{-i}^*) \geq u_i(s_i^*, s_{-i}^*)$, for every feasible strategy$s_i \in S_1$.

9.2.3 Bidding Strategy

As previously mentioned, the power producers pursue the maximum profit according to several suitable strategies in the market. Since the connectivity of profit between all participants in the oligopolistic market, which is the actuality of electricity market in china, it is very important for the power producer to know the choices of other participants. As a self-financing economic entity, each power producer has the knowledge of its own payoffs and generation costs but could lack such information on the operator. The MAS have been built in Sect. 9.2, in which all entities are assumed to be intelligent, so they can be aware of information of others according to corresponding approaches. At the same time, Nash equilibrium points can be got throughiterative process and the final result can be resolved by multi-objective planning based on the Nash equilibrium.

The formulas given in last section show that the elements of bidding game can be expressed as follows:

(1) Player: power producers, including the GAG and MSMAG in the upper-market; MAG is the player in the lower-market.

(2) Strategy: a strategy is a set of decision variables of each agent, which are different from each other.

(3) Payoff function: the revenue for every power producer, which have been modeled in last section.

In the iterative process, each bidding agent tries to obtain maximum profit in every step based on the assumption that the others retain current bidding. The flowchart of iterative process for searching Nash equilibrium will be proposed in detail in Sect. 9.4.

9.3 The Solution of Bidding Model Based on Game Theory

Associated with what is discussed above, the bidding variable for every bidding agent has its own quantity. Every agent maximizes its profit by changing the bidding quantity in the process of bidding. The agent of market collects these bidding information and take the corresponding information as the feedback to every bidding agent, based on which bidding agent will adjust its bidding quantity for the next round game. In upper-market, Nash equilibrium point is found by iterative search method in this chapter. The iterative process could be repeated several times by adjusting the quantity under the constraint condition. The end of iterative process can be affirmed as follows:

Assuming that the strategy of kth round game for every bidding agent is $\left\{q_{G1}^{k}, q_{G2}^{k}, \ldots q_{G(n-m)}^{k}, q_{MSM}^{k}\right\}$ then the result of $(k+1)$th round game which is described as $\left\{q_{G1}^{k+1}, q_{G2}^{k+1}, \ldots q_{G(n-m)}^{k+1}, q_{MSM}^{k+1}\right\}$ can be obtainedby the following formulas:

$$\begin{cases} q_{G1}^{k+1} = \underset{q_{G1}}{\text{argmax}} U_{G1}^{pf}\left(q_{G1}, q_{G2}^{k}, \ldots, q_{G(n-m)}^{k}, q_{MSM}^{k}\right) \\ q_{G2}^{k+1} = \underset{q_{G}}{\text{argmax}} U_{G2}^{pf}\left(q_{G1}^{k+1}, q_{G2}, \ldots, q_{G(n-m)}^{k}, q_{MSM}^{k}\right) \\ q_{G(n-m)}^{k+1} = \underset{q_{G2}}{\text{argmax}} U_{G(n-m)}^{pf}\left(q_{G1}^{k+1}, q_{G2}^{k+1}, \ldots, q_{G(n-m)}, q_{MSM}^{k}\right) \\ q_{MSM}^{k+1} = \underset{q_{MSM}}{\text{argmax}} U_{MSM}^{pf}\left(q_{G1}^{k+1}, q_{G2}^{k+1}, \ldots, q_{G(n-m)}^{k+1}, q_{MSM}\right) \end{cases} \tag{9.23}$$

where $q_{G1}^{k+1} = \text{argmax}_{q_{G1}} U_{G1}^{pf}\left(q_{G1}^{k+1}, \ldots, q_{Gi}, \ldots, q_{G(n-m)}^{k}, q_{MSM}^{k}\right)$ and $argmax_{q_{MSM}} U_{MSM}^{D}(q_{G1}^{k+1}, q_{G2}^{k+1} \ldots, q_{G(n-m)}^{k+1}, q_{MSM})$ are the maximum revenue of q_{Gi} and q_{MSM} based on that the frontagents fixed the bidding quantity of $(k+1)$th round game while the later agent fixed the quantity of kth round game.

If the revenue of two adjacent round games are equal for all bidding agents, no one will be willing to change the quantity of bidding, in other words, the strategy is Nash equilibrium, which can be described as follows:

$$\left\{q_{G1}^{k}, q_{G2}^{k}, \ldots, q_{G(n-m)}^{k}, q_{MSM}^{k}\right\} = \left\{q_{G1}^{k+1}, q_{G2}^{k+1}, \ldots, q_{G(n-m)}^{k+1}, q_{MSM}^{k+1}\right\} \tag{9.24}$$

As shown above, q_{Gi} is the variable of bidding strategy and the strategy set belongs to the purely strategic space. That there is a theorem of the existence of Nash equilibrium can be described as follows: In a game which involves n players, if the strategy set is a purely strategic space and belongs to Nonempty Bounded Closed Sets on Euclidean space, and the revenue function of the ith player is consecutive and belongs to quasiconcave function in terms of the strategy set of the ith player, then the game must have a Nash equilibrium. It is easy to prove that quadratic function which opens down is quasiconcave function among the real number field. However, the bidding game in electricity has some constraints, which would be not able to ensure that the revenue function is always quasiconcave function. Therefore, the equilibrium is not always possible, which is the sign of failure of the market bidding this time. Then the market transaction would be finished after one more bidding, which is feasible in engineering applications. Due to the contradiction between (9.19) and (9.20), the final result may not meet the constraint of power balance. Using the method of finding a compromising solution in multiobjective planning can solve the problem [5]. In order to make full use of renewable energy, the quantity of q_{MSM} will not be modified in (9.22), whereas the quantities of qGi will be modified by the method mentioned above, which can be described as follows:

(1) In the upper-market, quantity of q_{Gi} can be obtained by (9.19) with the constraint of (9.20). The modified model can be formulated as follows:

$$min \sum_{i=1}^{n-m} \left(q_{Gi} - \frac{P_{\max} - b_{Gi} - \alpha \cdot q_{all}}{a + 2c_{Gi}} \right)^2 \tag{9.25}$$

subject to

$$\left\{ \begin{array}{l} \sum_{i=1}^{n-m} q_{Gi} + q_{MSM} = q_{all} \\ q_{Gi}^{\min} \le q_{Gi} \le q_{Gi}^{\max} \end{array} \right\} \tag{9.26}$$

Then, solving (9.25) under the conditions (9.26)

$$\left\{ \begin{array}{l} L_1 = \left(q_{Gi} - \frac{P_{\max} - b_{Gi} - \alpha \cdot q_{all}}{a + 2c_{Gi}} \right) \\ L_2 = \sum_{i=1}^{n-m} q_{Gi} + q_{MSM} - q_{all} \end{array} \right\} \tag{9.27}$$

The Lagrange function L is constructed as follows:

$$L = \sum_{i=1}^{n-m} L_1 + \lambda \cdot L_2 \tag{9.28}$$

The optimizing solution to (9.25) can be gained through solving the problem as follows:

$$\min_{q_{Gi}} \max_{\lambda} L \tag{9.29}$$

The solution to (9.29) must satisfy

$$\left\{ \begin{array}{l} \frac{\partial L}{\partial \lambda} = 0 \\ \frac{\partial L}{\partial_{qGi}} = 0 \quad i = 1, 2, \ldots, n - m \end{array} \right. \tag{9.30}$$

$$\left\{ \begin{array}{l} 2 \left(q_{Gi} - \frac{P_{\max} - b_{Gi} - \alpha \cdot q_{all}}{a + 2c_{Gi}} \right) + \lambda = 0 \\ i = 1, 2, \ldots, n - m \\ \sum_{i=1}^{n=m} q_{Gi} + q_{MSM} = q_{all} \end{array} \right. \tag{9.31}$$

So

$$q_{Gi} = \frac{P_{\max} - b_{Gi} - \alpha \cdot q_{all}}{a + 2c_{Gi}} - \frac{\lambda}{2} \tag{9.32}$$

$$\lambda = \frac{2}{n-m} \times \left[\sum_{i=1}^{n-m} \frac{P_{\max} - b_{Gi} - \alpha \cdot q_{all}}{\alpha + 2c_{Gi}} + q_{MSM} - q_{all} \right] \tag{9.33}$$

$$\begin{aligned} q_{Gi} =& \frac{P_{\max} - b_{Gi} - \alpha \cdot q_{all}}{\alpha + 2c_{Gi}} - \frac{1}{n-m} \\ & \times \left[\sum_{i=1}^{n-m} \frac{P_{\max} - b_{Gi} - \alpha \cdot q_{all}}{a + 2c_{Gi}} + q_{MSM} - q_{all} \right] \end{aligned} \tag{9.34}$$

Formula (9.32) shows that the modified quantity of ith GAG is relevant to$q_{all}$and the number of bidding agents, which is the total quantity of bidding for all bidding agents and can be obtained by the interaction of MAS. Assuming the final quantity of ith GAG is q_{Gi}^*, which meets condition (9.19), the revenue of ith GAG can be calculated as follows:

$$U_{Gi}^{pf} = \left(p_{\max} - \alpha \cdot q_{\text{all}}^*\right) \cdot q_{Gi}^* - C_{Gi}^* \tag{9.35}$$

subject to

$$C_{Gi}^* = a_{Gi} + b_{Gi} \cdot q_{Gi}^* + c_{Gi} \cdot q_{Gi}^{*2} \tag{9.36}$$

The revenue of MSMAG can be calculated by (9.9).

(2) In the lower-market, the demand of load is fixed and equal to the quantity of MSMAG obtained from upper- market. The objective of optimization problem in lower-market is how to dispatch the quantity obtained from upper-market. The model can be described as follows:

$$U_{Mi}^{pf} = p_{MCP} \cdot q_{Mi} - C_{Mi} \tag{9.37}$$

$$C_{Mi} = a_{Mi} + b_{Mi} \cdot q_{Mi} + c_{Mi} \cdot q_{Mi}^2 \tag{9.38}$$

$$q_{Mi} = f\left(p_{MCP}\right) = f\left(p_{\max} - \alpha \cdot q_{all}^*\right) \tag{9.39}$$

Where $\{U_{Mi}^{pf}, C_{Mi}\}$ are the revenue and cost function of the ith MAG; $\{a_{Mi}, b_{Mi}, and c_{Mi}\}$ are the coefficient of cost function of the ith MAG;q_{Mi} is quantity of the ith MAG for bidding in the lowermaket; p_{MCP} is the market clear price which is equal to the market clear price in upper-market due to uniform price. The function $f(p_{MCP})$ shows q_{Mi} is a variable related to the market clear price. The objective function of optimization problem for ith MAG can be described as follows:

$$maxU_{mi}^{pf} \tag{9.40}$$

Associated with (9.37)–(9.39), the revenue of ith GAG can be described as:

$$\begin{aligned} U_{Mi}^{pf} =& \left(p_{\max} - \alpha q_{\text{all}}^*\right) \cdot q_{Mi} \\ & - \left(a_{Mi} + b_{Mi} \cdot q_{Mi} + c_{Mi} \cdot q_{Mi}^2\right) \end{aligned} \tag{9.41}$$

Similar to the condition in upper-market, the information every bidding agent can obtain is finite, and they need bid base on imperfect information. This chapter solves the problem of formula (9.40) by using mild increase quantity price strategy for maximum profit of bidding:

$$\frac{\partial U_{Mi}^{pf}}{\partial p_{MCP}} = 0 \tag{9.42}$$

$$q_{Mi} + p_{MCP} \cdot \frac{\partial q_{Mi}}{\partial p_{MCP}} - \left(b_{Mi} \cdot \frac{\partial q_{Mi}}{\partial p_{MCP}} + 2c_{Mi} \cdot q_{Mi} \cdot \frac{\partial q_{Mi}}{\partial p_{MCP}} \right) = 0 \tag{9.43}$$

Similar to the upper-market, the power supplied by all MAG must be equal to the load demand in the lower-market, which is known and equal to the power obtained by the MSMAG from the upper-market. So the sum of q_{Mi} is a constant and the following formula can be established:

$$\frac{\partial \left(q_{Mi} + \sum_{j=1, j \neq 1}^{m} q_{Mj} \right)}{\partial P_{MCP}} = 0 \tag{9.44}$$

So

$$\frac{\partial q_{Mi}}{\partial P_{MCP}} = -\frac{\sum_{j=1, j \neq i}^{m} q_{Mj}}{\partial P_{MCP}} \tag{9.45}$$

$$q_{Mi} = \frac{p_{MCP} - b_{Mi}}{1 + 2c_{Mi} \cdot G_i} \cdot G_i \tag{9.46}$$

$$\text{subject to} \begin{cases} \sum_{i=1}^{m} q_{Mi}^{*} = q_{MSM}^{*} \\ q_{Mi}^{\min} \leq q_{Mi} \leq q_{Mi}^{\max} \\ G_i = \frac{\partial \sum_{j \neq 1} q_{Mj}}{\partial p_{MCP}} \end{cases} \tag{9.47}$$

Where $\{U_{Mi}^{pf}, C_{Mi}\}$ are respectively the final quantity of q_{Gi} and the quantity of MSMAG which are obtained from the upper-market; the formula of Gi represents the response variable of the ith agent to the total increment of the other power generators in the market.

To sum up, the coordination between the upper and lower market can be described as follows:

(a) In the upper-market, MSMAG is only considered as a power generator to participate in market competition with other traditional electricity generators. The role of MSMAG at this time is to collect the information of all MAG, such as cost, the constraint of bidding capacity, etc. And then MSMAG calculates its own parameter information through mathematical methods.

When the transaction in the upper-market is finished, MSMAG obtains the final quantity, which will be considered as the load demand of the lower-market due to the goal of optimization problem in the lower-market being to dispatch the quantity of MSMAG well.

(b) In the lower-market, the responsibility of MSMAG is similar to the EMAG in the upper-market, that is, collecting the bidding information which all the MAG has submitted for bidding. And at the same time, the market clear price in the lower-market is equal to PMCP due to the electricity pricing mechanism of uniform price. In the lower-market, the ith MAG submits its own bidding quantity based on PMCP and the response variable to the total increment of the other MAG.

Assuming the final quantity of ith MAG is q_{Mi}^{*}, which meets condition (9.19), the revenue of the ith MAG can be calculated as follows:

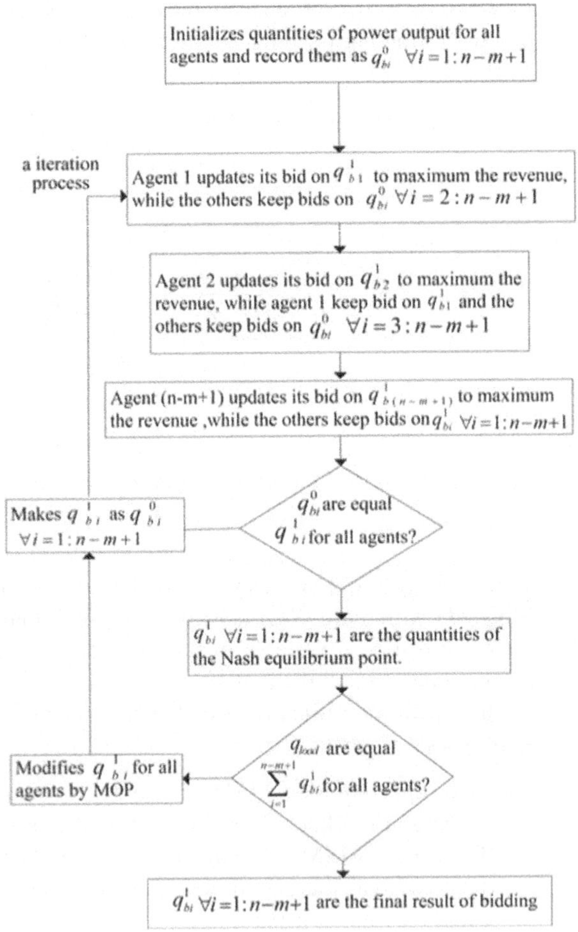

Fig. 9.4 Flowchart of bidding based game theory

$$U_{Mi}^{pf} = p_{MCP} \cdot q_{Mi}^{*} - C_{Mi}^{*} \tag{9.48}$$

subject to

$$C_{Mi}^{*} = a_{Mi} + b_{Mi} \cdot q_{Mi}^{*} + c_{Mi} \cdot q_{Mi}^{*2} \tag{9.49}$$

(3) According to MAS, Nash equilibrium point can be found by iterative process, and the final result should be modified by multi-objective planning (MOP) to meet the constraint conditions. Considering the bidding in upper-market as an example, the process of solution to bidding model based on game theory can be described by the flow chart shown in Fig. 9.4. For convenience, all agents of bidding in upper-market are numbered sequentially and the ith agent is recorded as agenti, in which all MAG are considered as a bidding agent.

In every process of iterative, quantities of bidding for all agents must meet the constraint conditions, which mainly refer to the upper and lower limits. For example, if the optimal quantity of ith agent in an iterative process is out of the constraint condition, it will take the upper or lower limit value as a result of the current round of iteration, which will be modified in the next round of iteration process.

9.4 Case Study and Analysis

In this section, a concrete example is adduced to analyze the model. To highlight the superiority of hierarchical bidding, the result will be compared with the result of monolayer market bidding, in which the MAG alone participates in the auction market.

As discussed above, this chapter assumes all the power generations-bid for accessing the electricity networks and there are no issues of large user bilateral trade, and then the simulation environment is described as in Fig. 9.5. In the electricity market, there are $(n - m)$ traditional generation companies (GENCOS), m microgrids (MG), and n load. For simplicity, this chapter takes five GENCOS and six MG in the electricity market as an example. Since the casestudy aims at demonstrating the superiority of the strategy proposed, the number of load is not able to affect the result of case study. So the case has a hypothetical value for the load in the market. The index of safety margin is introduced to highlight the necessity of the introduction of the method of MOP The power generations are recorded as player 1 to player 11, in which the first six players represent MG. The parameters of their generating cost are listed in Table 9.1, and the cost curves of every bidding agent are showed in Fig. 9.6. Without losing generality, the demand and power data in the case study is based on the information provided by NREL.

As mentioned above, the cost coefficient of MSMAG can be calculated by the method of undetermined coeffi- cients. Figure 9.6 shows that the cost of every bidding agent increases with the increase of power generation in the form of parabola, and the increase in the speed of cost is mainly determined by the coefficient of the

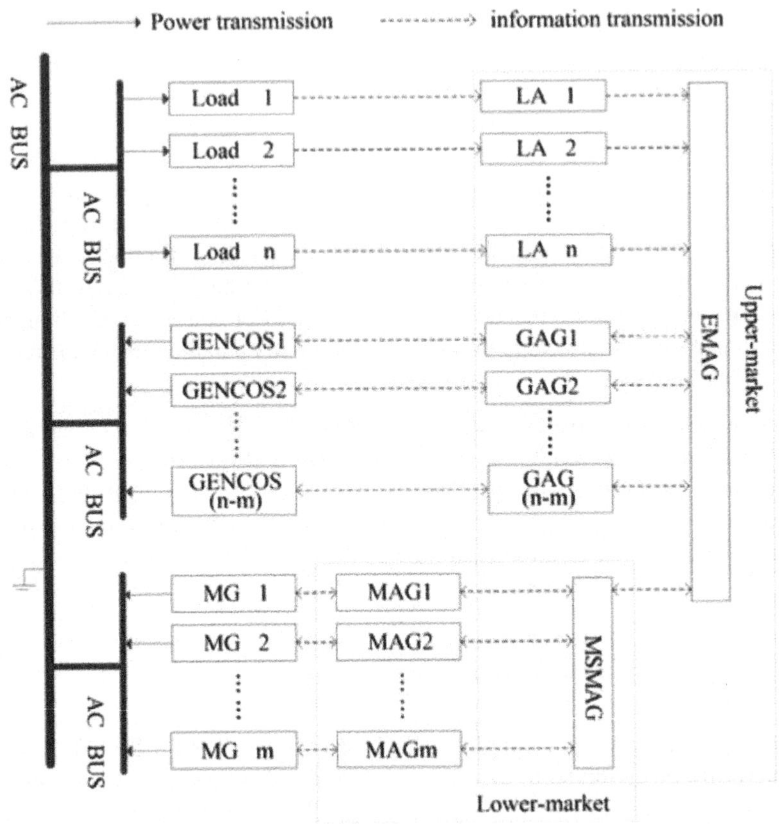

Fig. 9.5 The simulation environment

quadratic term. Table 9.1 shows that the maximum quantity of MAG is much smaller than GAG. The iterative process mentioned above shows that this is very bad for bidding to obtain more quantity in the market. So this chapter adopts allying MAG to overcome this disadvantage. Assuming the load demand is 900 MW at a certain time interval and the time interval is one hour. In order to calculate conveniently, the price at which the grid company purchases power can be defined as:

$$p_{MCP} = 70 - 0.015q_{\text{all}} \tag{9.50}$$

In this section, the simulation will be divided into A, B, and C parts. In part A, Case I calculates the market scheduling results for microgrids in the case of non-cooperation and cooperation respectively, and the load demands are both 900 MWH, which are shown in Tables 9.2 and 9.3, respectively. Without losing generality, in order to analyze the influence of production cost on the bidding results, and to study the superiority of microgrid joint bidding, in part B, Case II study the

Table 9.1 Players: power limits and cost factor

Player	a	b	c	q_{min} (MCH)	q_{max} (MWH)
1	30.2	24.30	0.143	10	55
2	50.1	25.26	0.189	5	45
3	22.7	29.47	0.145	12	60
4	47.2	24.60	0.176	6	45
5	29.7	24.20	0.143	10	65
6	23.2	28.70	0.142	7	48
7	40.2	16.80	0.166	50	165
8	38.1	21.05	0.168	45	150
9	32.2	19.63	0.185	35	155
10	39.8	16.7	0.165	40	145
11	38.6	16.2	0.161	47	161
MSMAG	30.8	10.47	0.136	5	318

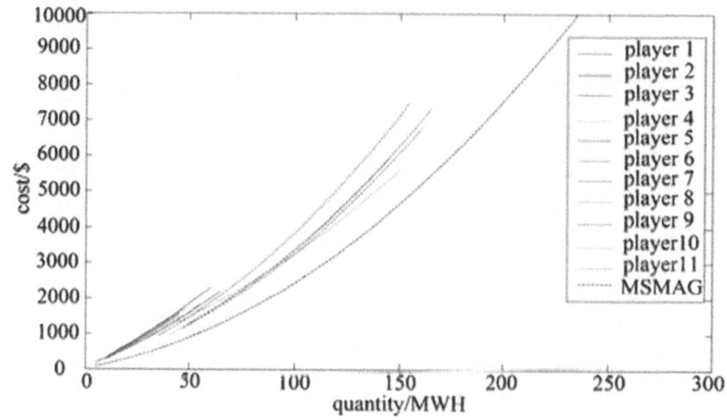

Fig. 9.6 The curve of cost for all players

market scheduling results under different load demand. Table 9.4 shows the final scheduling quantity of bidding agents in the case of non-cooperation with the different load demands, and Table 9.5 shows the final scheduling quantity of MSMAG and GAG in the case of cooperation with different load demands which are equal to the corresponding demands in Table 9.4. In part C, the main objective is to calculate the dispatched quantity of every MAG under the fixed total quantity and market clear price, which are acquired in the upper- market, and to show the final revenue of MAG in the case of non-cooperation and cooperation, respectively.

A. Case I: bidding in upper-market under a single load demand

In Table 9.2, the total quantity of the MAG is equal to the sum of the quantity of the six MAG, i.e., 228.9, which is smaller than the quantity of MSMAG in Table 9.3. Though the renewable energy obtains more market share, the corresponding total

Table 9.2 Players: The final quantity of scheduling for non-cooperation

Player	Quantity/MWH	MCP ($/MWH)	Revenue ($/MWH)
1	41.7	56.47	1062.63
2	27.4	56.47	663.16
3	42.9	56.47	868.74
4	31.9	56.47	790.35
5	51.6	56.47	1254.69
6	33.4	56.47	745.91
7	139.7	56.47	2262.03
8	123.4	56.47	177449
9	122.7	56.47	1702.84
10	140.6	56.47	2290.08
11	146.5	56.47	2405.3

Table 9.3 Players: The final quantity of scheduling for cooperation

Player	Quantity/MWH	MCP ($/MWH)	Revenue ($/MWH)
7	135.3	56.51	2293.75
8	119.3	56.51	1801.22
9	118.2	56.51	1742.34
10	138.8	56.51	2307.03
11	142.9	5651	2434.01
MSMAG	244.8	56.51	3089.71

Table 9.4 Agents: Bidding under different load demands in the way of non-cooperation

Bidding agents	Quantity (MWH)		
	Load demands (MWH)		
	300 MWH	600 MWH	900 MWH
Player1	12.8	27.1	41.7
Player2	6.7	16.2	27.4
Player3	13.7	28.2	42.9
Player4	9.2	20.6	31.9
Player5	16.8	33.8	51.6
Player6	10.9	21.8	33.4
Player7	48.2	93.7	139.7
Player8	42.4	84.2	123.4
Player9	41.8	83.5	122.7
Player10	48.5	93.8	140.6
Player11	49.4	97.7	146.5

Table 9.5 Player: Bidding under different load demands in the way of cooperation

Bidding agents	Quantity (MWH)		
	Load demands (MWH)		
	300 MWH	600 MWH	900 MWH
player 7	44.9	90.7	135.3
player 8	39.1	81.1	119.3
player 9	38.2	80.8	118.2
player 10	45.1	91.0	138.8
player 11	46.3	93.9	142.9
MAMAG	85.2	161.9	244.8

revenue of the three MAG may be greater than the MSMAG. The reason for this is that some players such as MAG2 have a very high cost, which would result in the decrease of revenue. However, the quantity obtained from the market is determined not only by the cost but also by the capacity of player, which can be seen from the contrast of the quantity between the player 5 and the player 1 in Table 9.2. In addition, although the cost of player 6 is slightly smaller than that of player 3, the capacity of player 6 is much smaller than that of player 3, which leads to the quantity of player 6 less than that of player 3. Obviously the capacity is the main factor of the final quantity for MAG. However, the contrast of the quantity between player 7 and player 10 shows that the cost is the main factor of the final quantity for GAG. So, with the decrease of capacity, the capacity of power producers becomes more and more important for the final quantity of bidder, which is also the reason why this chapter proposes a structure of bi-level bidding.

The market clear price in Table 9.3 is higher than that in Table 9.2, which demonstrates that the fierce competition in the market can reduce market clear prices, which will increase the consumer's income finally. The market clear price in Tables 9.2 and 9.3 shows the method of MOP can be as much as possible in the balance of final quantity and load demand. However, it may not be able to fully realize balance, since the MSMAG doesn't modify the final quantity by the method of MOP To highlight the necessity of the introduction of MOP, the index of safety margin marked qs is introduced in the case, which can be formulated as follows:

$$q_s = \left| \sum_{i=1}^{n-m} q_{Gi}^* + q_{MSM}^* - q_{\text{load}} \right| \tag{9.51}$$

B. Case II: bidding under different load demands

As shown in Table 9.4, the quantity of the 2th MAG is very close to its minimum when the load demand is 300 MWH. In addition, although the generation cost of the 3th MAG is higher than that of the 2th MAG, the final quantity obtained from the electricity market is less. So, as the load demand decreases, the upper limit of power producers becomes an important restraint to the quantity obtained from the

Table 9.6 Player: The quantity without modified by MOP

Bidding agents	Quantity (MWH)		
	Load demands (MWH)		
	300 MWH	600 MWH	900 MWH
player 7	44.76	90.58	135.21
player 8	38.98	80.97	119.22
player 9	38.07	80.49	118.11
player 10	45.01	90.86	138.73
player 11	46.47	93.79	142.79
MAMAG	85.2	161.9	244.8

market. From the comprehensive analysis of the data in Tables 9.4 and 9.5, it is easy to conclude that no matter what the load demand is, the MSMAG in Table 9.5 can obtain more power than the sum of the individual MAG auctions in Table 9.4.

The result shown in Table 9.5 is the final result of the GAG and MSMAG, which has been modified by the method of MOP The safety margin qs can be calculated easily by the formula (9.51). To highlight the necessity of introducing the method of MOP, Table 9.6 lists the quantity of every GAG in the case of cooperation, which hasn't been modified by the method of MOP.

It is easy to calculate the safety margin qs by the formula (9.51) in the case that the result hasn't been modified by the method of MOP, which is shown in Table 9.7.

Obviously, the introduction of MOP can increase the safety margin of the bidding market, which is beneficial for the stability of the electricity.

C. Case III: bidding in the lower-market

After the alliance of MG obtains the final scheduling quantity from upper-market, its agent will become a market agent from a bidding agent. Then, its main object receives the bidding of every MAG and feedback information corresponding to the bidding agents under certain constraints. At the same time, the total quantity of MAG and market clear price obtained from upper-market are considered as the load demand and price in lower-market. Table 9.8 lists the results of the competition between the MAG when the load is 85.2 MWH, 161.9 MWH, and 244.8 MWH, respectively.

The final quantity of every MAG in Table 9.8 is higher than that in Table 9.4, respectively. Thus MAG could obtain more quantity by way of cooperation, which

Table 9.7 The result of the safety margin in market

Case	qs(MWH)		
	Load demands (MWH)		
	300 MWH	600 MWH	900 MWH
Before modification	1.65	1.43	1.21
After modification	0.9	0.8	0.7

Table 9.8 Agents: The result of quantity allocation between MAG in lower-market

MAG	Quantity (MWH)		
	Load demands (MWH)		
	85.2 MWH	161.9 MWH	244.8 MWH
player 1	15.1	29.2	44.3
player 2	10.5	19.4	31.1
player 3	15.6	29.9	44.8
player 4	11.6	22.8	34.2
player 5	18.6	35.9	53.9
player 6	13.8	24.7	36.7

is beneficial to make full use of renewable resources. The total profit comparison of each MAG in the way of cooperation and non-cooperation is shown in Fig. 9.7. To highlight the increase of the total quantity of MAG, the quantity of the sum of all MAG obtained by the way of non-cooperation and the quantity of MSMAG obtained by the way of cooperation are also given in Fig. 9.7.

As shown in Fig. 9.7, the revenue of every MAG obtained by the way of cooperation is higher than that obtained by the way of non-cooperation, which is also the impetus of cooperating between the MAG. Especially for some MG who have a very high cost and low capacity MAG, such as the 2th MAG, the effect is more obvious. Simultaneously, the quantity of the microgrid obtained by way of cooperation from the electricity market is more than by way of non-cooperation. And when the load demand is 900 MWH, 600 MWH, and 30 MWH, the proportion of increasing quantity is 6.9%, 9.6%, and 21.5%, which shows the smaller the load demand, the more the advantage of cooperative biding.

9.5 Conclusion

This chapter studies the bidding problem of microgrid in the electricity market based on a bi-level game model. The MAS is used to build the platform for the interaction of bidder to achieve the goal of market dynamic equilibrium through the market game. Due to the market being an oligopolistic market, the bidder who has a larger capacity will have greater market power, which is beneficial to obtain more profit by strategic bidding. Therefore, the microgrids ally to improve the overall competitiveness and obtain more electricity in the market. Game theory as a power tool is widely used to address the solution of optimization problem and it has been introduced in the bi-level optimization strategy. Additionally, iterative search approach is used to find the Nash equilibrium point of market game in this chapter. Due to the contradiction between constraint conditions, some constraint may not be satisfied, and the approach of MOP is introduced in the model to address this problem. In lower-market, mild

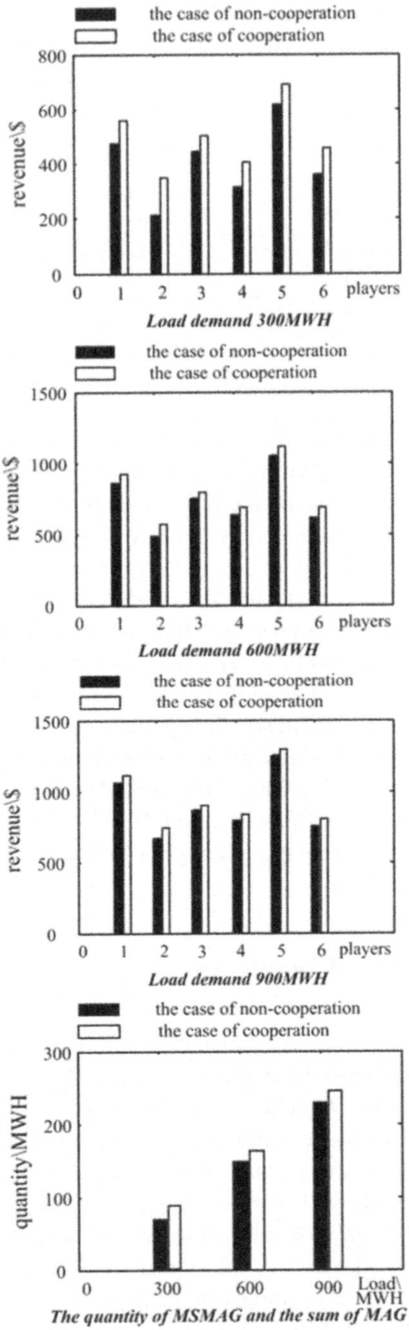

Fig. 9.7 The comparison of revenue for MAG between the ways of cooperation and non-cooperation

increase quantity price strategy is used in maximizing profit of MAG. Simulation case shows that the bidding strategy can improve the proportion of new energy in the electricity market and the introduction of MOP can improve the stability of electricity market. However, the hierarchical structure of the market proposed in this chapter will influence the real-time performance of the market. This study is not elaborate in this chapter due to space limitations, which will be focused on in the future studies.

References

1. M.A. Lopez, S. Martin, J. Aguado, S. D. La Torre, Market-oriented operation in microgrids using multi-agent systems, in *Proceedings of 2011 International Conference Power Engineering Energy Electrical* (2011), pp. 1–6
2. D. Pozo, E.E. Sauma, J. Contreras, A three-level static milp model for generation and transmission expansion planning. IEEE Trans. Power Syst. **28**(1), 202–210 (2013)
3. G. Zhang, G. Zhang, Y. Gao, J. Lu, Competitive strategic bidding optimization in electricity markets using bilevel programming and swarm technique. IEEE Trans. Ind. Electron. **58**(6), 2138–2146 (2011)
4. S. Soleymani, A.M. Ranjbar, A.R. Shirani, Strategic bidding of generating units in competitive electricity market with considering their reliability. Int. J. Electr. Power Energy Systems **30**(3), 193–201 (2008)
5. Z. Tan, X. Li, Y. Liu, Y. Han, Game models and solutions in the competitive electricity market. Trans. China Electrotech. Soc. **19**(6), 42–50 (2004)

Part IV
Distributed Cooperative Control of DC Microgrid

Chapter 10
Multiagent System-Based Distributed Coordinated Control for Radial DC Microgrid Considering Transmission Time Delays

This chapter focuses on a multiagent-based distributed coordinated control for radial DC microgrid considering transmission time delays. First, a two-level multiagent system is constructed, where local control is formulated based on local states and executed by means of the first-level agent, and distributed coordinated control law is formulated based on widearea information and executed by means of the secondary-level agent in order to improve the voltage control performances. Afterward, the research mainly focuses on designing the local controller and the distributed coordinated controller. For the purpose of robust stability, the local control is designed as local state-feedback-based $H\infty$ robust controller. It is worth mentioning that the distributed coordinated control consists of local state feedback control and decoupling coordinated control law that only come from adjacent distributed energy resource units. Moreover, the distributed coordinated controller is designed by means of delay-dependent $H\infty$ robust control method taking into account transmission time delays. Finally, the validity of the proposed control scheme is demonstrated by means of simulation results.

10.1 The MAS Based Distributed Coordinated Control Scheme of DC MG

The proposed MAS based distributed coordinated control scheme for radial DC MG is depicted in Fig. 10.1. The radial DC MG contains multiple DER units connected with each other in proper order through a DC line. And each DER unit supplies a local load connected on the point of common coupling (PCC) node. Here, a hierarchical MAS that consists of first-level unit control agents and secondary-level distributed coordinated control agents, is proposed to implement distributed coordinated control for the radial DC MG.

D. Yue et al., *Cooperative Optimal Control of Hybrid Energy Systems*,
https://doi.org/10.1007/978-981-33-6722-7_10

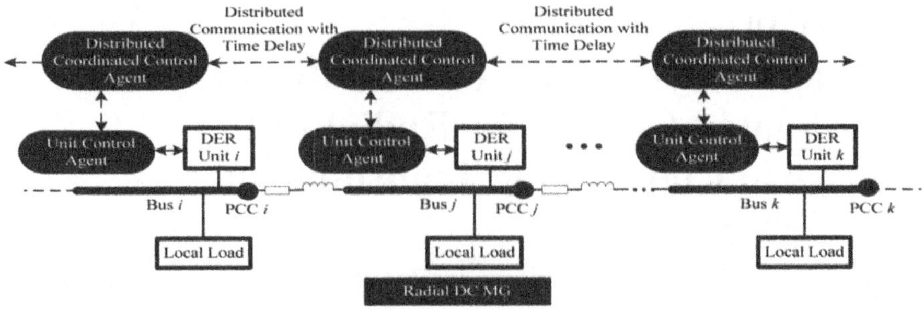

Fig. 10.1 MAS based distributed coordinated control scheme

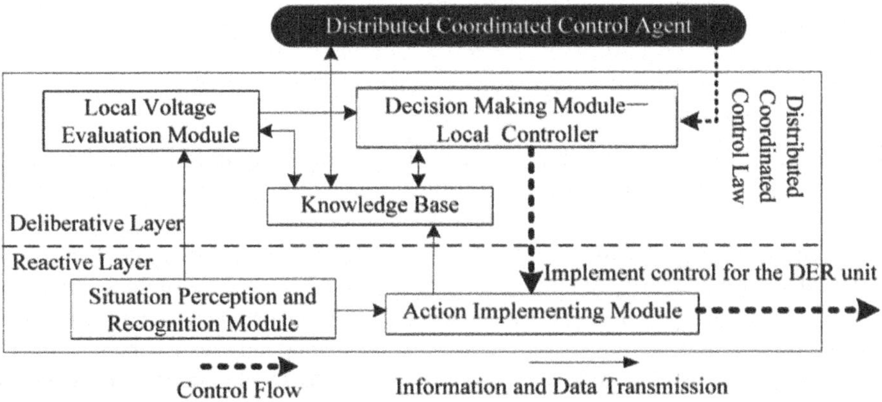

Fig. 10.2 Structure of the first-level unit control agent

Corresponding to Fig. 10.1, each first-level unit control agent is mainly responsible for formulating and implementing the local control for its DER unit. It is designed as hybrid agent, which is composed of reactive layer and deliberative layer as shown in Fig. 10.2. The reactive layer is defined as "recognition, perception and action" and has priority to respond quickly to emergencies of environment. The deliberative layer is defined as "belief, desire and intention", and has high intelligence to control the dynamic behavior of its DER unit.

Each secondary-level distributed coordinated control agent is mainly responsible determining what kind of control mode and implementing the distribution coordinated control law for its DER unit to improve control performance as shown in Fig. 10.3. The distributed coordinated control law is formulated by using wide-area states that only come from the adjacent DER unit. Each secondary-level agent is designed as a deliberative agent. The voltage deviation between any two adjacent DER units is estimated in voltage assessment module. The distributed coordinated control law is formulated in decision-making module. When the voltage deviation is larger than a specified threshed, the distributed coordinated control law is executed through action implementation module.

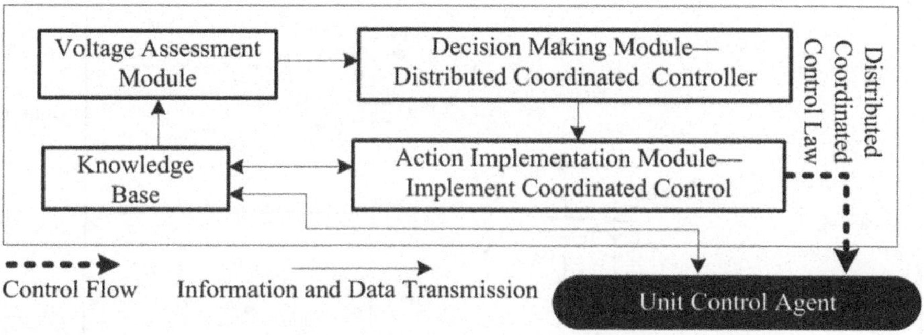

Fig. 10.3 Structure of the secondary-level distributed coordinated control agent

In the hierarchical MAS, the interactive manner among agents divides into two categories [1]: no fixed master-slave mode and master-slave mode. For example, when the request of one unit agent is responded by another unit agent, this kind of interaction is no fixed master-slave mode. While when the distributed coordinated control agent requests to send the coordinated control law to the unit agent, the request must be responded by the unit agent, in this case this kind of interaction is master-slave mode. The request of the coordinated control agent has the higher priority than that of any unit agent.

10.2 Control Mode and Dynamic Modeling

Corresponding to Fig. 10.1, the dynamic model of any three adjacent DER units is built as shown in Fig. 10.4. In Fig. 10.4, the pth DER is connected among both qth and rth DERs through DC lines with different impedances specified by parameters $Rpq = Rqp > 0$, $Lpq > 0$, $Rpr = Rrp > 0$ and $Lpr > 0$. In each DER unit, a Buck converter is presented to supply a local DC load connected to the PCC through a LC filter. The local load is unknown and is treated as current disturbance.

Corresponding to Fig. 10.4, according to Kirchoff's voltage law and Kirchoff's current law, the following set of equations is obtained

$$\text{DER } p \begin{cases} \dfrac{du_p}{dt} = \dfrac{1}{C_i} i_{tp} - \dfrac{1}{C_i} i_{Lp} + \dfrac{1}{C_i}\left(i_{pq} + i_{pr}\right) \\ \dfrac{di_{tp}}{dt} = -\dfrac{1}{L_{tp}} u_p - \dfrac{R_{tp}}{L_{tp}} i_{tp} + \dfrac{1}{L_{tp}} u_{tp} \end{cases} \tag{10.1}$$

$$\text{DER } k \in \{q, r\} \begin{cases} \dfrac{du_k}{dt} = \dfrac{1}{C_k} i_{tk} - \dfrac{1}{C_k} i_{Lk} + \dfrac{1}{C_k} i_{kp} \\ \dfrac{di_{tk}}{dt} = -\dfrac{1}{L_{tk}} u_k - \dfrac{R_{tk}}{L_{tk}} i_{tk} + \dfrac{1}{L_{tk}} u_{tk} \end{cases} \tag{10.2}$$

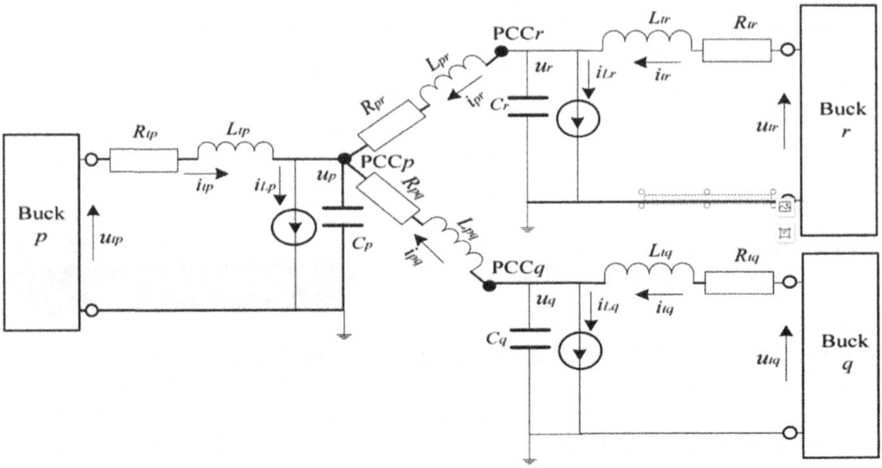

Fig. 10.4 Dynamic model of three adjacent DER units

As [1], i_{pq} and p_{pr} are currents on DC transmission line, $di_{pq}/d_t = -di_{qp} = 0$ and $di_{pr}/d_t = -di_{rp} = o$. Therefore, the following equations are given

$$i_{pq} = -i_{qp} = \left(u_q - u_p\right)/R_{pq} \tag{10.3}$$

$$i_{pr} = -i_{rp} = \left(u_r - u_p\right)/R_{pr} \tag{10.4}$$

Remark 10.1 According to Eqs. (10.3) and (10.4), when $\left|u_q - u_p\right| \leq \varepsilon_{pq}$ and $\left|u_r - u_p\right| \leq \varepsilon_{pr}$, where ε_{pq} and ε_{qr} are small specified thresholds, $\left|i_{pq}\right|$ and $\left|i_{pr}\right|$ are very small, since R_{pq} and R_{pr} are constant. In this case, $i_{pq} + i_{pr}, i_{pq}\ and\ i_{pr}$ are regarded as current disturbances, then the three DER units can be described as follows

$$\text{DER } i : \dot{\boldsymbol{x}}_i(t) = \boldsymbol{A}_i \boldsymbol{x}_i(t) + \boldsymbol{B}_i v_i(t) + \boldsymbol{D}_i \omega_i(t) i \in \{p, q, r\} \tag{10.5}$$

where $\boldsymbol{x}_i(t) = [u_i(t), i_{ti}(t)]^{\mathrm{T}}$ is state vectors; $v_i(t) = u_{ti}(t)$ is input variables; $\omega_p(t) = \left[i_{Lp}(t)\left(i_{pq}(t) + i_{pr}(t)\right)\right]^T$, $\omega_q(t) = \left[i_{Lq}(t) i_{qp}(t)\right]^T$ and $\omega_r(t) = \left[i_{Lr}(t) i_{rp}(t)\right]^T$ are disturbance vectors of the three DER units, respectively; the coefficient matrices are as follows

$$\boldsymbol{A}_i = \begin{bmatrix} 0 & 1/C_i \\ -1/L_{ti} - R_{ti}/L_{ti} & \end{bmatrix}; \boldsymbol{B}_i = \begin{bmatrix} 0; \\ 1/L_{ti} \end{bmatrix}$$

$$\boldsymbol{D}_i = \begin{bmatrix} -1/C_i & 1/C_i \\ 0 & 0 \end{bmatrix}$$

Since there is not coupling in Eq. (10.5), the three DER units can be respectively controlled only by means of their own local controller in the first-level unit agent. The local controller design will be discussed in Sect. 10.4.1. According to Remark 10.1, the control mode of the three DER units is defined as control mode 1.

Remark 10.2 When $|u_q - u_p| > \varepsilon_{pq}$ and $|u_r - u_p| \le \varepsilon_{pr}$, $|i_{pq}|$ is no longer very small, thus $i_{pq} + i_{pr}$ and i_{pq} are also no longer deemed as current disturbances. In this case, taking into account transmission time delay, both pth and qth DER units are described in the following form

$$\text{DER}\, i : \dot{x}_i(t) = \tilde{A}_{ij} x_i(t) + \boldsymbol{B}_i v_i(t) + \boldsymbol{A}_{ij} \boldsymbol{x}_j\left(t - \tau_{ij}\right) + \tilde{\boldsymbol{D}}_i \omega_i(t) \tag{10.6}$$

where $i \in \{p, q\}$, $j \in \{p, q\}$, $i \ne j$; $\tau_{pq} = \tau_{qp} \le \bar{\tau}(\bar{\tau} > 0)$ is the transmission time delay between pth DER and qth DER, $\boldsymbol{x}_j\left(t - \tau_{ij}\right) = \left[u_j\left(t - \tau_{ij}\right), i_{tj}\left(t - \tau_{ij}\right)\right]^{\mathrm{T}}$, $\boldsymbol{B}_i$ is in the same form of Eq. (10.5),

$$\tilde{\boldsymbol{A}}_{ij} = \begin{bmatrix} -1/R_{ij}C_i & 1/C_i \\ -1/L_{ii} & -R_{ti}/L_{ii} \end{bmatrix}, \boldsymbol{A}_{ij} = \begin{bmatrix} 1/R_{ij}C_i & 0 \\ 0 & 0 \end{bmatrix}$$
$$\tilde{\boldsymbol{D}}_i = \begin{bmatrix} -1/C_i & 0 \\ 0 & 0 \end{bmatrix}$$

In Eq. (10.6), since there is coupling between pth DER and qth DER, the two DER units need to be controlled by means of the local controller combined with the decoupling coordinated control law. The MAS based delay-dependent distributed coordinated controller will be studied in Sect. 10.4.2. While the rth DER units is still described in the same form of Eq. (10.5), where only $i = r$. Therefore, the rth DER unit can be controlled only by means of the local controller in the first-level unit agent. According to Remark 10.2, the control mode of the three DER units is defined as control mode 2.

Remark 10.3 When $|u_r - u_p| > \varepsilon_{pr}$ and $|u_q - u_p| \le \varepsilon_{pq}$, for the same reason, both pth and kth DER units are described in a similar form of Eq. (10.6), where only $i \in \{p, r\}$, $j \in \{p, r\}$, $i \ne j$, $\tau_{pr} = \tau_{rp} \le \bar{\tau}(\bar{\tau} > 0)$ is the transmission time delay between the pth DER and the rth DER. In this case, the two DER units need to be controlled by means of the MAS based delay-dependent distributed coordinated controller. While the qth DER units can be described in the same form of Eq. (10.5), where only $i = q$. Therefore, the qth DER unit can be controlled only by means of the local controller in the first-level unit agent. According to Remark 10.3, the control mode of the three DER units is defined as control mode 3.

Remark 10.4 When $|u_q - u_p| > \varepsilon_{pq}$ and $|u_r - u_p| > \varepsilon_{pr}$, all currents including $i_{pq} + i_{pr}$, i_{pq} and i_{pr} are no longer dealt as current disturbances. In this case, both qth and kth DER units are described in a similar form of Eq. (10.6) where only $i \in \{q, k\}$, $j = p$. While the pth DER unit is described as

$$\begin{aligned} \text{ER}\, p : \dot{x}_p(t) =& \tilde{A}_{pqr} x_p(t) + B_p v_p(t) \\ &+ A_{pq} x_q\left(t - \tau_{pq}\right) + A_{pr} x_r\left(t - \tau_{pr}\right) + \tilde{D}_p \omega_p(t) \end{aligned} \tag{10.7}$$

where, $\tilde{A}_{pqr} = \begin{bmatrix} -1/R_{pq}C_p - 1/R_{pr}C_p & 1/C_p \\ -1/L_{tp} & -R_{tp}/L_{tp} \end{bmatrix}$, other coefficient matrices are in a similar form of Eq. (10.6) only $i = p, i = pj \in \{q, r\}$.

The three DER units need to be controlled by means of the MAS based delay-dependent distributed coordinated controller. According to Remark 10.4, the control mode of the three DER units is defined as control mode 4.

No matter the local controller or the distributed coordinated controller, their control goal is to guarantee the state variables (voltage and current) to track their reference values. Therefore, here, the tracking reference model is given for the pth, qth and rth DER units as follows

$$\dot{x}_{ri}(t) = A_{ri}x_{ri}(t) + r_i(t) i \in \{p, q, r\} \tag{10.8}$$

where $x_{ri}()$t represents a desired trajectory for $x_i(t)$ to follow. A_{ri} is a specific asymptotically stable matrix, and $r_i(t)$ denotes bounded reference input.

10.3 The Controller Design

10.3.1 Local Controller Design in First-Level Unit Agent

When one DER unit (including each DER unit in Remark 10.1, the rth DER unit in Remark 10.2, and the qth DER unit in Remark 10.3) is modeled in the form of Eq. (10.5), it can be controlled by means of only local controller in the first-level unit control agent. The local controller is designed as

$$v_i(t) = K_i [x_i(t) - x_{ri}(t)] i \in \{p, q, r\} \tag{10.9}$$

where K_i is local controller parameter matrix of ith DER unit.

Then, by using Eqs. (10.5) and (10.8), the ith DER tracking control system under the local controller (10.9) is written in the following form

$$\text{DER } i : \dot{\hat{x}}_i(t) = \hat{A}_i\hat{x}_i(t) + \hat{D}_i\hat{\omega}_i(t) \tag{10.10}$$

where, $\hat{x}_i(t) = \left[x_i^T(t) x_{ri}^T(t)\right]^{\mathrm{T}}$, $\hat{\omega}_i(t) = \left[\omega_i^T(t) r_i^T(t)\right]^T$, and

$$\hat{A}_i = \begin{bmatrix} A_i + B_iK_i & -B_iK_i \\ 0 & A_{ri} \end{bmatrix}, \hat{D}_i = \begin{bmatrix} D_i & 0 \\ 0 & I \end{bmatrix}$$

H_∞ control performance regarding the tracking error is given in the following form

$$\frac{\int_0^{t_f} \left[(x_i(t) - x_{ri}(t))^T Q_i (x_i(t) - x_{ri}(t))\right] dt}{\int_0^{t_f} \hat{\omega}_i(t)^T \hat{\omega}_i(t) dt} \le \rho_i^2 \tag{10.11}$$

where t_f denotes terminal time of control, and the weighting matrix $\boldsymbol{Q}_i = \boldsymbol{Q}_i^T > 0$ Physical meaning of (10.11) is that effect of any $\hat{\omega}_i(t)$ on tracking error $\boldsymbol{x}_i(t) - \boldsymbol{x}_{ri}(t)$ must be attenuated below a desire level ρ_i from viewpoint of energy. H_∞ control performance with a prescribed attenuation level is useful for a robust control design without knowledge of $\omega_i(t)$ and $ri(t)$.

Considering the initial condition, H_∞ control performance can be rewritten as follows:

$$\int_0^{t_f} \hat{x}_i^T(t)\hat{\boldsymbol{Q}}_i\hat{x}_i(t)dt \le \rho_i^2 \int_0^{t_f} \hat{\omega}_i^T(t)\hat{\omega}_i(t)dt + \hat{\boldsymbol{x}}_i^T(0)\boldsymbol{P}_i\hat{\boldsymbol{x}}_i(0) \tag{10.12}$$

where $\hat{\boldsymbol{Q}}_i = \begin{bmatrix} \boldsymbol{Q}_i & -\boldsymbol{Q}_i \\ -\boldsymbol{Q}_i & \boldsymbol{Q}_i \end{bmatrix}$, and weighting matrix $\boldsymbol{P}_i = \boldsymbol{P}_i^T > 0$.

According to the following Theorem 10.1, the first-level unit control agent can determine the local controller (10.9) for the ith tracking system (10.10) with the guaranteed H_∞ control performance (10.12) for $\forall\hat{\omega}_i(t)$.

Theorem 10.1 *The ith DER unit tracking control system (10.10) is asymptotically stable with the guaranteed H_∞ control performance in (10.12) for $\forall\hat{\omega}_i(t)$ if there exists $\boldsymbol{P}_i = \boldsymbol{P}_i^T > 0$, satisfying the following matrix inequality*

$$\begin{bmatrix} \hat{\boldsymbol{A}}_i^T\boldsymbol{P}_i + \boldsymbol{P}_i\hat{\boldsymbol{A}}_i + \hat{\boldsymbol{Q}}_i & \boldsymbol{P}_i\hat{\boldsymbol{D}}_i \\ \hat{\boldsymbol{D}}_i^T\boldsymbol{P}_i & -\rho_i^2\boldsymbol{I} \end{bmatrix} \le 0 \tag{10.13}$$

The proof is given as follows

Define a Lyapunov function for the tracking system (10.10) as

$$\mathrm{V}_1(t) = \hat{x}_i^T(t)P_i\hat{x}_i(t), \text{ where } P_i = P_i^T > 0$$

Then, it is easy to obtain

$$\begin{aligned}
&\int_0^{t_f}\left[(x_i(t) - x_{ri}(t))^T Q_i\,(x_i(t) - x_{ri}(t))\right]dt \\
&= \int_0^{t_f}\hat{x}_i^T(t)\hat{Q}_i\hat{x}_i(t)dt = \hat{x}_i^T(0)P_i\hat{x}_i(0) - \hat{x}_i^T(t_f)P_i\hat{x}_i(t_f) \\
&\quad + \int_0^{t_f}\left\{\hat{x}_i^T(t)\hat{Q}_i\hat{x}_i(t) + \tfrac{d}{dt}\left(\hat{x}_i^T(t)P_i\hat{x}_i(t)\right)\right\}dt \\
&\le \hat{x}_i^T(0)P_i\hat{x}_i(0) + \int_0^{t_f}\left\{\hat{x}_i^T(t)\hat{Q}_i\hat{x}_i(t) + \dot{\hat{x}}_i^T(t)P_i\hat{x}_i(t)\right. \\
&\qquad \left. + \hat{x}_i^T(t)P_i\dot{\hat{x}}_i(t)\right\}dt \\
&= \hat{x}_i^T(0)P\hat{x}_i(0) \\
&\quad + \int_0^{t_f}\left\{\begin{bmatrix} \hat{\boldsymbol{x}}_i(t) \\ \hat{\omega}_i(t) \end{bmatrix}^T \begin{bmatrix} \hat{\boldsymbol{A}}_i^T\boldsymbol{P}_i + \boldsymbol{P}_i\hat{\boldsymbol{A}}_i + \hat{\boldsymbol{Q}}_i & \boldsymbol{P}_i\hat{\boldsymbol{D}}_i \\ \hat{\boldsymbol{D}}_i^T\boldsymbol{P}_i & -\rho_i^2\boldsymbol{I} \end{bmatrix} \begin{bmatrix} \hat{\boldsymbol{x}}_i(t) \\ \hat{\omega}_i(t) \end{bmatrix}\right. \\
&\qquad \left. + \rho_i^2\hat{\omega}_i(t)^T\hat{\omega}_i(t)\right\}dt
\end{aligned}$$

By the above inequality, it can be implied that if the inequality (10.13) is satisfied, the ith augmented DER unit system (10.10) is asymptotically stable with the guaranteed H_∞ control performance in (10.12) for $\forall\hat{\omega}_i(t)$. The proof is completed.

The inequality (10.13) can be transformed into linear matrix inequality (LMI) according to the following procedures

(1) Denote a new matrix

$$\boldsymbol{W}_i = \begin{bmatrix} \bar{\boldsymbol{W}}_i & 0 \\ 0 & \boldsymbol{I} \end{bmatrix} = \begin{bmatrix} \boldsymbol{P}_i^{-1} & 0 \\ 0 & \boldsymbol{I} \end{bmatrix}$$

where $\bar{\boldsymbol{W}}_i^T = \bar{\boldsymbol{W}}_i = \boldsymbol{P}_i^{-1} > 0$, left and right multiplying it into (10.14), it can be obtained

$$\begin{bmatrix} \bar{\boldsymbol{W}}_i\hat{\boldsymbol{A}}_i^T + \hat{\boldsymbol{A}}_i\bar{\boldsymbol{W}}_i + \bar{\boldsymbol{W}}_i\hat{\boldsymbol{Q}}_i\bar{\boldsymbol{W}}_i & \hat{\boldsymbol{D}}_i \\ \hat{\boldsymbol{D}}_i^T & -\rho_i^2\boldsymbol{I} \end{bmatrix} \le 0 \tag{10.14}$$

(2) Define

$$\bar{W}_i = \begin{bmatrix} \bar{W}_{i1} & 0 \\ 0 & \bar{W}_{i1} \end{bmatrix} = \begin{bmatrix} \boldsymbol{P}_{i1}^{-1} & 0 \\ 0 & \boldsymbol{P}_{i1}^{-1} \end{bmatrix}, \hat{\boldsymbol{K}}_i = \boldsymbol{K}_i\bar{\boldsymbol{W}}_{i1}, \check{\boldsymbol{Q}} = \bar{\boldsymbol{W}}_1\boldsymbol{Q}\bar{\boldsymbol{W}}_1,$$

it is easy to gain

$$\hat{\boldsymbol{A}}_i\bar{\boldsymbol{W}}_1 = \begin{bmatrix} \boldsymbol{A}_i\bar{\boldsymbol{W}}_{i1} + \boldsymbol{B}_i\hat{\boldsymbol{K}}_i & -\boldsymbol{B}_i\hat{\boldsymbol{K}}_i \\ 0 & \boldsymbol{A}_{ri}\bar{\boldsymbol{W}}_{i1} \end{bmatrix}$$

Hence, the inequality (10.14) is equivalent to LMI.

Remark 10.5 The local controller design in (10.9) is transformed into the following LMI convex optimization problem

$$\min_{\bar{W}_i\hat{K}_i} \rho_i^2$$
$$\text{subject to } \bar{\boldsymbol{W}}_i = \bar{\boldsymbol{W}}_i^T > 0 \text{ and (10.14)} \tag{10.15}$$

10.3.2 MAS Based Distributed Coordinated Controller Design

(1) Distributed Coordinated Controller Design Between Two DER Units: When two DER units (including both pth and qth DER units in Remark 10.2, as well as both pth and rth DER units in Remark 10.3) are modeled in the form of Eq. (10.6), they need to be controlled by means of the two-level MAS. The distributed coordinated controller is designed as

$$v_i(t) = \boldsymbol{K}_i\left[\boldsymbol{x}_i(t) - \boldsymbol{x}_{ni}(t)\right] + \boldsymbol{K}_{ij}\boldsymbol{x}_j\left(t - \tau_{ij}\right) \tag{10.16}$$

where $i \in \{p,q\}, j \in \{p,q\}, i \neq j$; or $i \in \{p,r\}, j \in \{p,r\}, i \neq j$; $\boldsymbol{K}_i$ is local controller parameter matrix of the ith DER unit, $\boldsymbol{K}_{ij}$ is decoupling coordinated control law that comes form jth DER unit.

By Eqs. (10.6) and (10.8), the ith tracking control system under the controller (10.16) is given as follows

$$\text{DER}\, i: \dot{\hat{x}}_i(t) = \bar{A}_{ij}\hat{x}_i(t) + \bar{A}_{ij}\hat{x}_j\left(t - \tau_{ij}\right) + \bar{D}_i\hat{\omega}_i(t) \tag{10.17}$$

where $i \in \{p,q\}, j \in \{p,q\}, i \neq j$; or $i \in \{p,r\}, j \in \{p,r\}, i \neq j$,

$$\overline{\boldsymbol{A}}_{ij} = \begin{bmatrix} \tilde{\boldsymbol{A}}_{ij} + \boldsymbol{B}_i\boldsymbol{K}_i & -\boldsymbol{B}_i\boldsymbol{K}_i \\ 0 & \boldsymbol{A}_{ni} \end{bmatrix}, \overline{\boldsymbol{A}}_{ij} = \begin{bmatrix} \boldsymbol{A}_{ij} + \boldsymbol{B}_i\boldsymbol{K}_{ij} & 0 \\ 0 & 0 \end{bmatrix}$$

$$\overline{\boldsymbol{D}}_i = \begin{bmatrix} \tilde{\boldsymbol{D}}_i & 0 \\ 0 & \boldsymbol{I} \end{bmatrix}$$

Then augmented system by integrating the ith DER with the jth DER is described as

$$\text{DER}\, ij: \dot{\tilde{x}}(t) = \tilde{A}\tilde{x}(t) + \widehat{A}\tilde{x}\left(t - \tau_{ij}\right) + \tilde{D}\tilde{\omega}(t) \tag{10.18}$$

where $i = p, j \in \{q,r\}$; $\tilde{\boldsymbol{x}}(t) = \left[\boldsymbol{x}_i^T(t), \boldsymbol{x}_{ri}^T(t), \boldsymbol{x}_j^T(t), \boldsymbol{x}_{rj}^T(t)\right]^{\mathrm{T}}$ is the state vector of the augmented system, $\tilde{v}(t) = \left[v_i(t), v_j(t)\right]^{\mathrm{T}}$ is the input vector and $\tilde{\omega}(t) = \left[\omega_i^T(t)\omega_j^T(t)\right]^T$ is the disturbance vector of the augmented system. Moreover,

$$\tilde{\boldsymbol{A}} = \begin{bmatrix} \bar{\boldsymbol{A}}_{ij} & 0 \\ 0 & -\bar{\boldsymbol{A}}_{ji} \end{bmatrix}, \widehat{\boldsymbol{A}} = \begin{bmatrix} 0 & \bar{\boldsymbol{A}}_{ij} \\ \bar{\boldsymbol{A}}_{ji} & 0 \end{bmatrix}, \tilde{\boldsymbol{D}} = \begin{bmatrix} \bar{\boldsymbol{D}}_i & 0 \\ 0 & \bar{\boldsymbol{D}}_j \end{bmatrix}.$$

Corresponding to the augmented system (10.18), H_∞ control performance can be given as follows:

$$\int_0^{t_f} \tilde{\boldsymbol{x}}^T(t)\tilde{\boldsymbol{Q}}\tilde{\boldsymbol{x}}(t)dt \leq \rho^2 \int_0^{t_f} \tilde{\omega}^T(t)\tilde{\omega}(t)dt + \mathrm{V}(0) \tag{10.19}$$

where $\tilde{\boldsymbol{Q}} = \begin{bmatrix} \boldsymbol{Q}_i & -\boldsymbol{Q}_i & 0 & 0 \\ -\boldsymbol{Q}_i & \boldsymbol{Q}_i & 0 & 0 \\ 0 & 0 & \boldsymbol{Q}_j & -\boldsymbol{Q}_j \\ 0 & 0 & -\boldsymbol{Q}_j & \boldsymbol{Q}_j \end{bmatrix}$, V(0) is Lyapunov function initial value.

By mans of the following Theorem 10.2, the two-level agents can determine the distributed coordinated controller (10.16) for the augmented system (10.18) with the guaranteed H_∞ control performance (10.19) for $\forall\tilde{\omega}(t)$.

Theorem 10.2 *Given allowable upper bound $\bar{\tau}$ of the time delays, the controlled system (10.18) is asymptotically stable for all time-delays satisfying $\tau_{ij} \in [0, \bar{\tau}]$ with the H_∞ control performance in (10.19), if there exist symmetric positive definite matrices* **P**, **S**, **Z**, **X** *satisfying the following matrix inequalities*

$$\begin{bmatrix} \Theta & P\widehat{A} - Y & P\tilde{D} & \bar{\tau}\tilde{A}^T Z \\ * & -S & 0 & \bar{\tau}\widehat{A}^T Z \\ * & * & -\rho^2 & \bar{\tau}\tilde{D}^T Z \\ * & * & * & -\bar{\tau}Z \end{bmatrix} \leq 0 \text{ and } \begin{bmatrix} X & Y \\ Y^T & Z \end{bmatrix} \geq 0 \tag{10.20}$$

where $i = p$, $j \in \{q, r\}$, $\Theta = \tilde{Q} + S + P\tilde{A} + \tilde{A}^T P + \bar{\tau}X + Y + Y^T$

The proof is given as follows.

In order to prove the Theorem 10.2, firstly give a Lemma 10.1.

Lemma 10.1 ([2]) *For all vectors* **a** *and* **b**, *matrices* **N**, **X**, **Y**, **Z** *with appropriate dimensions, if* $\begin{bmatrix} X & Y \\ Y^T & Z \end{bmatrix} \geq 0$, *then*

$$-2a^T Nb \leq \inf_{X,Y,Z} \left\{ \begin{bmatrix} a \\ b \end{bmatrix}^T \begin{bmatrix} X & Y - N \\ Y^T - N^T & Z \end{bmatrix} \begin{bmatrix} a \\ b \end{bmatrix} \right\}$$

where, **X** *and* **Z** *are symmetrical positive matrices.*

Define a delay-dependent Lyapunov function for the system (10.18) as

$$\mathrm{V}(t) = \mathrm{V}_1(t) + \mathrm{V}_2(t) + \mathrm{V}_3(t) \tag{(1B)}$$

where $\mathrm{V}_1(t) = \tilde{x}^T(t)P\tilde{x}(t)\mathrm{V}_2(t) = \int_{t-\bar{\tau}}^{t} \tilde{x}^T(\tau)S\tilde{x}(\tau)d\tau$

$$\mathrm{V}_3(t) = \int_{-\bar{\tau}}^{0}\int_{t+\beta}^{t} \dot{\tilde{x}}^T(\alpha)\dot{\tilde{x}}(\alpha)d\alpha d\beta$$

And **P**, **S**, **Z** are symmetric positive definite weighting matrices. The derivative of $V_1(t)$ along the trajectory of system (10.18) satisfies that

$$\begin{aligned} \dot{\mathrm{V}}_{i1}(t) =& 2\tilde{x}^T(t)P(\tilde{A} + \widehat{A})\overline{x}_i(t) \\ &- 2\tilde{x}^T(t)P\widehat{A}\int_{t-\bar{\tau}}^{t} \dot{\tilde{x}}(\alpha)d\alpha + 2\tilde{x}^T(t)P\tilde{D}\tilde{\omega}(t) \end{aligned}$$

By using Lemma 10.1, it is very easy to obtain

$$\begin{aligned} -2\tilde{x}^T(t)P\widehat{A}\dot{\tilde{x}}(\alpha) \leq& \tilde{x}^T(t)X\tilde{x}(t) + \dot{\tilde{x}}^T(\alpha)\left[Y^T - (P\widehat{A})^T\right]\tilde{x}(t) \\ &+ \tilde{x}^T(t)(Y - P\widehat{A})\dot{\tilde{x}}(\alpha) + \dot{\tilde{x}}^T(\alpha)Z\dot{\tilde{x}}(\alpha) \end{aligned}$$

then

$$\dot{V}_1(t) \leq \left\{ \tilde{x}^T(t) \left(P \left(\tilde{A} + \tilde{A}^T P + \bar{\tau} X + Y + Y^T \tilde{x}(t) \right.\right.\right. \\ -2\tilde{x}^T(t)(Y - P\widehat{A})\tilde{x}(t - \bar{\tau}) \\ \left. + \int_{t-\bar{\tau}}^{t} \dot{\tilde{x}}^T(\alpha)\dot{x}(\alpha)d\alpha + 2\tilde{x}^T(t)P\tilde{D}\tilde{\omega}(t) \right\} \tag{2B}$$

The derivative of other Lyapunov functions along the trajectory of system (10.18) satisfy that

$$\dot{V}_2(t) = \tilde{x}^T(t)S\tilde{x}(t) - \tilde{x}^T(t - \bar{\tau})S\tilde{x}(t - \bar{\tau}) \tag{3B}$$

$$\dot{V}_3(t) = \bar{\tau}\dot{\tilde{x}}^T(t)Z\dot{\tilde{x}}(t) - \int_{t-\bar{\tau}}^{t} \dot{\tilde{x}}^T(\alpha)Z\dot{\tilde{x}}(\alpha)d\alpha \tag{4B}$$

Then, it can be obtained

$$\int_0^t \tilde{x}^T(t)\tilde{Q}\tilde{x}(t)dt = V(0) - V(t_f) + \int_0^{t_f} \left\{ \tilde{x}^T(t)\tilde{Q}\tilde{x}(t) + \dot{V}(t) \right\} dt$$
$$\leq V(0) + \int_0^{t_f} \left\{ \tilde{x}^T(t) \left(\tilde{Q} + S + P\tilde{A} + \tilde{A}^T P + \bar{\tau}X + Y + Y^T \right) \tilde{x}(t) \right.$$
$$-2\tilde{x}^T(t)(Y - P\widehat{A})\tilde{x}(t - \bar{\tau}) + 2\tilde{x}^T(t)P\tilde{D}\tilde{\omega}(t) - \tilde{x}^T(t - \bar{\tau})S\tilde{x}(t - \bar{\tau}) + \bar{\tau}\dot{\tilde{x}}^T(t)Z\dot{\tilde{x}}(t)$$
$$\left. +\rho^2\tilde{\omega}^T(t)\tilde{\omega}(t) - \rho^2\tilde{\omega}^T(t)\tilde{\omega}(t) \right\} dt$$
$$= V(0) + \int_0^{t_f} \left\{ \begin{bmatrix} \tilde{x}(t) \\ \tilde{x}(t - \bar{\tau}) \\ \tilde{\omega}(t) \end{bmatrix}^T \left\{ \begin{bmatrix} \Theta & P\widehat{A} - Y & P\tilde{D} \\ * & -S & 0 \\ * & * & -\rho^2 \end{bmatrix} \right.\right.$$
$$\left.\left. +\bar{\tau} \begin{bmatrix} \tilde{A} \\ \widehat{A} \\ \tilde{D} \end{bmatrix}^T Z \begin{bmatrix} \tilde{A} \\ \widehat{A} \\ \tilde{D} \end{bmatrix} \right\} \begin{bmatrix} \tilde{x}(t) \\ \tilde{x}(t - \bar{\tau}) \\ \tilde{\omega}(t) \end{bmatrix} + \rho^2\tilde{\omega}^T(t)\tilde{\omega}(t) \right\} dt$$

According to the above inequality, by using Schur complement, it is easy to obtain that, if the inequality (10.20) hold, the system (10.18) is asymptotically stable with the $H\infty$ performance in (10.19). This completes the proof. The inequality (10.20) is not linear, so left multiply and right multiply matrix diag $\{P^{-1}, I, I, Z^{-1}\}$, then define

$$P^{-1} = \begin{bmatrix} P_1^{-1} & 0 & 0 & 0 \\ 0 & P_1^{-1} & 0 & 0 \\ 0 & 0 & P_1^{-1} & 0 \\ 0 & 0 & 0 & P_1^{-1} \end{bmatrix}, P_1 = P_1^T > 0$$

$\tilde{K}_i = K_i P_1^{-1}$, $\tilde{K}_j = K_j P_1^{-1}$, $\tilde{K}_{ij} = K_{ij} P_1^{-1}$, $\tilde{K}_{jl} = K_{ji} P_1^{-1}$, $\tilde{X} = P^{-1} X P^{-1}$, $\tilde{S} = P^{-1} S P^{-1}$,

$\hat{Q} = P^{-1}\tilde{Q}P^{-1}, \tilde{Y} = P^{-1}YP^{-1}$, it is easy to gain,

$$\tilde{A}^{-1} = \begin{bmatrix} \tilde{A}_{ij}P_1^{-1} + B_i\tilde{K}_i & -B_i\tilde{K}_i & 0 & 0 \\ 0 & A_{ri}P_1^{-1} & 0 & 0 \\ 0 & 0 & \tilde{A}_{ji}P_1^{-1} + B_j\tilde{K}_j & -B_j\tilde{K}_j \\ 0 & 0 & 0 & A_{rj}P_1^{-1} \end{bmatrix}$$

Then the inequality (10.20) is transformed to LMI (10.21).

$$\begin{bmatrix} \tilde{\Theta} & \widehat{A} - P^{-1}Y & \tilde{D} & \bar{\tau}P^{-1}\tilde{A}^T \\ * & -S & 0 & \bar{\tau}\widehat{A}^T \\ * & * & -\rho^2 & \bar{\tau}\tilde{D}^T \\ * & * & * & -\bar{\tau}Z^{-1} \end{bmatrix} \leq 0 \tag{10.21}$$

$$\text{where } \tilde{\Theta} = \widehat{Q} + \tilde{S} + \tilde{A}P^{-1} + P^{-1}\tilde{A}^T + \bar{\tau}\tilde{X} + \tilde{Y} + \tilde{Y}^T. \tag{10.22}$$

Remark 10.6 The distributed coordinated controller design in (10.16) is transformed into the following LMI convex optimization problem

$$\min_{P^{-1},\tilde{K}_i,\tilde{K}_j,\tilde{K}_{ij},\tilde{K}_{ji}} \rho^2$$

subject to $P^{-1} = P^{-T} > 0$, $\begin{bmatrix} X & Y \\ Y^T & Z \end{bmatrix} \geq 0$, and (10.21) (10.22) where $i = p, j \in \{q, r\}$

(1) Distributed Coordinated Controller Design Among Three DER Units: When three DER units are described as Remark 10.4, the two-level MAS needs to implement the distributed coordinated control among three DER units. In this case, the distributed coordinated controller is designed as

$$\begin{aligned} v_i(t) =& K_i\left[x_i(t) - x_{ri}(t)\right] \\ &+ K_{ij}x_j\left(t - \tau_{ij}\right), i \in \{q, r\}, j = p \end{aligned} \tag{10.23}$$

$$\begin{aligned} v_p(t) =& K_p\left[x_p(t) - x_{rp}(t)\right] \\ &+ K_{pq}x_q\left(t - \tau_{pq}\right) + K_{pr}x_r\left(t - \tau_{pr}\right) \end{aligned} \tag{10.24}$$

where the definition regarding controller parameters is similar to Eq. (10.16). By Eqs. (10.6) and (10.8), the ith tracking control system under the controller (10.23) is given as follows

$$\text{DER}\, i : \dot{\hat{x}}_i(t) = \bar{A}_{ij}\hat{x}_i(t) + \bar{A}_{ij}\hat{x}_j\left(t - \tau_{ij}\right) + \bar{D}_i\hat{\omega}_i(t), i \in \{q, r\}, j = p \tag{10.25}$$

where the coefficient matrices are in a similar form of Eq. (10.17), only $i \in \{q, r\}, j = p$.

By Eqs. (10.7) and (10.8), the pth tracking control system under the controller (10.24) is given as follows

$$\text{DER}\, p : \dot{\hat{x}}_p(t) = \bar{A}_{pq}\hat{x}_i(t) + \bar{A}_{pq}\hat{x}_q\left(t - \tau_{pq}\right) + \bar{A}_{pr}\hat{x}_r\left(t - \tau_{pr}\right) + \bar{D}_p\hat{\omega}_p(t) \tag{10.26}$$

where $\bar{\boldsymbol{A}}_{pqr} = \begin{bmatrix} \tilde{\boldsymbol{A}}_{pqr} + \boldsymbol{B}_p\boldsymbol{K}_p & -\boldsymbol{B}_p\boldsymbol{K}_p \\ 0 & \boldsymbol{A}_{rp} \end{bmatrix}$, other coefficient matrices are in a similar form of Eq. (10.17) only $i = p, j \in \{q, r\}$.

Then augmented system by integrating the three DER units is described as

$$\text{DER pqr} : \dot{\tilde{x}}(t) = \tilde{\boldsymbol{A}}_1\tilde{\boldsymbol{x}}(t) + \widehat{\boldsymbol{A}}_1\tilde{\boldsymbol{x}}(t - \tau) + \tilde{\boldsymbol{D}}_1\tilde{\omega}(t) \tag{10.27}$$

where $\tilde{\boldsymbol{x}}(t) = \left[\boldsymbol{x}_p^T(t), \boldsymbol{x}_{rp}^T(t), \boldsymbol{x}_q^T(t), \boldsymbol{x}_{rq}^T(t), \boldsymbol{x}_r^T(t), \boldsymbol{x}_{rr}^T(t)\right]^{\mathrm{T}}$ is the state vector, $\tilde{\boldsymbol{x}}(t - \tau) = \left[\boldsymbol{x}_p^T(t), \boldsymbol{x}_{rp}^T(t), \boldsymbol{x}_q^T\left(t - \tau_{pq}\right), \boldsymbol{x}_{rq}^T(t), \boldsymbol{x}_r^T\left(t - \tau_{pr}\right), \boldsymbol{x}_{rr}^T(t)\right]^{\mathrm{T}}$ $\tilde{v}(t) = \left[v_p(t), v_q(t), v_r(t)\right]^{\mathrm{T}}$ is the input vector of the augmented system, and $\tilde{\omega}(t) = \left[\omega_p^T(t)\omega_q^T(t)\omega_r^T(t)\right]^T$ is the disturbance vector. Moreover,

$$\tilde{\boldsymbol{A}}_1 = \begin{bmatrix} \overline{\boldsymbol{A}}_{pqr} & 0 & 0 \\ 0 & \overline{\boldsymbol{A}}_{qp} & 0 \\ 0 & 0 & \overline{\boldsymbol{A}}_{rp} \end{bmatrix}, \widehat{\boldsymbol{A}}_1 = \begin{bmatrix} 0 & \overline{\boldsymbol{A}}_{pq} & \overline{\boldsymbol{A}}_{pr} \\ \overline{\boldsymbol{A}}_{qp} & 0 & 0 \\ \overline{\boldsymbol{A}}_{rp} & 0 & 0 \end{bmatrix}.$$

$$.\tilde{\boldsymbol{D}}_1 = \begin{bmatrix} \boldsymbol{D}_p & 0 & 0 \\ 0 & \overline{\boldsymbol{D}}_q & 0 \\ 0 & 0 & \overline{\boldsymbol{D}}_r \end{bmatrix}$$

Remark 10.7 According to the similar design process of Theorem 10.2, the distributed coordinated controller design in (10.23) and (10.24) is transformed into the following LMI convex optimization problem

$$P^{-1}, \tilde{K}_p, \tilde{K}_q, \tilde{K}_r^{\min}\tilde{K}_{pq}, \tilde{K}_{qp}, \tilde{K}_{pr}, \tilde{K}_p^{\rho^2}$$

$$\text{subject to } \boldsymbol{P}^{-1} = \begin{bmatrix} \boldsymbol{P}_1^{-1} & 0 & \cdots & 0 \\ 0 & \boldsymbol{P}_1^{-1} & \cdots & 0 \\ \vdots & \vdots & \ddots & \vdots \\ 0 & 0 & \cdots & \boldsymbol{P}_1^{-1} \end{bmatrix}^{6\times 6},$$

$$\boldsymbol{P}_1 = \boldsymbol{P}_1^T > 0, \begin{bmatrix} XY \\ Y^T Z \end{bmatrix} \geq 0,$$

$$\text{and} \begin{bmatrix} \tilde{\Theta}_1 & \widehat{A}_1 - P^{-1}Y & \tilde{D}_1 & \bar{\tau}P^{-1}\tilde{A}_1^T \\ * & -S & 0 & \bar{\tau}\hat{A}_1^T \\ * & * & -\rho^2 & \bar{\tau}\tilde{D}_1^T \\ * & * & * & -\bar{\tau}Z^{-1} \end{bmatrix} \leq 0 \tag{10.28}$$

where

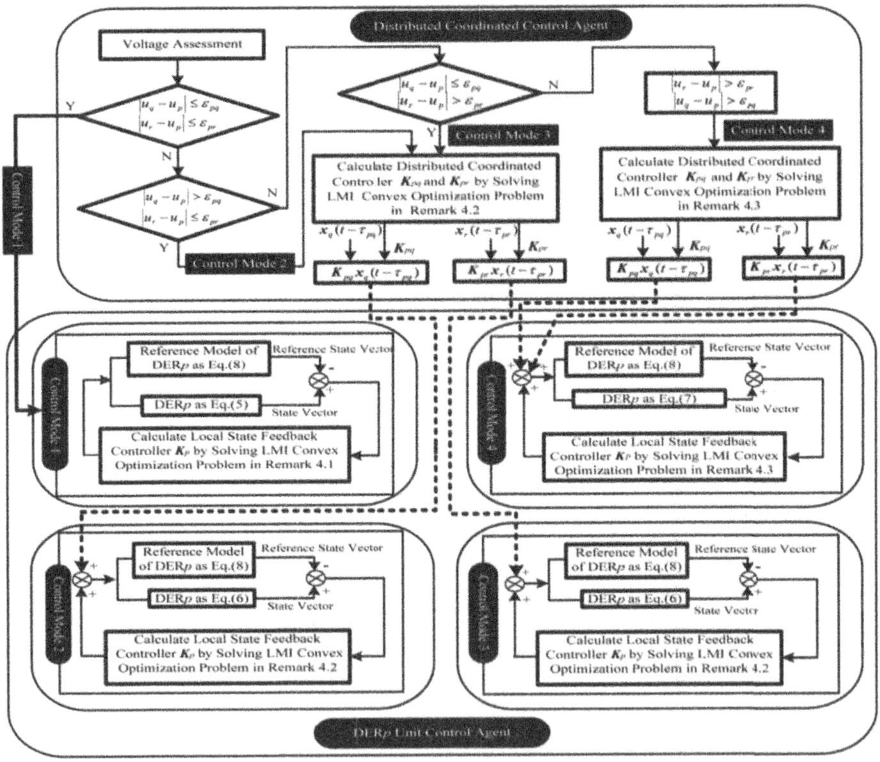

Fig. 10.5 Implementation flowchart of four kinds of control modes based on the MAS

$$\widehat{Q}_1 = \boldsymbol{P}^{-1}\begin{bmatrix} \hat{Q}_p & 0 & 0 \\ 0 & \hat{\boldsymbol{Q}}_q & 0 \\ 0 & 0 & \hat{\boldsymbol{Q}}_r \end{bmatrix}, \hat{\boldsymbol{Q}}_i = \begin{bmatrix} \boldsymbol{Q}_i & -\boldsymbol{Q}_i \\ -\boldsymbol{Q}_i & \boldsymbol{Q}_i \end{bmatrix}, i \in \{p, q, r\}$$

$\tilde{\Theta}_1 = \widehat{Q}_1 + \tilde{\boldsymbol{S}} + \tilde{\boldsymbol{A}}_1 \boldsymbol{P}^{-1} + \boldsymbol{P}^{-1}\tilde{\boldsymbol{A}}_1^T + \bar{\tau}\tilde{\boldsymbol{X}} + \tilde{\boldsymbol{Y}} + \tilde{\boldsymbol{Y}}^T$, the definition regarding $\tilde{\boldsymbol{K}}_p$, $\tilde{\boldsymbol{K}}_q$, $\tilde{\boldsymbol{K}}_r$, $\tilde{\boldsymbol{K}}_{pq}$, $\tilde{\boldsymbol{K}}_{pr}$, $\tilde{\boldsymbol{K}}_{qp}$, $\tilde{\boldsymbol{K}}_{rp}$, $\tilde{\boldsymbol{S}}$, $\tilde{\boldsymbol{X}}$, $\tilde{\boldsymbol{Y}}$ is similar to Remark 10.6.

10.3.3 Implementation of the Distributed Coordinated Control

The pth DER unit in DC MG can be controlled in four kinds of modes. The first mode is that the DER unit is controlled only by the local controller in the first-level unit agent. The second mode is that the DER unit is controlled by the distributed coordinated controller between both pth and qth DERs based on MAS. The third mode is by the distributed coordinated controller between both pth and rth DERs. The fourth mode is by the distributed coordinated controller among the pth, qth and rth DERs. The implementation flowchart of four kinds of control modes is given in Fig. 10.5.

10.4 Experiment Studies

To evaluate the performance of the distributed coordinated control in a radial DC MG, two kinds of cases are designed in order to test different scenarios. The two cases mainly focus on two aspects: the control performance in response to different load changes, as well as the control performance in different time delays. The simulation parameters are given in Table 10.1.

10.4.1 Case 1: The Load Demand Doubled in the DERj

The load demand in the DER_q increases twice at $t = 3$ s instant as shown in Fig. 10.6a. The load current disturbance leads to a big bus voltage deviation between both pth and qth DER units, i.e., $|u_q - u_p| > \varepsilon_{pq}$ and $|u_r - u_p| \le \varepsilon_{pr}$. In this case, the secondary-level agent determines to execute the distributed coordinated control mode in both pth and qth DER units after $t = 3$ s, while the rth DER unit is still controlled by local control mode 1. By solving LMI convex optimization problem in Remark 10.6, we find that the integrated system of both pth and qth DER units is robust stable for anytime delays satisfying $0 \le \bar{\tau} \le 2.2848$ s. For the comparison purpose, in Fig. 10.6b, c, we firstly give the PCC voltage performance of both pth

Table 10.1 Simulation parameters

Parameter	Symbol	Value
DC power supply	U_{DC}	1 kV
Output capacitance	C_p	2.2 mF
Output capacitance	C_q	2.0 mF
Output capacitance	C_r	1.8 mF
Converter inductance	L_{tp}	2.0 mH
Converter inductance	L_{tq}	1.8 mH
Converter inductance	L_{tr}	1.6 mH
Converter resistance	R_{tp}	0.2 Ω
Converter resistance	R_{tq}	0.2 Ω
Converter resistance	R_{tr}	0.2 Ω
Transmission line inductance	L_{pq}	2.0 μH
Transmission line resistance	R_{pq}	0.05 Ω
Transmission line inductance	L_{pr}	2.2 μH
Transmission line resistance	R_{pr}	0.06 Ω

and qth DER units still in local control mode 1 after $t = 3$ s. Then the PCC voltage performance in the proposed distributed coordinated control mode 2 is displayed in Fig. 10.6d–g taking into account different time delays.

From Fig. 10.6b, c, it can be shown that if the two DER units are still controlled by means of the local control mode 1 after $t = 3$ s, the PCC bus voltages, especially the PCC_q voltage, have larger fluctuation. It is because that, only by local controller in each unit control agent, no decoupling control signal eliminates the coupling effect between both pth and qth DER units, so that the bus voltage deviation can not be controlled effectively. From Fig. 10.6d–g, it can be implied that, by means of the proposed distributed coordinated control mode 2, the PCC voltages of both pth and qth DER units can be stabilized within the secure range with less bus voltage deviation. Moreover, although taking into account different time delays, the PCC voltage performance in each DER is almost same. It is because that, by solving LMI convex optimization problem in Remark 10.6, as long as $\tau_{pq} \in [0, \bar{\tau}]$, the designed distributed coordinated controller can guarantee the system optimal control performance. The above simulation results imply that the MAS based distributed coordinated control can improve bus voltage performance in response to the load change and different transmission time delays.

Also for comparison purpose, Fig. 10.7 gives the voltage performance by means of the proposed method of Ref. [3] in Case 1.

When $\tau = 20$ ms, the voltage performance in Fig. 10.7a, b by means of the proposed method of Ref. [3] is similar to that in Fig. 10.6d, e, only there is a bit larger deviation between both pth and qth bus voltages than in Fig. 10.6d, e. While when $\tau = 2$ s, compared with Fig. 10.6f, g, there are larger voltage fluctuations in Fig. 10.7c, d by means of the proposed method of Ref. [3] after $t = 3$ s. These voltage fluctu-

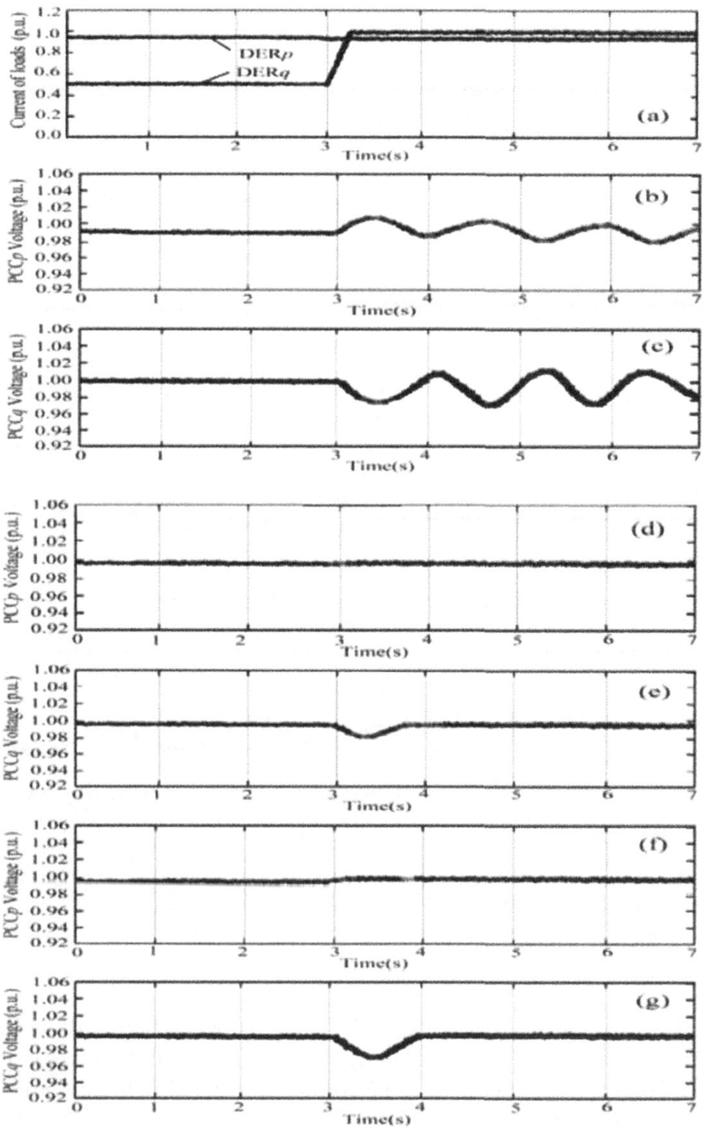

Fig. 10.6 PCC control performance in Case 1. **a** Load currents in the DERp and DERq; **b** PCC voltage of the DERp in the local control mode 1 when $\tau = 20\,\text{ms}$; **c** PCC voltage of the DERq in the local control mode 1 when $\tau = 20\,\text{ms}$; **d** PCC voltage of the DERp in the distributed coordinated control mode 2 when $\tau = 20\,\text{ms}$; **e** PCC voltage of the DERq in the distributed coordinated control mode 2 when $\tau = 20\,\text{ms}$; **f** PCC voltage of the DERp in the distributed coordinated control mode 2 when $\tau = 2\,\text{s}$; **g** PCC voltage of the DERq in the distributed coordinated control mode 2 when $\tau = 2\,\text{s}$

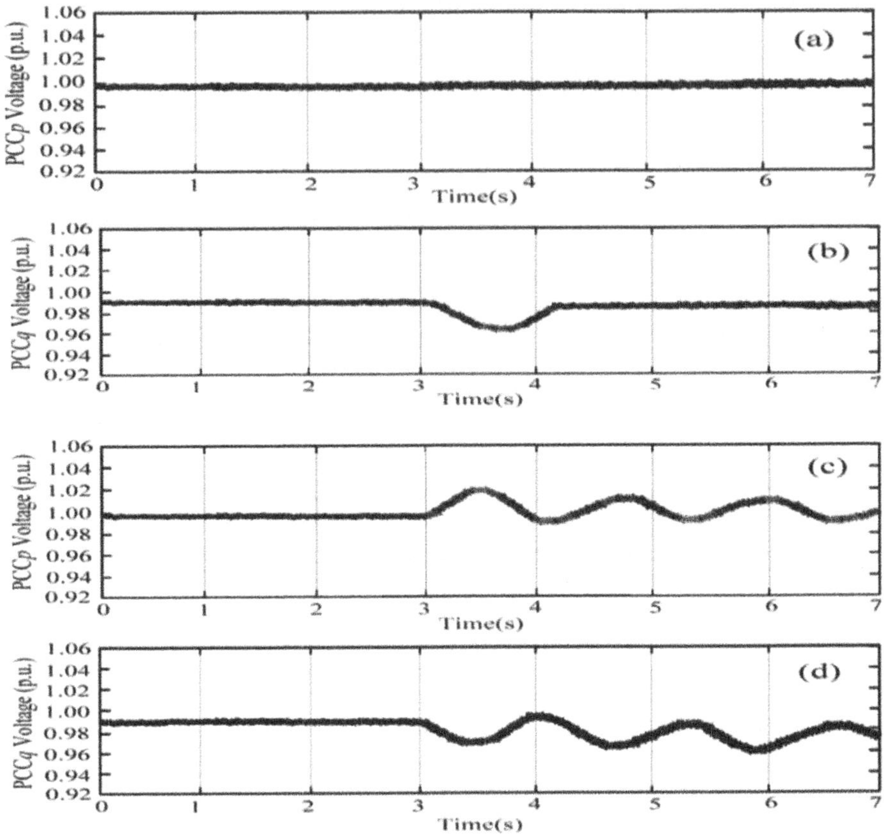

Fig. 10.7 PCC control performance by means of the proposed method of Ref. [3] in Case 1. **a** PCC voltage of the DERp when $\tau = 20\,\text{ms}$; **b** PCC voltage of the DERq when $\tau = 20\,\text{ms}$; **c** PCC voltage of the DERp when $\tau = 2\,\text{s}$; **d** PCC voltage of the DERq when $\tau = 2\,\text{s}$

ations can lead to much larger voltage deviation between both pth and qth DER units. The comparative results above imply that the proposed MAS based distributed coordinated control has better control performance especially in large time delays.

10.4.2 Case 2: The Load Changes in Both DERp and DERq

The load demand in the DERq increases twice at $t = 3\,\text{s}$ instant, at the same time, the load demand in the DERp decreases half as shown in Fig. 10.7a. The load changes in the two DER units result in a larger bus voltage deviation among three adjacent DER units, i.e., $\left|u_q - u_p\right| > \varepsilon_{pq}$ and $\left|u_r - u_p\right| > \varepsilon_{pr}$. Therefore, the secondary-level agent determines to execute the distributed coordinated control mode 4 for the pth, qth and rth DER units after $t = 3$ s. By using LMI convex optimization technique in Remark 10.7, it can be found that the augmented system by integrating the pth, qth and rth DER units is asymptotically stable for any time delays satisfying $0 \leq \bar{\tau} \leq 2.1774$ s. To evaluate the performance of the distributed coordinated control by comparison with the local control, the PCC voltage performance in the local control mode 1 is firstly given in Fig. 10.7b–d. Then taking into account different time delays, the PCC voltage performance in the distributed coordinated control mode 4 is displayed in Fig. 10.7e–j.

From Fig. 10.8b–d, it can seen that the local control is not able to maintain the PCC bus voltage stabilization in Case 2. There are very large voltage deviations especially between both pth and qth DER units since there are larger load changes in the two units. From Fig. 10.8e–j, it can shown that by means of the proposed distributed coordinated control mode 4, at the $t = 3$ s instant, the larger load disturbances only lead to a bit bus voltage fluctuations in the three DER units, afterwards the PCC voltages are rapidly stabilized down almost without any deviation in different time delays. The simulation results above indicate that the proposed MAS based distributed coordinated control still can ensure the bus voltages performance better in the face of larger load disturbances.

In order to testify the performance regarding current sharing by comparison with the proposed method of Ref. [3], the current performance is given in Fig. 10.9 by using two kinds of methods taking into account different time delays in Case 2.

From Fig. 10.9a–f, it can be shown that when $\tau = 200$ ms, the current sharing in the three DER units by means of the proposed control mode 4 is a bit better than that in the proposed method of Ref. [3], since after $t = 4.5$ s there are a bit current deviation among the three DER units by means of the proposed method of Ref. [3]. When $\tau = 2$ s, the current sharing is almost similar to that when $\tau = 200$ ms in the each DER unit by means of the proposed control mode 4 as shown in Fig. 10.9g–i. While by means of the proposed method of Ref. [3], when $\tau = 2$ s, after $t = 3$ s there are very large current fluctuations in the three DER units, which leads to much larger current sharing deviation among the three DER units. The results above imply that when there is a large time delay, the proposed method in this chapter can achieve better performance regarding current sharing.

In order to illustrate the robust performance of the proposed method, we assume that there is a failure of the communication link between both pth and qth secondary-level agents. Still in Case 2, the voltage performance of the three DER units is shown in Fig. 10.10.

If there is no failure of the communication link between both pth and qth secondary-level agents, in case 2 the voltage performance in the three DER units should be the same as Fig. 10.8e–g. Therefore, compared with Fig. 10.8e–g, the PCC voltage fluctuations in the three DER units as shown in Fig. 10.10a–c are a bit larger. However, the voltage fluctuations can damp down after $t = 4.5$ s. It is because that, though there is a failure of the communication link between both pth and qth secondary-level agents, but based on the MAS platform, through the interactions between first-level and secondary-level agents of pth DER unit and of qth DER units, as well as the interaction between both pth and qth first-level agents, the

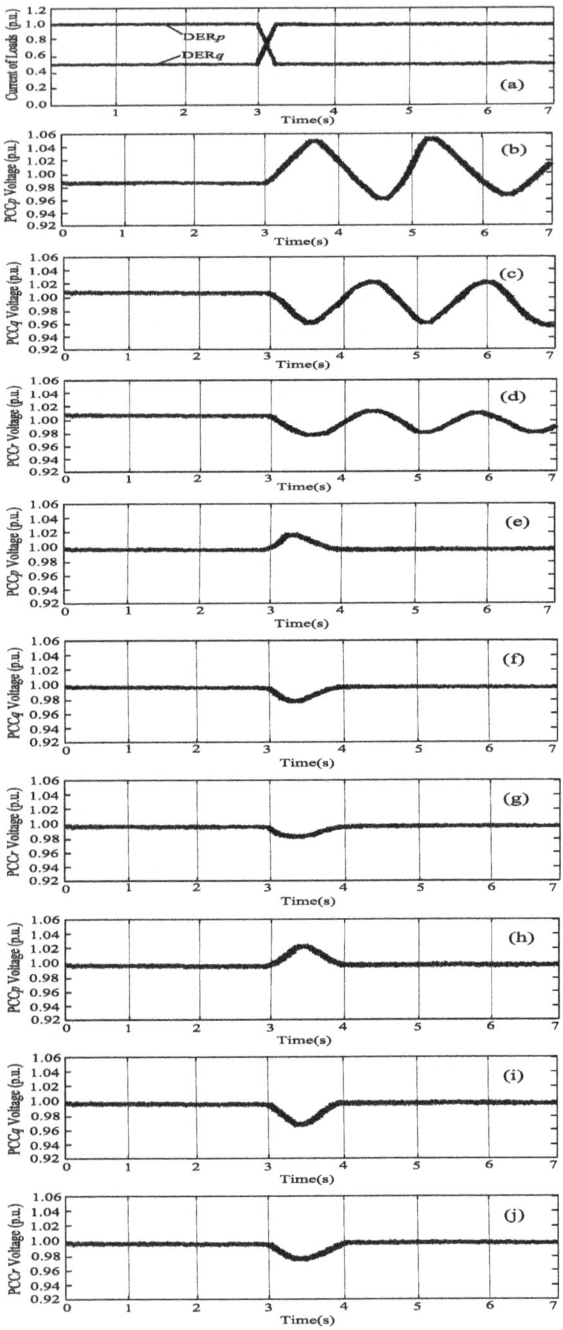

Fig. 10.8 PCC voltage control performance in Case 2. **a** Load currents in the DERp and DERq; **b** PCC voltage of the DERp in the local control mode 1 when $\tau = 200$ ms; **c** PCC voltage of the DERq in the local control mode 1 when $\tau = 200$ ms; **d** PCC voltage of the DERr in the local control mode 1 when $\tau = 200$ ms; **e** PCC voltage of the DERp in the distributed coordinated control mode 4 when $\tau = 200$ ms; **f** PCC voltage of the DERq in the distributed coordinated control mode 4 when $\tau = 200$ ms; **g** PCC voltage of the DERr in the distributed coordinated control mode 4 when $\tau = 200$ ms; **h** PCC voltage of the DERp in the distributed coordinated control mode 4 when $\tau = 2$ s; **i** PCC voltage of the DERq in the distributed coordinated control mode 4 when $\tau = 2$ s; **j** PCC voltage of the DERr in the distributed coordinated control mode 4 when $\tau = 2$ s

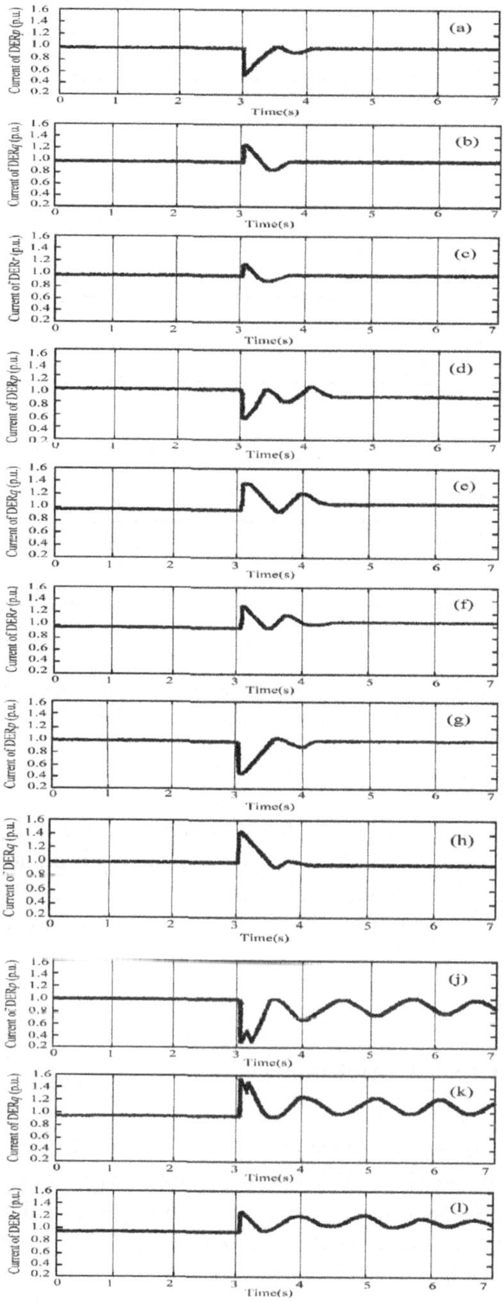

Fig. 10.9 Current control performance in Case 2. **a** Current of the DERp in the proposed control mode 4 when $\tau = 200\,\text{ms}$; **b** Current of the DERq in the proposed control mode 4 when $\tau = 200\,\text{ms}$; **c** Current of the DERr in the proposed control mode 4 when $\tau = 200\,\text{ms}$; **d** Current of the DERp in the proposed method of Ref. [3] when $\tau = 200\,\text{ms}$; **e** Current of the DERq in the proposed method of Ref. [3] when $\tau = 200\,\text{ms}$; **f** Current of the DERr in the proposed method of Ref. [3] when $\tau = 200\,\text{ms}$; **g** Current of the DERp in the proposed control mode 4 when $\tau = 2\,\text{s}$; **h** Current of the DERq in the proposed control mode 4 when $\tau = 2\,\text{s}$; **i** Current of the DERr in the proposed control mode 4 when $\tau = 2\,\text{s}$; **j** Current of the DERp in the proposed method of Ref. [3] when $\tau = 2\,\text{s}$; **k** Current of the DERq in the proposed method of Ref. [3] when $\tau = 2\,\text{s}$; **l** Current of the DERr in the proposed method of Ref. [3] when $\tau = 2\,\text{s}$

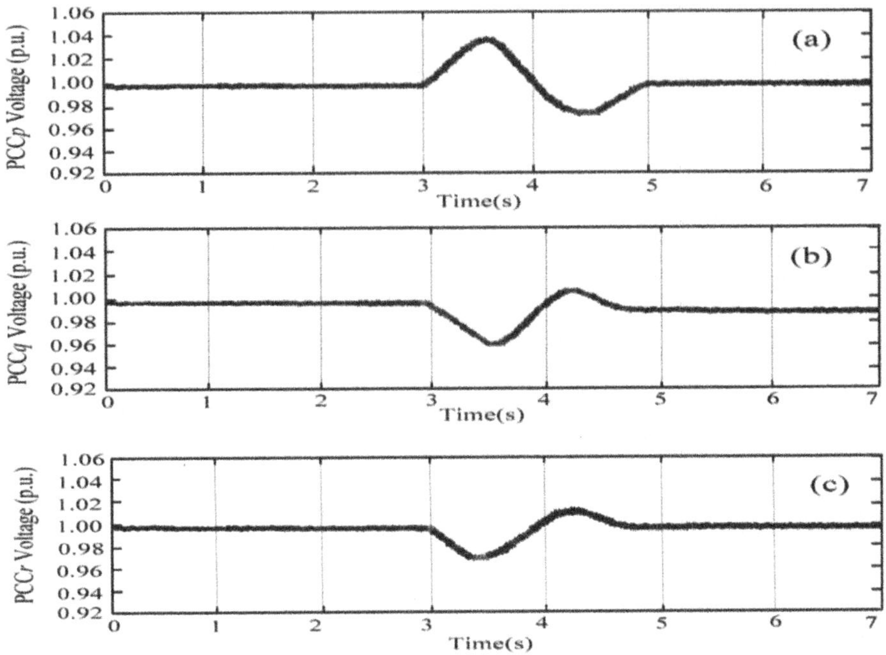

Fig. 10.10 Voltage control performance when there is a failure of the communication in Case 2 when $\tau = 200\,\text{ms}$ **a** PCC voltage of the DER_p; **b** PCC voltage of the DER_q; **c** PCC voltage of the DERr

coordinated control laws between both pth and qth DER units can be still sent. Therefore, the proposed MAS distributed coordinated control method still can guarantee the PCC voltage stabilization. Only since it takes a larger time when the coordinated control laws are sent than that without failure, PCC voltage fluctuations are a bit larger.

10.5 Conclusion

This chapter develops a delay-dependent distributed coordinated control based on the two-level MAS, so that the DC MG can ensure voltage and current sharing performance better in response to large load disturbances and different time delays. Compared with the relevant research results, this chapter contributes the following original works: The MAS based distributed coordinated control is proposed, where beside local-statefeed-back control in unit control agent, the remote states that only come from adjacent DER based on low bandwidth communication are used to synthesize decoupling coordinated control law by means of the distributed coordinated control agent. Moreover, taking into account transmission time delays, the distributed coordinated control is designed by means of delay-dependent H_∞ robust control method. Finally, the better control performance has been demonstrated by means of simulation results.

However, the MAS based distributed coordinated control scheme still a bit relies on communication technology. Based on communication technology, the scheme can not only be applied into DC MG, but also be feasible for controlling any kind of smart grids with multiple DERs by extending the control function of the agents and creating additional agents in the system. The future development of this study will focus on extending application.

References

1. S. Riverso, F. Sarzo, G. Ferraritrecate, Plug-and-play voltage and frequency control of islanded microgrids with meshed topology. IEEE Trans. Smart Grid **6**(3), 1176–1184 (2015)
2. C. Dou, Z. Duan, X. Jia, Delay-dependent $h\infty$ robust control for large power systems based on two-level hierarchical decentralised coordinated control structure. Int. J. Syst. Sci. **44**(2), 329–345 (2013)
3. S. Anand, B.G. Fernandes, J.M. Guerrero, Distributed control to ensure proportional load sharing and improve voltage regulation in low-voltage dc microgrids. IEEE Trans. Power Electron. **28**(4), 1900–1913 (2013)

Chapter 11
MAS-Based Distributed Cooperative Control for DC Microgrid Through Switching Topology Communication Network with Time-Varying Delays

A multiagent system based distributed cooperative control is proposed for a DC microgrid. First, the multiagent system based control scheme is built, where each distributed energy resource unit is associated with a first-level unit control agent to deal with primary control and a secondary-level distributed cooperative control agents to implement secondary control. For the secondary control purposes of global voltage regulation and proportional current sharing, both voltage and current regulators are installed in each secondary-level agent. Their consensus secondary control protocols are designed by using the state information exchanged among only adjacent secondary-level control agents based on a switching topology communication network with time-vary delays. Then these protocols are applied into the associated first-level agent to structure an improved outer-loop droop controller. According to the reference set points corrected by the improved droop controller, an inner-loop voltage/current robust controller is designed in place of the conventional PI control to regulate the voltage and current of distributed energy resource unit. Finally, the validity of the proposed control is testified by means of simulation results.

11.1 MAS-Based Control Structure

An islanded low-voltage DC MG consists of N DER units, where each DER unit supplies a public load connected nearby the point of common coupling (PCC) on its bus. To implement primary/secondary distributed cooperative control, a two-level MAS is proposed as shown in Fig. 11.1.

Each first-level unit control agent is in charge of local primary control for its DER unit. It is designed as hybrid agent composed of "*reactive layer*" and "*deliberative layer*," as shown in Fig. 11.2. The reactive layer defined as "recognition, perception, and action" has priority to respond quickly to emergencies of operation status. The

D. Yue et al., *Cooperative Optimal Control of Hybrid Energy Systems*,
https://doi.org/10.1007/978-981-33-6722-7_11

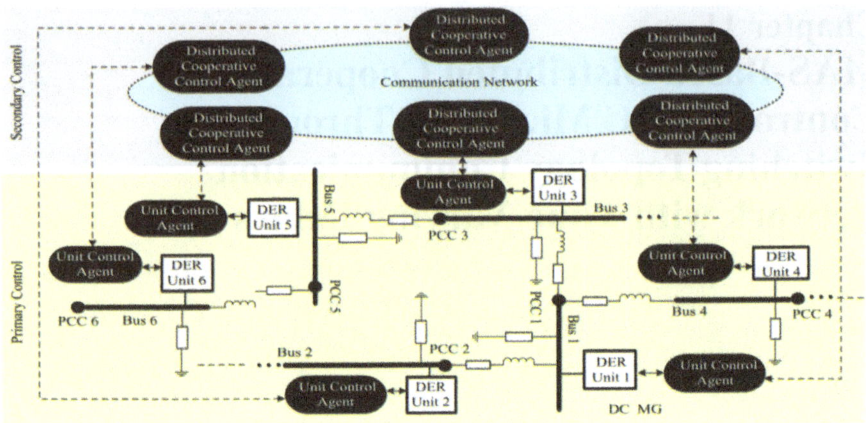

Fig. 11.1 MAS-based distributed cooperative control structure

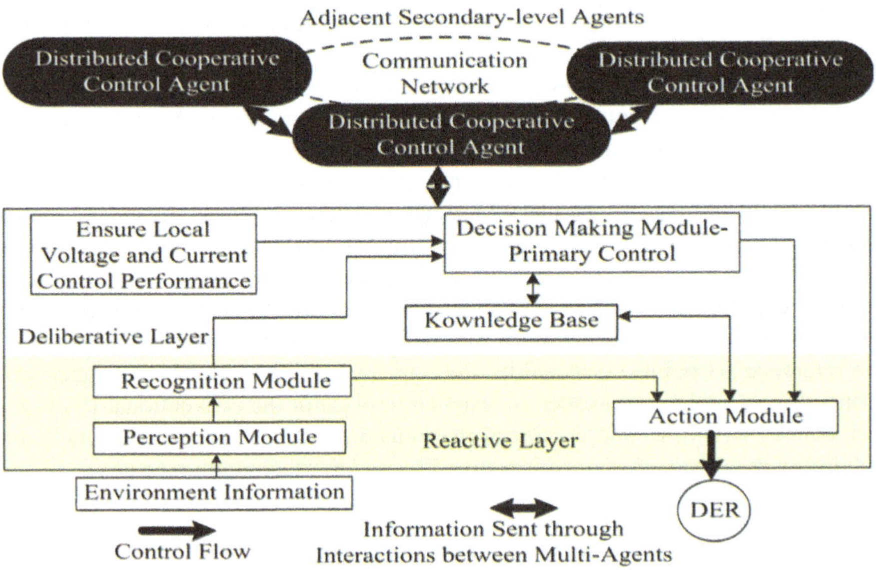

Fig. 11.2 Structure of the first-level unit control agent

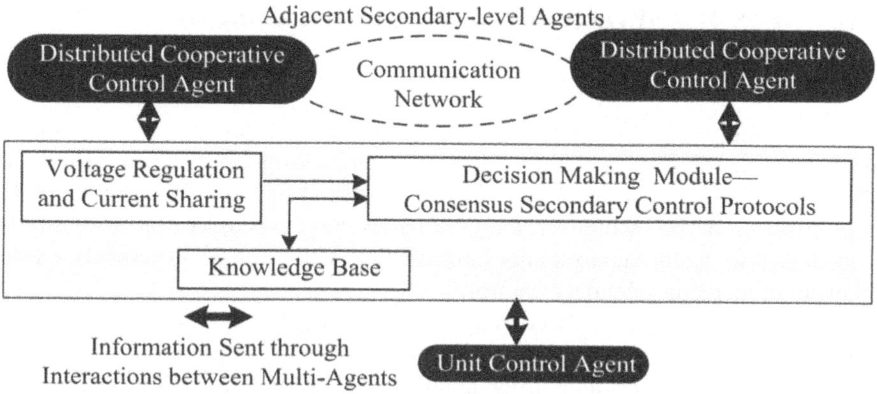

Fig. 11.3 Structure of the first-level unit control agent

deliberative layer defined as "belief, desire, and intent (BDI)" has high intelligence to control the dynamic behaviors of its DER unit. In detail, the primary control (i.e., intent) is determined in decision making module by using both outer-loop droop controller and inter-loop voltage/current controller based on knowledge and recognition information as well as consensus secondary control protocols (i.e., belief), and then is implemented by means of action module to regulate local voltage and current dynamic performance (i.e., *desire*).

Each secondary-level agent is designed as a BDI agent, as shown in Fig. 11.3. The agent is responsible for global voltage regulation and equal per-unit current sharing (i.e., *desire*) to implement secondary control. The both voltage and current regulators (i.e., *intent*) are installed and designed in decision making module based on knowledge information and the states exchanged between adjacent secondary level agents (i.e., *belief*). Accordingly, their consensus secondary control protocols are applied to the first-level primary control through interactions between two-level agents.

The interactive manner among agents is designed the following two types: (1) master–slave mode between different levels of agents, which is defined that the request of secondary-level agent must be responded by the asked first-level agent, that is, the secondary-level agent has priority over the first-level one and (2) nonmaster–slave mode between same level of agents, which means that same level of agents interact in an equal way.

11.2 MAS-Based Distributed Cooperative Control Strategies

Corresponding to Fig. 11.1, with respect to the jth DER unit($j \in 1, 2, \ldots N$), the MAS-based distributed cooperative control scheme is designed as shown in Fig. 11.4. The jth DER unit isassociated with the jth first-level unit control agent to implement primary control. Moreover, the jth first-level agent is associated with the jth secondary-level agent that exchanges information with its adjacent secondary-level agents to implement secondary control.

In detailed, the consensus secondary control protocols from the jth and its adjacent secondary-level agents are sent to the jth first-level agent, accordingly constituting an improved outer- loop droop controller to tune the inner-loop reference set point. According to the corrected reference set values, the inner-loop current/voltage controller dynamically regulates the current and voltage of the jth DER unit.

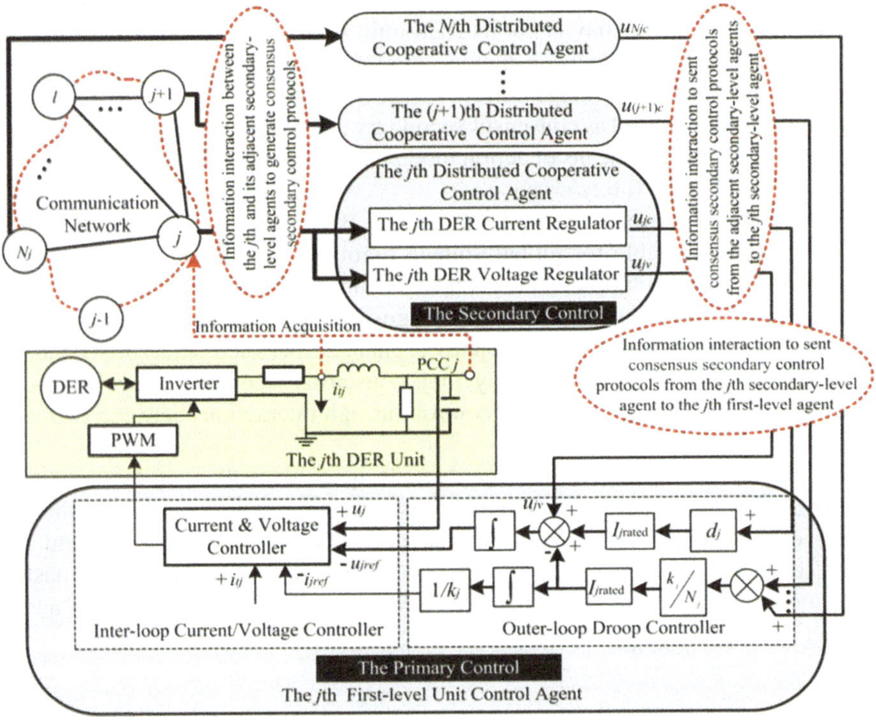

Fig. 11.4 MAS-based distributed cooperative control scheme of the jth DER unit

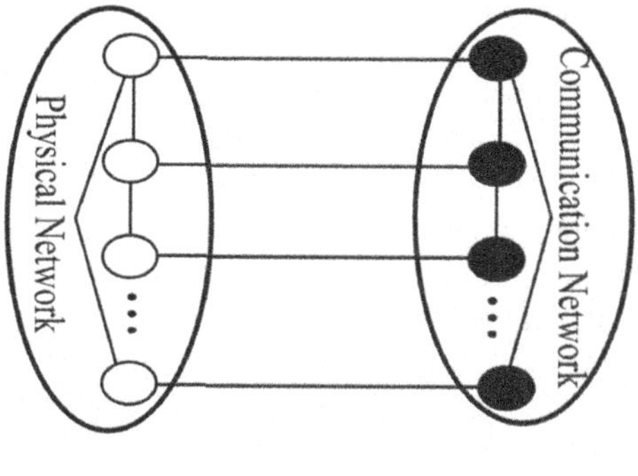

Fig. 11.5 Cyber-physical networks

11.2.1 Secondary Control

In this chapter, the consensus secondary control protocols are proposed based on cyber-physical "one-to-one correspondence" interdependent network model, as shown in Fig. 11.5, where each communication node is corresponding to a physical node of DER unit.

A brief introduction about algebraic graph theory is given as follows [1, 2]. An undirected graph is connected if any two distinct nodes can be connected via a path that follows the lines of the graph. Let $\psi = \{\boldsymbol{V}, \boldsymbol{E}, \boldsymbol{W}\}$ describes the undirected graph of communication network. $\boldsymbol{V} = \{1, \ldots, N\}$ denotes node set of communication network, $\boldsymbol{E} \subseteq \boldsymbol{V} \times \boldsymbol{V}$ represents the transmission lines. $\boldsymbol{W} = \left[w_{jk}\right] N \times N$ is a nonnegative weighted adjacency matrix. Any line of $\psi = \{\boldsymbol{V}, \boldsymbol{E}, \boldsymbol{W}\}$ is denoted as $e_{jk} = (j, k)$. Self-line (j, j) is not allowed, i.e., $e_{jk} = (j, k) \in \boldsymbol{E}$ if and only if $\omega_{jk=1}$. The degree of nodes is defined as $\deg(j) = \sum_{k=1}^{N} w_{jk}$, and then the Laplacian matrix is defined as $\boldsymbol{L}_\psi = \boldsymbol{D} - \boldsymbol{W}$, where $\boldsymbol{D} = \text{diag}\{\deg(1), \ldots, \deg(N)\}$. So every row sum of $\boldsymbol{L}_\psi$ is zero, i.e., $\boldsymbol{L}_\psi 1_N = 0$, where $\boldsymbol{1}_N = [1, 1, \ldots, 1]^T_{1\times N}$ is a right eigenvector corresponding to eigenvalue 0. If $\psi = \{\boldsymbol{V}, \boldsymbol{E}, \boldsymbol{W}\}$ is connected, all the eigenvalues of $\boldsymbol{L}_\psi$ are positive except one 0 eigenvalue. The leader adjacency matrix is defined as $\boldsymbol{F} = \left[f_j\right]_{N\times 1}$ with $f_j = 1$ if the jth node can receive information from the reference voltage, otherwise $f_j = 0$.

It is worth mentioning that the "plug and play" of any DER unit might change the topology of physical network, accordingly resulting in the change of topology of "one-to-one correspondence" interdependent communication network. For this reason, the consensus secondary control protocols are discussed based on a time-varying topology graph. The dependence of the graph upon time can be characterized by a switching signal $\sigma(t)$ of piecewise constant, and thus the time-varying topology graph is denoted as $\psi_\sigma(t)$. All the switching graphs only need to be jointly connected across each interval.

(1) Consensus Secondary Control for Global Voltage Regulation: For the purpose of global voltage regulation, the containment-based consensus control protocol is designed in the voltage regulator.

The dynamic of the jth bus voltage under consensus secondary control is described as

$$\dot{v}_j(t) = \eta_\Delta u_{jv}(t), \quad j \in \{1, 2, \ldots, N\} \tag{11.1}$$

where $v_j(t)$ is the jth bus voltage (per-unit value); $\eta_\Delta u_{jv}(t)$ is the consensus control protocol subject to actuator saturation. For a positive scalar Δ, $\eta_\Delta : \mathbf{R} \to \mathbf{R}$ is a scalar valued saturation defined as

$$\eta_\Delta u_{jv}(t) = \Gamma_j(t) u_{jv}(t) = \begin{cases} \operatorname{sign}\left[u_{jv}(t)\right]\Delta, & \left|u_{jv}(t)\right| \geq \Delta \\ u_{jv}(t), & \left|u_{jv}(t)\right| < \Delta \end{cases} \tag{11.2}$$

It is easy to deduce that $\Gamma_j(t)$ can be expressed as

$$\Gamma_j(t) = \begin{cases} \Delta / \left|u_{jv}(t)\right|, & \left|u_{jv}(t)\right| \geq \Delta \\ 1, & \left|u_{jv}(t)\right| < \Delta \end{cases} \tag{11.3}$$

Considering the communication delay in the MG, we denote the time-varying delay $0 \leq \tau(t) \leq \tau, 0 \leq t \leq \infty$, where τ is the upper bound. In this case, it is available to design the control protocol $u_{jv}(t)$ using relative states $u_{jv}(t - \tau(t))$. Through the switching topology communication network with time-varying delays, the jth secondary-level agent exchanges information with its adjacent secondary-level agents, constituting the consensus secondary control protocol as

$$\begin{aligned} u_{jv}(t) = &- K_{j,\sigma(t)} \\ &\times \left\{ \sum_{i \in \Omega_j, i \neq j} w_{ji,\sigma(t)} \left[v_i(t - \tau(t)) - v_j(t - \tau(t))\right] \right. \\ &\left. + f_{j,\sigma(t)} \left[v_{\text{ref}}(t) - v_j(t - \tau(t))\right] \right\} \end{aligned} \tag{11.4}$$

where $\Omega_j = \left\{j, j+1, \ldots, l, \ldots, N_j\right\}$ is the sequence number set of the jth and its adjacent secondary-level agents, and N_j is the sum total of the jth and its adjacent DER units; $v_{i \in \Omega_j}(t) = 0$, if $t < 0$; $K_{j,\sigma(t)}$; $K_{j,\sigma(t)}$ is voltage control gain; $v_{\text{ref}}(t)$ is the global voltage reference set point. In the ω_jth region (composed of the jth and its adjacent nodes) it is required that at least one node can receive information from the reference voltage.

The error between the jth bus voltage and global voltage reference set point is denoted as $\bar{v}_j(t) = v_j(t) - v_{\text{ref}}(t)$, then in the ω_jth region the global voltage regulation dynamics can be written as

$$\dot{\bar{\boldsymbol{v}}}_{\Omega_j}(t) = \boldsymbol{\Pi}_{\Omega_j \sigma(t)} \bar{\boldsymbol{v}}_{\Omega_j}(t - \tau(t)) \tag{11.5}$$

where $\bar{\mathbf{v}}_{\Omega_j}(t) = \left[\bar{v}_j(t), \bar{v}_{j+1}(t), \ldots, \bar{v}_{N_j}(t)\right]^{\mathrm{T}}$; $\boldsymbol{L}_{\Omega_j \sigma(t)}$ is the Laplacian matrix in the ω_jth region of switching communication network; $\Pi_{\Omega_j \sigma(t)} = K_{\Omega_j \sigma(t)} \Gamma_{\Omega_j}(t)$

$\left(\boldsymbol{L}_{\Omega_j\sigma(t)} + \boldsymbol{F}_{\Omega_j\sigma(t)}\right)$; $\mathbf{K}_{\Omega_j\sigma(t)} = \text{diag}\left[K_{j,\sigma(t)}, K_{j+1,\sigma(t)}, \ldots, K_{N_j,\sigma(t)}\right]$; $_{\Omega_j}(t) = \text{diag}\left[\Gamma_j(t), \Gamma_{j+1}(t), \ldots, \Gamma_{N_j}(t)\right]$; $\mathbf{F}_{\Omega_j\sigma(t)} = \text{diag}\left[f_{j,\sigma(t)}, f_{j+1,\sigma(t)}, \ldots, f_{N_j,\sigma(t)}\right]$.

Theorem 11.1 *With respect to the dynamics (11.5), if there exist matrix $Y \in R^{N\times N}$ and symmetrical positive definite matrices $\boldsymbol{P}, \boldsymbol{Q}, \boldsymbol{X}, \boldsymbol{Z} \in R^{N\times N}$ satisfying*

$$\begin{bmatrix} \boldsymbol{X} & \boldsymbol{Y} \\ \boldsymbol{Y}^T & \boldsymbol{Z} \end{bmatrix} \geq 0 \quad \text{and} \quad \begin{bmatrix} \Phi_{11} & \Phi_{12} \\ * & \Phi_{22} \end{bmatrix} < 0 \tag{11.6}$$

where $\Phi_{11} = 2P\Pi_{\Omega_j\sigma(t)} + \tau X + Y + Y^T + Q$, $\Phi_{12} = \Phi_{21} = P\Pi_{\Omega_j\sigma(t)} - Y$, $\Phi_{22} = -Q + \tau\Pi^T_{\Omega_j\sigma(t)}\Pi_{\Omega_j\sigma(t)}$, *then global voltage consensus in the sense of* $\lim_{t\to\infty} v_k(t) \equiv 0$ *can be achieved, where* $k \in \Omega_j = \{j, j+1, \ldots, N_j\}$.

Considering limitation of chapter length, the proof regarding Theorem 11.1 is not presented here. If necessary, it can be presented in a separate file.

Remark 11.1 According to Theorem 11.1, when the switching topology is fixed, corresponding to certain upper bound of the time-varying delay, the voltage control gain $\boldsymbol{K}_{\Omega_j\sigma(t)}$ can be achieved.

(2) Consensus Secondary Control for Current Sharing: The consensus secondary control for current sharing is designed based on the same switching topology communication network as the global voltage regulation.

The dynamic of current of the jth DER unit under consensus secondary control is described as

$$\dot{i}_{jpu}(t) = \eta_\Delta u_{jc}(t), \quad j \in \{1, 2, \ldots, N\} \tag{11.7}$$

where $i_{jpu}(t)$ is the per-unit current of the jth DER unit; $\eta_\Delta u_{jc}(t)$ is consensus control protocol subject to actuator saturation, and defined in a similar way of (11.2). Considering the communication delay, the consensus secondary control protocol of the jth current regulator is designed as

$$\begin{aligned} u_{jc}(t) = -K^c_{j,\sigma(t)} \textstyle\sum_{i\in\Omega_j, i\neq j} w_{ji,\sigma(t)} \big[& i_{ipu}(t-\tau(t)) \\ & -i_{jpu}(t-\tau(t))\big] \end{aligned} \tag{11.8}$$

where $K^c_{j,\sigma(t)}$ is current sharing control gain.

Remark 11.2 Theorem 11.1 is still suitable for current sharing consensus control, but in the linear matrix inequalities (11.6), $\Pi_{\Omega_j\sigma(t)} = \boldsymbol{K}^c_{\Omega_j\sigma(t)}\boldsymbol{\Gamma}_{\Omega_j}(t)\boldsymbol{L}_{\Omega_j\sigma(t)}$ where $\mathbf{K}^c_{\sigma(t)} = \text{diag}\left[K^c_{j,\sigma(t)}, K^c_{j+1,\sigma(t)}, \ldots, K^c_{N_j,\sigma(t)}\right]$. According to Theorem 11.1, corresponding to the same upper bound of the time-varying delay as the global voltage regulation, the current sharing control gain $K^c_{\sigma(t)}$ can be also achieved.

11.2.2 Primary Control

The first-level primary control consists of the outer-loop improved droop controller and inner-loop voltage/current robust controller. The improved droop controller provides the corrected reference values for the voltage/current controller.

(1) Improved Outer-Loop Droop Controller: The unequal nominal voltages and load distribution among **DER** units result in that the currents deviate significantly from their ideal values, and then equal per-unit current sharing cannot be guaranteed. Despite these problems can be solved by increasing the droop gains higher than the line resistance, the voltage varies significantly from no-load to full-load condition. Therefore, an improved droop controller is required to overcome the limitation. In this chapter, it is designed as follows:

$$v_j = u_{j\,\text{ref}} - d_j I_{j\,\text{rated}\,i,pu} + k_j I_{j\,\text{rated}} \sum_{k\in\Omega_j} i_{kpu}/N_j \tag{11.9}$$

where d_j is the droop gain of the jth DER unit; u_{jref} is the nominal voltage, i.e., the reference voltage of inner-loop controller; k_j is the gain of shift controller; and I_{jrated} is the rated current of the jth DER unit.

Compared with conventional droop characteristics, the improved droop controller (11.9) has an additional voltage shift term as $\Delta v_j^0 = k_j I_{jrated} \sum_{k\in\Omega_j} i_{kpu}/N_j$, which also called shift controller. The shift controller is related to the per-unit currents of the jth and its adjacent DER units, as well as the rated current of the jth DER unit.

The gain k_j of shift controller is determined based on the following criteria.

(1) The shift controller Δv_j^0 should be as close as possible to the product of droop gain and DER current, i.e., $d_j I_{j\,\text{rated}}\, i_{jpu}$. This is to ensure the voltage drop to be compensated by the appropriate shift, and making operating voltage close to the nominal voltage.
(2) Droop characteristics of all DERs in the ω_jth region should be shifted by the same amount to ensure that the shift controller does not affect the load sharing, that is,

$$k_j I_{j\,\text{rated}} \sum_{k\in\Omega_j} i_{kpu}/N_j = \cdots = k_l I_{l\,\text{rated}} \sum_{k\in\Omega_j} i_{kpu}/N_j \tag{11.10}$$

(3) The shift gain should be smaller than the droop gain for stable operation, that is,

$$k_j \le d_j \tag{11.11}$$

Based on the above criteria, the shift controller can appropriately lift the operation voltage when faced with an increased load, making the voltage close to its nominal voltage. In other word, although a large droop gain is used to ensure equal load sharing, the operation voltage can be close to its nominal value. According to the improved droop controller, the reference voltage set point of the inner-loop voltage/current controller is

$$u_{j\mathrm{ref}} = \int_0^t \left(\dot{v}_j + d_j I_{j\mathrm{rated}} \dot{i}_{jpu} - k_j I_{j\mathrm{rated}} \sum_{k \in \Omega_j} i_{kpu} / N_j \right) \tag{11.12}$$

Substitute (11.1) into (11.12), the reference voltage set point is rewritten in the following form:

$$\begin{aligned} u_{\mathrm{jref}} = & \int_0^t \left(\eta_\Delta u_{jv} + d_j I_{j\ \mathrm{rated}} \eta_\Delta u_{jc} \right. \\ & \left. - k_j I_{j\ \mathrm{rated}} \sum_{k \in \Omega_j} \eta_\Delta u_{kc} / N_j \right) \end{aligned} \tag{11.13}$$

In addition, the reference current set point of the inner-loop voltage/current controller is given as follows:

$$i_{j\ \mathrm{ref}} = I_{j\ \mathrm{rated}} \sum_{k \in \Omega_j} i_{kpu} / N_j = \int_0^t I_{j\ \mathrm{rated}} \sum_{k \in \Omega_j} \eta_\Delta u_{kc} / N_j \tag{11.14}$$

From (11.13) and (11.14), it implies that the unit control agent needs to collect the consensus secondary control protocols from the jth and its adjacent secondary-level agents to adjust the reference voltage/current set points, and accordingly the innerloop controller dynamically regulate the voltage/current of its DER unit, which is so-called distributed cooperative control.

(2) Inner-Loop Voltage/Current Controller: The jth DER unit is connected with $N_j - 1$ DER units through DC lines with different impedances specified by parameters $R_{jk} \geq 0$ and $L_{jk} > 0$, $k \in \Omega_j$, $k \neq j$, as shown in Fig. 11.6. In the DER unit, a Buck converter is presented to supply a local DC load connected to the PCC through an LC filter. In accordance with the circuit structure of the jth DER unit, a set of dynamic equations is obtained as follows:

$$\text{DER } j \begin{cases} \frac{du_j}{dt} = -\frac{1}{C_j R_{Lj}} \Delta u_j + \frac{1}{C_j} \Delta i_{tj} + \frac{1}{C_j} \Delta i_j \\ \frac{di_{tj}}{dt} = -\frac{1}{L_{tj}} \Delta u_j - \frac{R_{tj}}{L_{tj}} \Delta i_{tj} + \frac{1}{L_{tj}} u_{tj} \end{cases} \tag{11.15}$$

where all parameters and variables are described as shown in Fig. 11.6; $\Delta u_j = u_j - u_{j\ \mathrm{ref}}$ and $\Delta i_{tj} = i_{tj} - i_{j\ \mathrm{ref}}$. As [3], due to $i_{jk}\left(k \in \Omega_j, k \neq j\right)$ is the DC current between the jth and the kth buses, $di_{jk}/dt = -di_{kj}/dt = 0$, and then the following equations are obtained:

$$i_{jk} = -i_{kj} = \left(u_k - u_j\right) / R_{jk} \tag{11.16}$$

$$\Delta i_j = \sum_{k \notin \Omega_j, k \neq j} i_{jk} = \sum_{k \in \Omega_j, k \neq j} \left(u_k - u_j\right) / R_{jk} \tag{11.17}$$

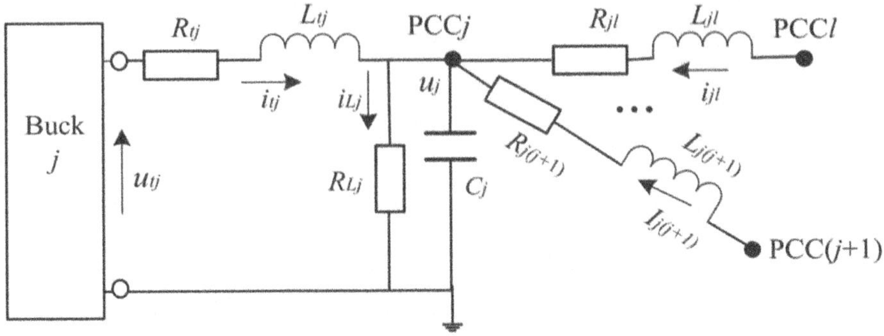

Fig. 11.6 Circuit structure of the jth DER unit

Remark 11.3 The global voltage regulation is guaranteed by the secondary-level agents, which results in Δi_j very small. In this case, Δi_j in (11.12) can be treated as a current external disturbance, so that the jth DER unit can be described as follows:

$$\text{DER } j : \dot{\mathbf{x}}_j(t) = \mathbf{A}_j \mathbf{x}_j(t) + \mathbf{B}_j u_j^{vc}(t) + \mathbf{D}_j \omega_j(t) \tag{11.18}$$

where $\boldsymbol{x}_j(t) = \left[\Delta u_j(t), \Delta i_{tj}(t)\right]^T$ is the state vector of the jth DER; $u_j^{vc}(t) = u_{tj}(t)$ is the input variable; $\omega_j(t) = \Delta i_j$ is the disturbance vector; and the coefficient matrices are given as follows:

$$\boldsymbol{A}_j = \begin{bmatrix} -1/C_j R_{Lj} & 1/C_j \\ -1/L_{tj} & -R_{tj}/L_{tj} \end{bmatrix}$$
$$\boldsymbol{B}_j = \begin{bmatrix} 0 \\ 1/L_{tj} \end{bmatrix}$$
$$\boldsymbol{D}_j = \begin{bmatrix} 1/C_j \\ 0 \end{bmatrix}$$

The current/voltage controller is designed as

$$u_j^{vc}(t) = \mathbf{G}_j \mathbf{x}_j(t) \tag{11.19}$$

where $G_j \in R^{1\times 2}$ is the control gain of the inner-loop controller.

The controlled system in (11.18) under the controller in (11.19) is written in the following form:

$$\text{DER } j : \dot{\boldsymbol{x}}_j(t) = \hat{\boldsymbol{A}}_j \boldsymbol{x}_j(t) + \boldsymbol{D}_j \omega_j(t) \tag{11.20}$$

where $\hat{\mathbf{A}}_j = \mathbf{A}_j + \mathbf{B}_j \mathbf{G}_j$.

For the robust control purpose, considering the initial condition, the H_∞ control performance is given as follows:

$$\int_0^{t_f} \mathbf{x}_j^T(t)\mathbf{Q}_j\mathbf{x}_j(t)dt \leq \rho_j^2 \int_0^{t_f} \omega_j^T(t)\omega_j(t)dt + \mathbf{x}_j^T(0)\mathbf{P}_j\mathbf{x}_j(0) \tag{11.21}$$

where t_f denotes terminal time of control, and the weighting matrix $\boldsymbol{Q}_j = \boldsymbol{Q}_j^T > 0$, $\boldsymbol{P}_j = \boldsymbol{P}_j^T > 0$.

The first-level unit control agent can determine the control gain of inner-loop voltage/current controller based on the following theorem.

Theorem 11.2 *The jth DER unit controlled system (11.20) is asymptotically stable with the guaranteed H_∞ control performance (11.21) for $\forall\omega_j(t)$, if there exists $\boldsymbol{P}_j = \boldsymbol{P}_j^T > 0$ satisfying the following linear matrix inequality (LMI):*

$$\begin{bmatrix} \hat{\mathbf{A}}_j^T\mathbf{P}_j + \mathbf{P}_j\hat{\mathbf{A}}_j + \mathbf{Q}_j & \mathbf{P}_j\mathbf{D}_j \\ \mathbf{D}_j^T\mathbf{P}_j & -\rho_j^2\mathbf{I} \end{bmatrix} \leq 0 \tag{11.22}$$

Consider limitation of chapter length, the proof is not presented. If necessary, it can be also given in a separate file.

Remark 11.4 The inner-loop voltage/current controller design can be transformed into the following LMI convex optimization problem:

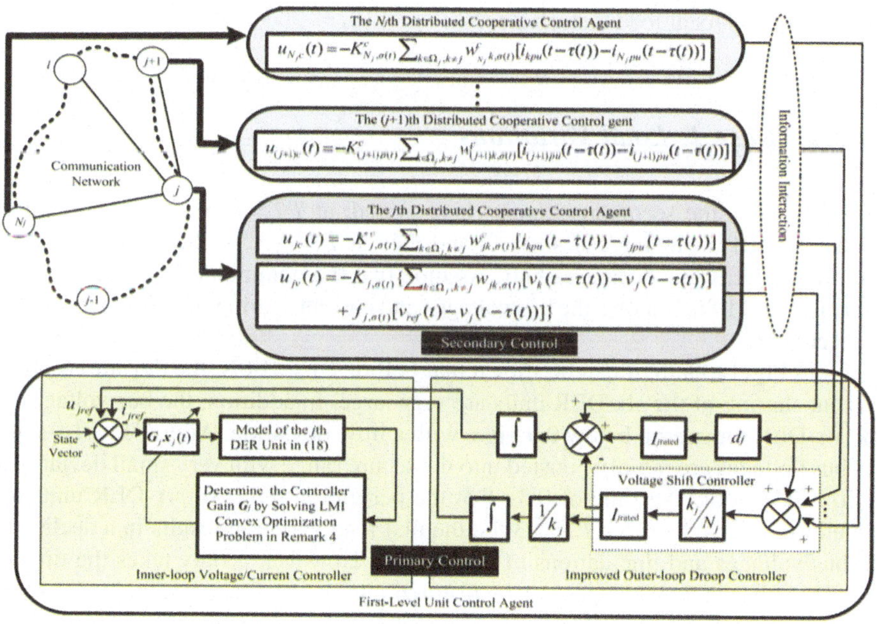

Fig. 11.7 Implementation strategies of the distributed cooperative control for the jth DER unit based on two-level MAS

$$\min_{P_j}, G_j\rho_j^2$$
$$Subject\ to\ \mathrm{P}_j = \mathrm{P}_j^T > 0\ and\ (22) \tag{11.23}$$

By solving the LMI convex optimization problem above, the inner-loop controller gain can be obtained.

11.2.3 Implementation Strategies of Distributed Cooperative Control Based on the MAS

By using the designed control strategies, the distributed cooperative control for the jth DER unit is implemented based on the two-level MAS, as shown in Fig. 11.7 which gives the detailed design of primary and secondary controllers.

11.3 Simulation Studies

To evaluate the performance regarding the MAS-based distributed cooperative control, the simulation results are performed in the following three kinds of cases: (1) public load variations; (2) "plug and play" of a DER; (3) different communication delays; and (4) local load disturbance.

11.3.1 Case 1: Load Variations

At $T1 = 1.0$ s, the secondary control is activated; at $T2 = 3.5$ s, the load demand on the first DER bus increases twice; and then at $T3 = 5.5$ s, the load demand on the second DER bus decreases half. Assume that the communication delay $\tau(t) \leq 200$ ms, Fig. 11.8a, b shows the bus voltages and current sharing of the six DER units, respectively.

From Fig. 11.8a, b, it can be shown that before $T1 = 1.0$ s, the deviations about current sharing of the six DER units are very large. In addition, the bus voltages of the six DER units drop below 0.95 p.u. with a little deviation. After $T1 = 1.0$ s, the six bus voltages are rapidly boosted into the secure range with very small deviations. Also it takes only the time of 0.3 s that the per-unit currents of six DER units are equally shared well. After $T2 = 3.5$ s, the first load variation results in a decline of all bus voltages and fluctuations of all currents. However, it only takes the time of 0.3 s that, the six bus voltages can be restored into the secure range with very small deviations. And the per-unit currents of six DER units are equally shared well again. Similarly, after $T3 = 5.5$ s, the twice load variation results in transient disturbances of all the voltages and currents. But it also takes the time of about 0.3 s that the bus voltages and per-unit.

To evaluate the control performance by means of comparative results, by using the method in [4], the bus voltages and current sharing of the six DER units are, respectively, shown in Fig. 11.8c, d, where the load demand on the first DER bus increases twice at $T1 = 2.0\,\mathrm{s}$ and then the load demand on the second DER bus decreases half at $T2 = 5.0\,\mathrm{s}$. From Fig. 11.8c, d, it can be shown that the deviations regarding bus voltages and current sharing are larger.

All the results imply that the proposed MAS-based distributed cooperative control presents better control performance regarding global voltage regulation and equal per-unit current sharing when faced with load variations.

11.3.2 Case 2: "Plug and Play" of a der Unit

Similarly, at $T1 = 1.0\,\mathrm{s}$ the secondary control is activated; at $T2 = 3.5\,\mathrm{s}$ the third DER unit quits running, and at $T3 = 5.5\,\mathrm{s}$ it accesses again. By using the proposed control, the bus voltages and current sharing of the six DER units are shown in Fig. 11.9a, b, respectively. In addition, Fig. 11.9c, d shows the comparative results by using the method proposed in [4], where the third DER unit quits running at $T1 = 2.0\,\mathrm{s}$ and accesses again at $T3 = 5.0\,\mathrm{s}$. In Fig. 11.9a, b, it can be shown that the activated secondary control can ensure the six bus voltages to boost rapidly into the secure range with very small deviations. Also the per-unit current sharing of six DER units is equalized well. After $T2 = 3.5\,\mathrm{s}$, quit running of the third DER unit results in a decline of other five bus voltages, as well as transient fluctuations of five DER currents. Afterward, it takes the time of about 0.3 s that the five bus voltages are rapidly restored. Also corresponding five DER currents are equally shared well. After $T3 = 5.5\,\mathrm{s}$, the third DER unit accesses again, after smaller transient fluctuations, all bus voltages and current sharing are rapidly restored again. The comparison results shown in Fig. 11.9c, d imply that the deviations of bus voltages and per-unit current sharing are larger when face with "plug and play" of the third DER unit.

11.3.3 Case 3: Different Communication Delays

Assume that $\tau(t) \leq 2.0\,\mathrm{s}$ with the same load variations as Case 1, by using the proposed control, the bus voltages and current sharing of the six DER units are shown in Fig. 11.10a, b, respectively.

Compared with the simulation results shown in Fig. 11.7a, b with $\tau = 200\,\mathrm{s}$, the six bus voltages and current sharing in Fig. 11.10a, b have almost no obvious change. The above simulation results imply that the MAS-based distributed cooperative control has strong adaptability to the different time-varying communication delays.

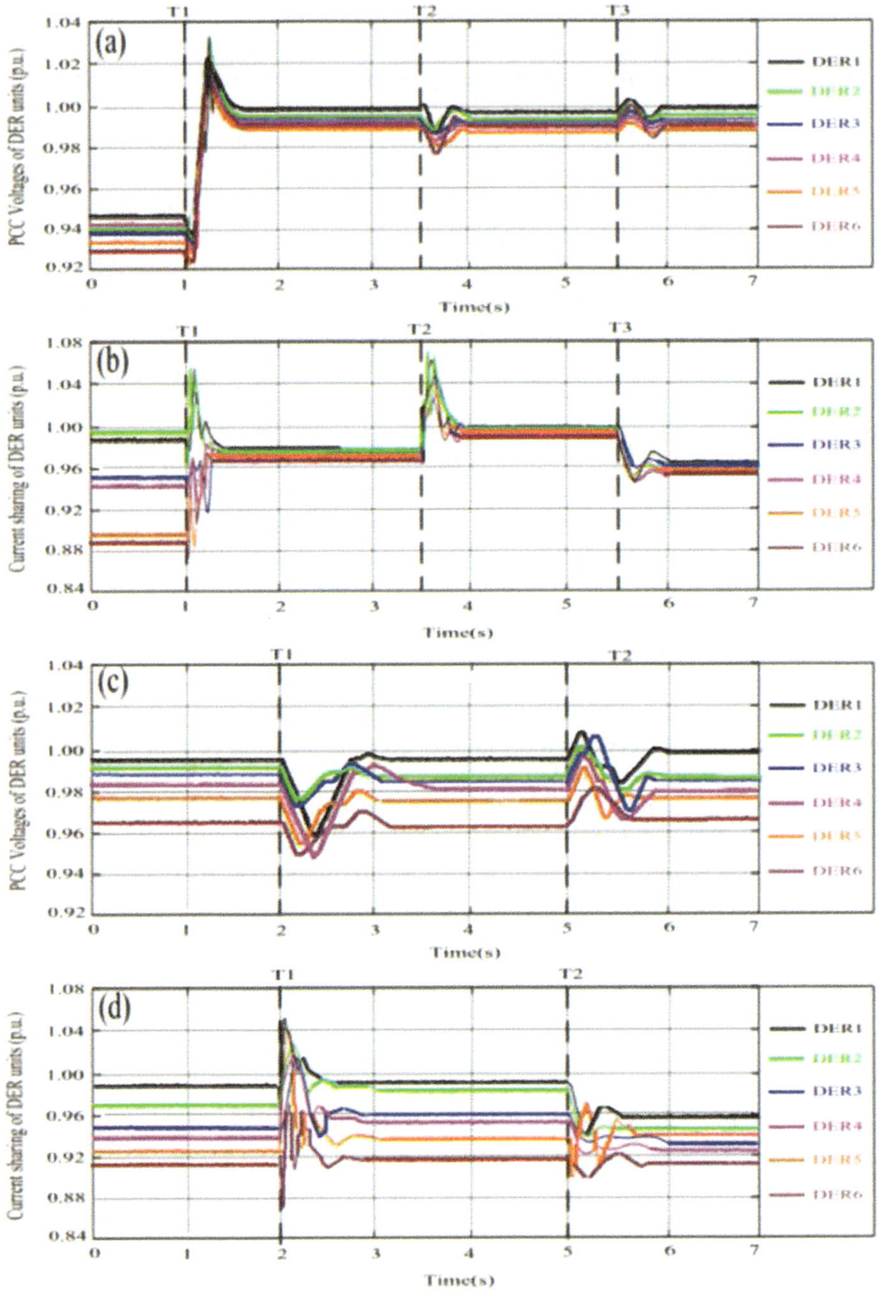

Fig. 11.8 Control performance in Case 1. **a** Bus voltages under the proposed control, **b** current sharing under the proposed control, **c** bus voltages by using the method in [4], and **d** current sharing by using the method in [4]

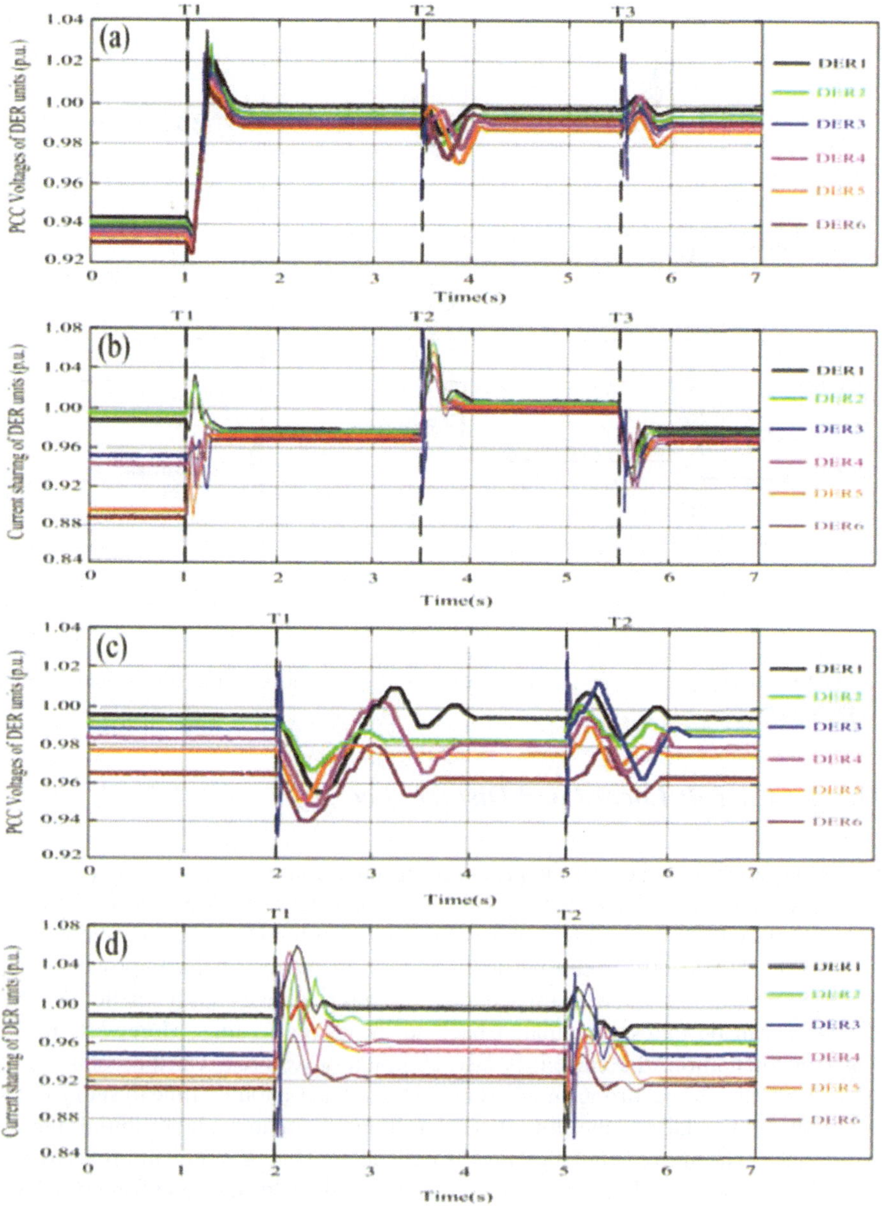

Fig. 11.9 Control performance in Case 2. **a** Bus voltages by using the proposed control, **b** current sharing by using the proposed control, **c** bus voltages by using the method in [4], and **d** current sharing by using the method in [4]

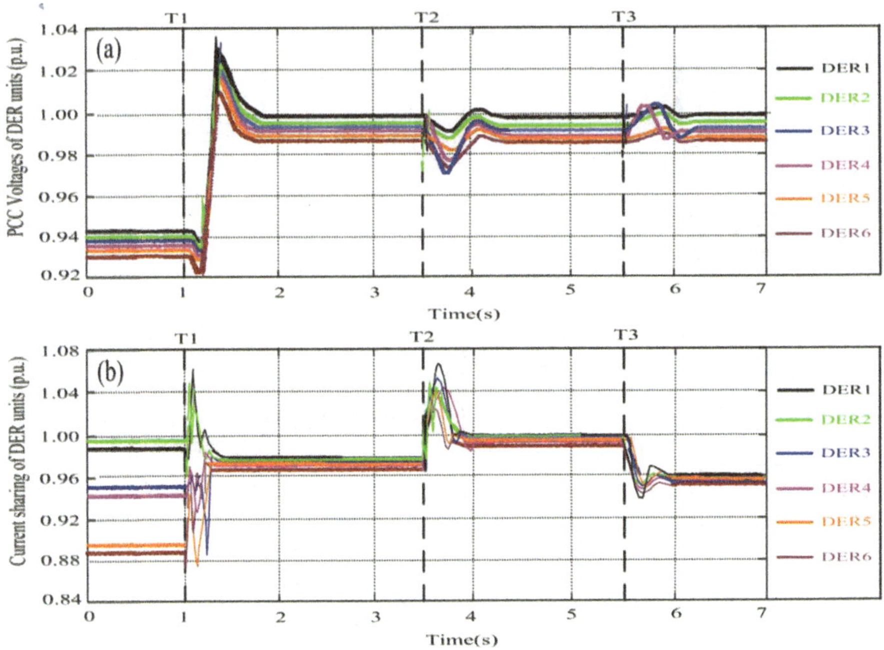

Fig. 11.10 Control performance in Case 3. **a** Six DER bus voltages and **b** current sharing of the six DER units

11.3.4 Case 4: Local Load Disturbance

To verify the effectiveness of the proposed primary control, at $t = 2.0$ s, a transient local load disturbance occurs in the DER1 unit. After 0.3 s, the disturbance is cleared. Figure 11.11a, b shows the bus voltage and current of the DER1, respectively.

From Fig. 11.11a, b, it can be deduced that by means of the improved droop and $H\infty$ robust controllers in this chapter, the bus1 voltages has a smaller decline, afterward can be quickly regulated back to the desired value. Moreover, after a transient fluctuation, the current is also completely restored. On the contrary, by using the conventional droop and PI controllers, it takes a longer time to stabilize the bus1 voltage. In addition, the current also has larger fluctuations in the initial stage of disturbance. The above simulation results indicate that the proposed improved droop and $H\infty$ robust controllers can ensure better control performance in front of a local disturbance.

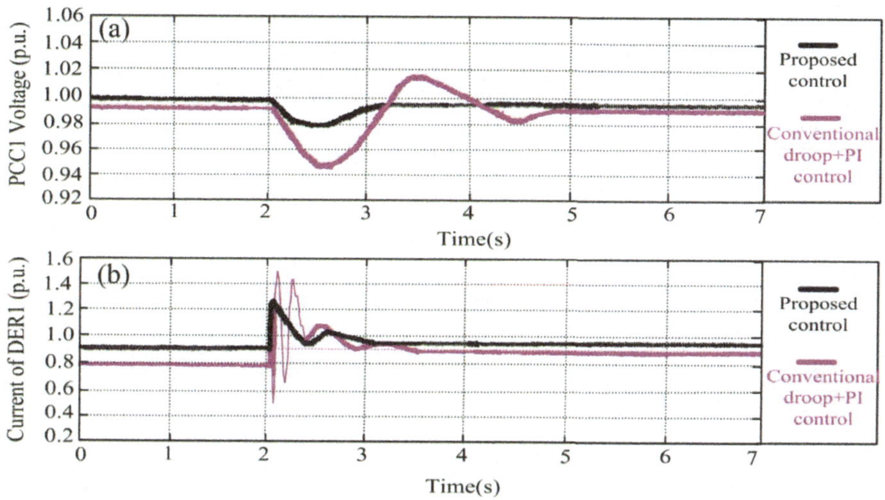

Fig. 11.11 Control performance in Case 4. **a** Bus1 voltage and **b** current of DER1

11.4 Conclusion

This chapter develops a distributed cooperative control based on the two-level MAS, so that the DC MG ensures fine global voltage regulation and equal per-unit current sharing. The original works are as follows.

1. Contributing the MAS-based distributed cooperative control scheme.
2. Developing the voltage/current consensus secondary control protocols in the secondary-level agents, taking account into actuator saturation, switching topologies, and time-varying communication delays.
3. Constituting an improving droop controller and a robust local controller in the first-level unit control agent. The better control performance has been demonstrated by means of the simulation results. The MAS-based distributed cooperative control scheme can be applied into any DC MG by extending the control function of the agents and creating additional agents in the future.

References

1. Z. Wang, J. Xu, H. Zhang, Consensusability of multi-agent systems with time-varying communication delay. Syst. Control Lett. **65**, 37–42 (2014)
2. V. Nasirian, S. Moayedi, A. Davoudi, F.L. Lewis, Distributed cooperative control of dc microgrids. IEEE Trans. Power Electron. **30**(4), 2288–2303 (2015)

3. S. Riverso, F. Sarzo, G. Ferraritrecate, Plug-and-play voltage and frequency control of islanded microgrids with meshed topology. IEEE Trans. Smart Grid **6**(3), 1176–1184 (2015)
4. H. Liang, B.J. Choi, W. Zhuang, X. Shen, A.S.A. Awad, A. Abdr, Multiagent coordination in microgrids via wireless networks. IEEE Wirel. Commun. **19**(3), 14–22 (2012)

Chapter 12
Multiagent System-Based Integrated Design of Security Control and Economic Dispatch for Interconnected Microgrid Systems

Hybrid and intermittent characteristics of the distributed energy resources (DERs) bring great challenges to the security control and economic dispatch (ED) of the microgrids. To bypass these hurdles, this chapter proposes a multiagent system-based integrated design of security control and ED to guarantee the effective and economical operation of the interconnected microgrids. First, a hierarchical control scheme is constructed by two-level unit agents, in which the switching control and dynamic regulation are fully implemented with the corresponding hybrid behaviors based on the differential hybrid Petri-net (DHPN) model. Based on the DHPN model, a novel dynamic ED integrated with security control is proposed to overcome the issues that cannot be solved in conventional models. Furthermore, to reduce the computational complexity and unified the mathematical model of the DERs, the inverter-based power control strategy is converted to a predictive control model which can be decomposed into several subsystems. In the optimization process, all the subsystems are implemented in a fully distributed, communication free and rolling optimization manner based on the distributed model predictive control (DMPC). The validity of the proposed design is demonstrated according to the simulation results in case studies.

12.1 MAS-Based Hierarchical Control Scheme

In sake of capturing the hybrid characteristics of the DER and load units in the IMS, a hierarchical control scheme constructed by the two-level unit agents is proposed in this section. A typical paradigm of the IMS is given in Fig. 12.1, which consists of two island microgrids:

(1) microgrid 1 compromises storage units, photovoltaic (PV) units, combined heat and power (CHP) units and a group of load units;

D. Yue et al., *Cooperative Optimal Control of Hybrid Energy Systems*,
https://doi.org/10.1007/978-981-33-6722-7_12

(2) microgrid 2 compromises PV units, storage units, as well as load units.

As shown in Fig. 12.1, the DER and load unit agents comprise the following features:

(1) In the lower level, each unit agent is responsible for local switching control (LSC) and distributed dynamic control;
(2) In the upper level, the hierarchical control center agent (HCCA) is in charge of global optimization and coordinating the switching control of all lower-level agents;
(3) All the lower-level unit agents are implemented in a communication free manner among peers, which only exchange information with the HCCA for saving the investment.

Remark 12.1 In order to achieve a better performance in an island operation environment, a particular unit is usually running at f-V droop control strategy to regulate the reference frequency/voltage of the IMS. For example, the storage unit in microgrid 1. Furthermore, the other units like the PV units and the storage units in microgrid 2 running at PQ control strategies are used to adjust the power output to maintain a generation supply and demand balance. Besides, the communication free manner among peers means that the information exchange is needed between the heterogeneous lower-level agents so that the investment of additional communication links between the homogeneous ones can be avoided. The details are given in the following sections.

12.1.1 Lower Level Unit Agent

The lower level unit agent is designed as a multi time-scale hybrid agent, which compromises *perception*, *deliberate* and *reaction* features as shown in Fig. 12.2. The *perception* feature is composed of "*perception* and *recognition*" modules that are responsible for delivering acceptable information of the operation states of the DERs, or directly respond quickly to the emergencies cooperating with the *action* module. The *deliberate* feature consists of "*desire*, *intent* and *knowledge base*" modules. The *knowledge base* drives from experience and professional knowledge is used to help the *intent* module to make decisions. The *desire* module capturing multi objectives can ensure the secure and economical performance of the IMS. The *intent* module is of making decisions including the LSC and distributed dynamic control based on *desire* and *knowledge base* module. The *reaction* feature constructed by the *action* module is to execute the control signals. Note that the time scale of the LSC is in the level of millisecond and the distributed optimization is in the level of minute. Besides, each lower-level unit agent connects to the upper-level HCCA is in parallel, which implies that all the unit agents have a similar structure and interactions through the coordination of the HCCA.

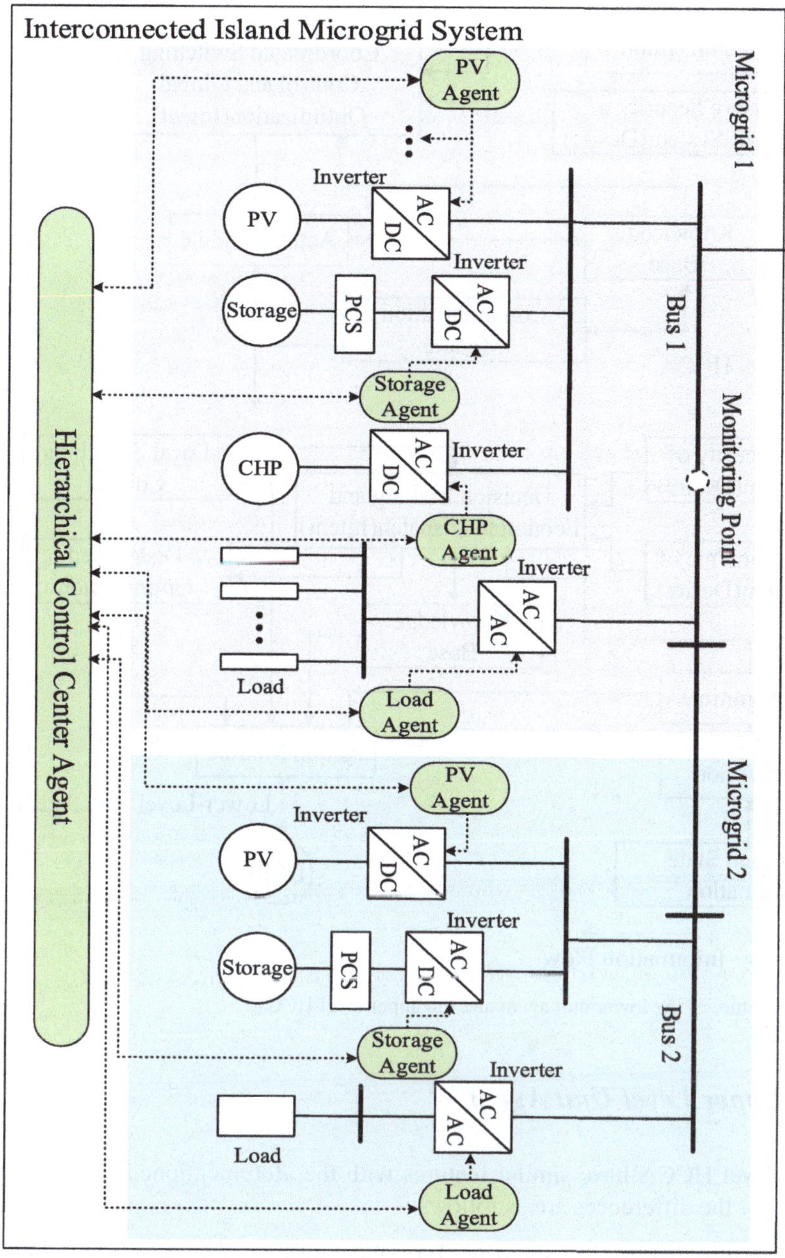

Fig. 12.1 MAS-based hierarchical control scheme of an IMS

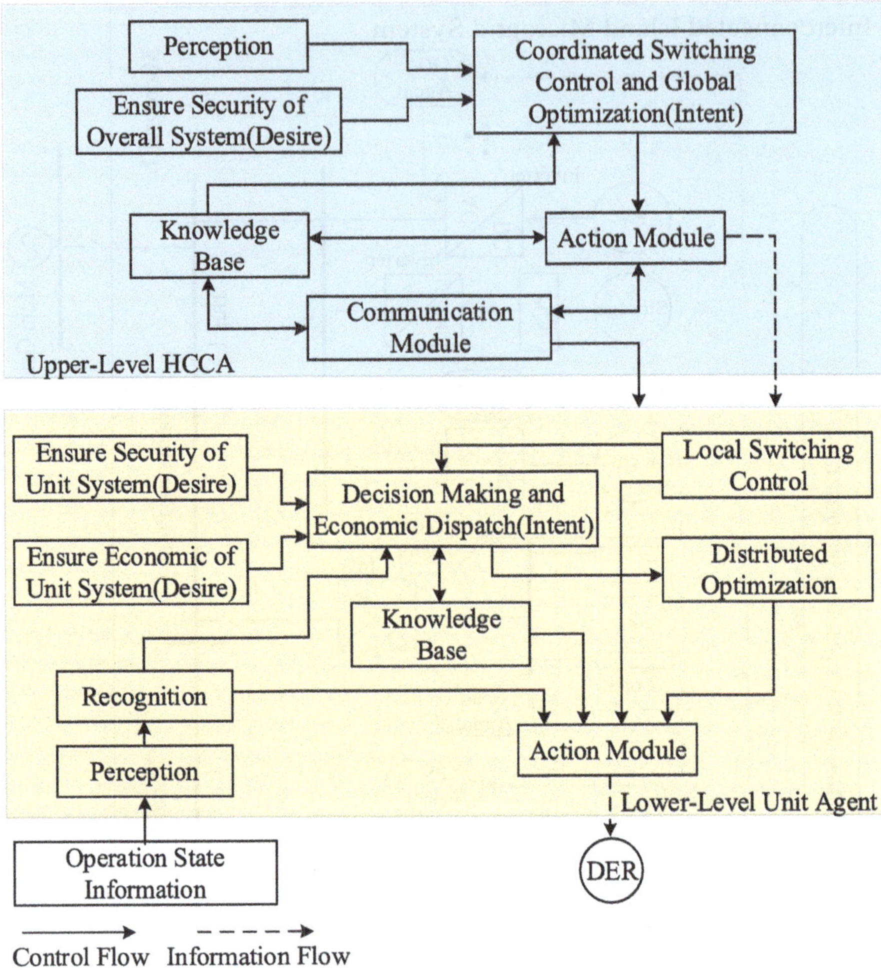

Fig. 12.2 Structure of the lower unit agent and the upper level HCCA

12.1.2 Upper Level Unit Agent

The upper-level HCCA have similar features with the aforementioned lower-level unit agent, but the differences are as follows:

(1) the *perception* feature is responsible for collecting the information come from all the unit agents in the lower level;
(2) the *deliberate* feature is determining the security and economy of the whole system and achieve coordinated switching control and global optimization based on the *desire* and *knowledge base* module.

In the MAS-based hierarchical control structure, the whole system is designed as an integration of ED and security control, which is also a hybrid system that can realize local and coordinated switching control, as well as global and distributed optimization in multi time-scale.

Remark 12.2 The interactions among agents can be implemented by the FIPA-ACL in BDI4JADE [1]. If a new agent joined the control scheme, the communication between the agent and the HCCA should be built so that the real-time state of the former can be observed by the later. Therefore, the security control and economic dispatch can be implemented. If an agent was about to exit the control scheme, the HCCA should receive this "message" and update all the control and optimization procedures relate to the exited agent. In this manner, the plug and play of the MAS can be implemented.

12.2 Insecurity-Events-Triggered Switching Controls

In this section, the DHPN is used to build the model of the hybrid switching control in the IMS, which has been proved to be effective for the hybrid system.

12.2.1 DHPN Model of the IMS

As shown in Fig. 12.3, the DHPN model is constructed by a six-tuple set $\mathcal{D} = (\mathcal{P}, \mathcal{T}_t, \mathcal{A}, \mathcal{M}, \mathcal{F}, \mathcal{T}_\tau)$ so that the event-triggered switching behaviors derived from the insecurity event judgement can be captured by means of the security assessment index. The entrics comprised in $\mathcal{D}$ are defined as:

$$\mathcal{P} = \mathcal{P}_0 \cup \mathcal{P}_1 = \left\{P_1, \ldots, P_{11}, P_{1f}, \ldots, P_{4f}\right\} \tag{12.1}$$

$$\mathcal{T}_t = \mathcal{T}_0 \cup \mathcal{T}_1 = \left\{T_1, \ldots, T_{15}, T_{1f}, \ldots, T_{6f}\right\} \tag{12.2}$$

$$\mathcal{A} = (\mathcal{P}_0 \times \mathcal{T}_0) \cup (\mathcal{T}_0 \times \mathcal{P}_0) \cup (\mathcal{P}_1 \times \mathcal{T}_1) \cup (\mathcal{T}_1 \times \mathcal{P}_1) \tag{12.3}$$

$$\mathcal{M} = \{M_{10}, M_{20}, \ldots, M_{40}\} \tag{12.4}$$

$$\mathcal{F} = \mathcal{F}_+ \cup \mathcal{F}_- \tag{12.5}$$

$$\mathcal{T}_\tau = \left\{\Delta t_{T_1}, \ldots, \Delta t_{T_{15}}, \Delta t_{T_{1f}}, \ldots, \Delta t_{T_{6f}}\right\} \tag{12.6}$$

$\mathcal{P}_0$ and $\mathcal{P}_1$ denote the set of discrete and differential places, which represent the operation modes and continuous dynamics of the DER and load units as it is shown in Tables 12.1 and 12.2, respectively. The set of discrete and differential transitions are denoted as $\mathcal{T}_0$ and $\mathcal{T}_1$, which describes the switching behaviors and distributed continuous controls of all units as it is shown in Tables 12.1 and 12.2, respectively. All the discrete and differential places are connected to the corresponding transitions through the arcs comprised in $\mathcal{A}$, which is associated with a predecessor or posterior

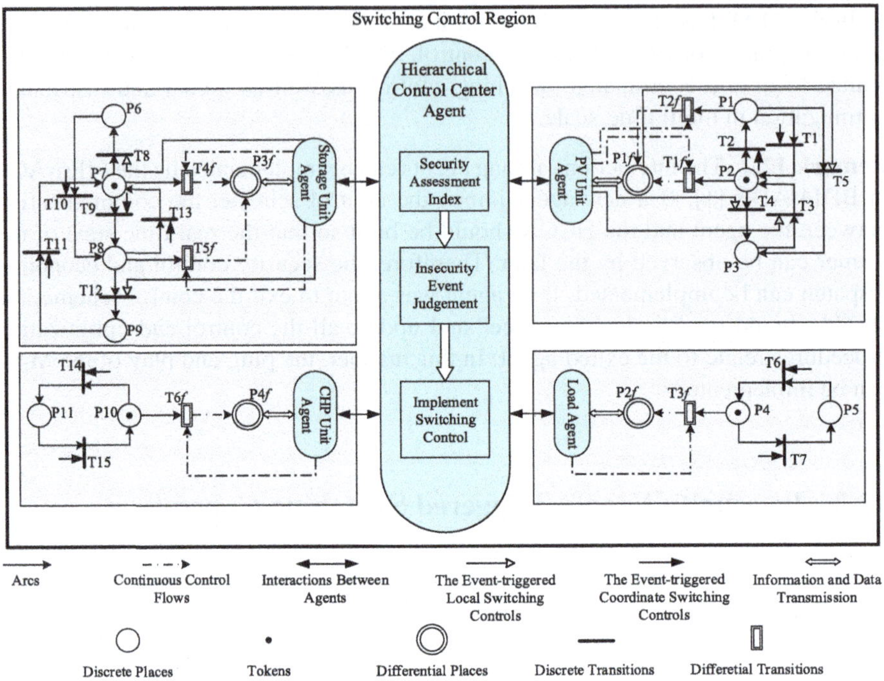

Fig. 12.3 Switching control method based on the DHPN

function defined in $\mathcal{F}_+$ and $\mathcal{F}_-$, respectively. $\mathcal{T}_\tau$ denotes the set of the triggered time of the transitions in $\mathcal{T}_t$.

The places marked with a token represents the initial operation mode of each unit, which can be represented by logic "1", otherwise, by logic "0". All the initial markings construct the set $\mathcal{M}$ which is defined as follows:

$$\begin{aligned} &M_{10}(P_1, P_2, P_3) = (0, 1, 0), \; M_{20}(P_4, P_5) = (1, 0) \\ &M_{30}(P_6, P_7, P_8, P_9) = (0, 1, 0, 0), \; M_{40}(P_{10}, P_{11}) = (1, 0) \end{aligned} \tag{12.7}$$

The switching principle depends on the transition of the tokens in $\mathcal{P}$. For example, if the transition T_2 is triggered, the token in P_2 would transmit into P_1 result in the switching of operation modes of PV unit. The details of the insecurity-events-triggered controls are presented in the following sections.

12.2.2 Local Switching Control

The LSC is responsible for switching the operation mode within the lower-level unit agent. In order to facilitate the following discussion, we introduced a unit step

Table 12.1 Description of discrete places

Discrete places	Description
P_1	PQ mode of the PV unit
P_2	MPPT mode of the PV unit
P_3	Stopping mode of the PV unit
P_4	Normal operation mode of the load unit
P_5	Load shedding mode of the load unit
P_6	Stopping mode with maximal SOC of the storage unit
P_7	Charging mode of the storage unit
P_8	Discharging mode of the storage unit
P_9	Stopping mode with minimal SOC of the storage unit
P_{10}	Normal operation mode of the CHP unit
P_{11}	Stopping mode of the CHP unit

Table 12.2 Description of differential places and transitions

Differential places and transitions	Description
$P_{1f} - P_{4f}$	Continuous dynamics of the corresponding units
$T_{1f} - T_{2f}$	Dynamic controls of the PV unit in MPPT and PQ mode respectively
T_{3f}	Dynamic controls of the load unit in normal operation mode
$T_{4f} - T_{5f}$	Dynamic controls of the storage unit in charging and discharging mode, respectively
T_{6f}	Dynamic controls of the CHP unit in normal operation mode

function:

$$S(t - t_0) = \begin{cases} 1, & t \geq t_0 \\ 0, & t < t_o \end{cases} \tag{12.8}$$

In the PV unit, the LSCs are designed as:
If $t = t_0$ and $G_{PV}(t)$ rises to $G_{PV}(t) > \underline{G}$, then

$$LSC(T_3) = S(t - t_0) - S(t - t_0 - \Delta t_{T_3}) \tag{12.9}$$

$$LSC(T_{1f}) = S(t - t_0) - S(t - t_0 - \Delta t_{T_{1f}}) \tag{12.10}$$

If $t = t_0$ and $G_{PV}(t)$ drops to $G_{PV}(t) \leq \underline{G}$, then

$$LSC(T_4) = S(t - t_0) - S(t - t_0 - \Delta t_{T_4}) \tag{12.11}$$

If $t = t_0$ and $G_{PV}(t)$ rises to $G_{PV}(t) > \overline{G}$, then

$$LSC(T_5) = S(t - t_0) - S(t - t_0 - \Delta t_{T_5}) \tag{12.12}$$

where $G_{PV}(t)$ denotes the intensity of irradiance at time t. $\underline{G}$ and $\overline{G}$ are the cut-in and cut-off threshold value, respectively. Function (12.9) and (12.10) imply that if $G_{PV}(t)$ rises to $G_{PV}(t) > \underline{G}$, then T_3 is triggered lasted for Δt_{T_3} result in that the operation mode is switched from P_3 to P_2. The other LSCs in each unit agent are designed in a similar way.

In the storage unit, the LSCs are designed as:

If $t = t_0$ and $SOC(t)$ rises to $SOC(t) > \overline{S}$, then

$$LSC(T_8) = S(t - t_0) - S(t - t_0 - \Delta t_{T_8}) \tag{12.13}$$

If $t = t_0$ and $SOC(t)$ drops to $SOC(t) \leq \underline{S}$,

$$LSC(T_{12}) = S(t - t_0) - S(t - t_0 - \Delta t_{T_{12}}) \tag{12.14}$$

where $SOC(t)$ denotes the state of charge (SOC) at time t, $\overline{S}$ and $\underline{S}$ are the maximum and minimum values of the SOC, respectively.

Remark 12.3 As mentioned above, the storage unit running at $f - V$ droop control strategy in microgrid 1 is used to provide the reference voltage and frequency for the whole system, and the other units are running at PQ control strategy. Furthermore, tradition PI controllers is used to enhance the robust capacity, other controllers like H∞ [2] etc., are designed to regulate the inner loop voltage/current. In this chapter, a DMPC-based distributed dynamic control is proposed to design the control scheme in each unit. More details about the control scheme are introduced in Sect. 12.4.2.

12.2.3 Coordinated Switching Control

The CSC is also defined as a set of event-triggered functions based on the triggered set $\mathcal{E} = \{E_1(t), E_2(t), \ldots, E_n(t)\}$, each element in $\mathcal{E}$ denotes the insecurity event which can be represented by logic "1" or "0". Furthermore, the voltage $U(t)$ of the PCC node which can be obtained by dynamic voltage sequence [3] or deep neural networks [4], etc., is used to design the insecurity events. In order to facilitate the illustration of CSC, two insecurity events $E_1(t)$ and $E_2(t)$ are designed as follows:

If $U(t) < \underline{U}$, then

$$E_1(t) = 1 \tag{12.15}$$

If $U(t) > \overline{U}$, then

$$E_2(t) = 1 \tag{12.16}$$

Function (12.5) and (12.6) imply that $E_1(t) = \{U(t) < \underline{U}\}$ and $E_2(t) = \{U(t) > \overline{U}\}$ are triggered under different conditions at time t, where $\underline{U}$ and $\overline{U}$ denote the lower and upper threshold of the voltage at the PCC node, respectively. Accordingly, the CSCs are designed as follows:

$$\begin{aligned} CSC(E_1) =& E_1(t)\{P_6(S(t-t_0) - S(t-t_0-\Delta t_{T10})) \\ &+ P_7(S(t-t_0) - S(t-t_0-\Delta t_{T9})) \\ &+ P_{11}(S(t-t_0) - S(t-t_0-\Delta t_{T15}))\} \\ &+ E_1(t+\Delta t)\{P_1(S(t-t_0-\Delta t) \\ &- S(t-t_0-\Delta t-\Delta t_{T1})) \\ &+ P_3(S(t-t_0-\Delta t) \\ &- S(t-t_0-\Delta t-\Delta t_{T3}))\} \\ &+ E_1(t+2\Delta t)\{P_4(S(t-t_0-2\Delta t) \\ &- S(t-t_0-2\Delta t-\Delta t_{T7}))\} \end{aligned} \tag{12.17}$$

$$\begin{aligned} CSC(E_2) =& E_2(t)\{P_8(S(t-t_0) - S(t-t_0-\Delta t_{T13})) \\ &+ P_9(S(t-t_0) - S(t-t_0-\Delta t_{T11})) \\ &+ P_{11}(S(t-t_0) - S(t-t_0-\Delta t_{T15}))\} \\ &+ E_2(t+\Delta t)\{P_2(S(t-t_0-\Delta t) \\ &- S(t-t_0-\Delta t-\Delta t_{T2}))\} \\ &+ E_2(t+2\Delta t)\{P_5(S(t-t_0-2\Delta t) \\ &- S(t-t_0-2\Delta t-\Delta t_{T6}))\} \end{aligned} \tag{12.18}$$

Function (12.17) implies that:

(1) when $t = t_0$, if $E_1(t) = 1$ and the operation modes of storage and CHP unit $\in \{P_6, P_7, P_{11}\}$, then $CSC(E_1)$ is activated;
(2) when $t = t_0 + \Delta t$, if $E_1(t) = 1$ and the operation modes of PV unit $\in \{P_1, P_3\}$, then $CSC(E_1)$ is activated again;
(3) when $t = t_0 + 2\Delta t$, if $E_1(t) = 1$ and the operation modes of load unit $\in \{P_4\}$, then $CSC(E_1)$ is activated again.

where Δt is the switching interval of the CSC. Function (12.18) is implemented in a similar way.

Function (12.17) and (12.18) define the CSCs in the IMS, which are implemented to each lower-level unit agent through the HCCA. As a result, the corresponding transitions is triggered according to the activation sequence of $CSC(E_1)$ and $CSC(E_2)$, of which the implementations of the CSCs are designed as follows:

$$CSC(E_1) = \{T_{10} \rightharpoonup T_9 \rightharpoonup T_{15} \rightharpoonup T_1 \rightharpoonup T_3 \rightharpoonup T_7\} \tag{12.19}$$

$$CSC(E_2) = \{T_{13} \rightharpoonup T_{11} \rightharpoonup T_{15} \rightharpoonup T_2 \rightharpoonup T_6\} \tag{12.20}$$

Remark 12.4 It's worth noting that the two insecurity events introduced as a paradigm in this chapter are used to illustrate the effectiveness of the switching control under large disturbance, but it is not limited in the scope of these two cases due to the flexibility of the proposed control scheme.

12.3 Dynamic Economic Dispatch Model

With the implementation of LSC and CSC, the conventional ED model is inappropriate to formulate the economy of DER and load units due to lacking consideration of the dynamics during switching the operation modes. Therefore, a dynamic ED model integrated with the security switching control is proposed, in which the hybrid characteristics of each DERs are embedded naturally. Furthermore, a DMPC-based method is used to derive several subsystems to achieve distributed optimal control after solving the dynamic ED model.

12.3.1 Insecurity-Event-Triggered Global Optimization

This section proposed a day-ahead ED model in the presence of insecurity events which can be expressed as the following formulation:

$$\min \sum_{t\in\mathcal{T}} \left\{ \sum_{v\in\mathcal{V}} z_{vt} + \sum_{s\in\mathcal{S}} z_{st} + \sum_{m\in\mathcal{M}} z_{mt} + \sum_{n\in\mathcal{N}} z_{nt} \right\} \tag{12.21}$$

$$s.t.\ \ z_{vt} = C_v \frac{p_{vt}}{\eta^p_{vt} L_v},\ \forall v \in \mathcal{V},\ \forall t \in \mathcal{T} \tag{12.22}$$

$$z_{st} = a_{st} + b_{st},\ \forall s \in \mathcal{S},\ \forall t \in \mathcal{T} \tag{12.23}$$

$$z_{mt} = C_m \Delta p_{mt},\ \forall m \in \mathcal{M},\ \forall t \in \mathcal{T} \tag{12.24}$$

$$z_{nt} = C_n \Delta p_{nt},\ \forall n \in \mathcal{N},\ \forall t \in \mathcal{T} \tag{12.25}$$

$$a_{st} = C_s[\mu_{st}\left(\frac{1}{\eta^d_{st}} - 1\right) p_{st} + (1 - \mu_{st})(1 - \eta^c_{st}) p_{st}] \quad \forall s \in \mathcal{S},\ \forall t \in \mathcal{T} \tag{12.26}$$

$$b_{st} = \lambda_s[(1 - D_{st})\overline{S} - S_{s(t-1)} + S_{st}],\ \forall s \in \mathcal{S},\ \forall t \in \mathcal{T} \tag{12.27}$$

$$S_{st} = \int_{t-1}^{t} \mu_{st} \frac{p_{st}}{\eta_{st}^{d}} dt - (1-\mu_{st})\eta_{st}^{c} \int_{t-1}^{t} p_{st} dt \quad \forall s \in \mathcal{S},\ \forall t \in \mathcal{T} \tag{12.28}$$

$$\overline{P} + \sum_{v \in \mathcal{V}} p_{vt} + \sum_{s \in \mathcal{S}} (\mu_{st} p_{st} + (\mu_{st} - 1) p_{st}) + \sum_{m \in \mathcal{M}} (\overline{p}_{mt} - \Delta p_{mt}) - \sum_{n \in \mathcal{N}} (\overline{p}_{nt} - \Delta p_{nt}) = 0,\ \forall t \in \mathcal{T} \tag{12.29}$$

$$\underline{P}_{v} \delta_{vt} \le p_{vt} \le \overline{P}_{v} \delta_{vt},\ \forall v \in \mathcal{V},\ \forall t \in \mathcal{T} \tag{12.30}$$

$$\mu_{st} \underline{P}_{s}^{d} + (1-\mu_{st}) \underline{P}_{s}^{c} \le p_{st} \le \mu_{st} \overline{P}_{s}^{d} + (1-\mu_{st}) \overline{P}_{s}^{c} \quad \forall s \in \mathcal{S},\ \forall t \in \mathcal{T} \tag{12.31}$$

$$0 \le \Delta p_{mt} \le \overline{p}_{mt} \alpha_{mt},\ \forall m \in \mathcal{M},\ \forall t \in \mathcal{T} \tag{12.32}$$

$$0 \le \Delta p_{nt} \le \overline{P}_{nt} \beta_{nt},\ \forall n \in \mathcal{N},\ \forall t \in \mathcal{T} \tag{12.33}$$

$$\overline{P}_{nt} \le \overline{p}_{nt},\ \forall n \in \mathcal{N},\ \forall t \in \mathcal{T} \tag{12.34}$$

$$\underline{S} \le S_{s(t-1)} - S_{st} \le \overline{S},\ \forall s \in \mathcal{S},\ \forall t \in \mathcal{T} \tag{12.35}$$

$$\delta \in \{0,1\}^{|\mathcal{V}| \times |\mathcal{T}|},\ \mu \in \{0,1\}^{|\mathcal{S}| \times |\mathcal{T}|},\ \alpha \in \{0,1\}^{|\mathcal{M}| \times |\mathcal{T}|},\ \beta \in \{0,1\}^{|\mathcal{N}| \times |\mathcal{T}|},\ \mathcal{V} = \mathcal{V}_0 \times \mathcal{R}_v,\ \mathcal{S} = \mathcal{S}_0 \times \mathcal{R}_s,\ \mathcal{M} = \mathcal{M}_0 \times \mathcal{R}_m,\ \mathcal{N} = \mathcal{N}_0 \times \mathcal{R}_n \tag{12.36}$$

In this formulation, the objective function (12.21) minimize the total costs of CHP, storage, and PV units, denoted by z_{vt}, z_{st} and z_{mt}, as well as the cost of load shedding z_{nt}. The generation outputs from CHP and storage units are denoted by p_{vt} and p_{st}, which confine to the maximum and the minimum capacity defined in constraints (12.30) and (12.31). $\overline{p}_{mt}$ and $\overline{p}_{nt}$ denote the forecast value of the PV and load units. The generation cost and life loss cost of the storage unit are presented in (12.26) and (12.27), respectively. Constraints (12.29) ensures the power balance between the generation supply and demand, where $\overline{P}$ is the supporting power output from the storage unit in microgird 1 which is running at $f - V$ droop control strategy. The down-regulation and load shedding limits of units are enforced by constraints (12.32) and (12.34), where $\bar{P}_{nt}$ denotes the maximum load shedding of load unit running in operation mode n during time t. Constraints (12.35) confine the SOC of the storage unit in the bound $[\underline{S}, \overline{S}]$. The upper and lower limits of p_{st} are defined in inequalities (12.31). $\mathcal{R}_v$, $\mathcal{R}_s$, $\mathcal{R}_m$ and $\mathcal{R}_n$ are the set of CHP, storage, PV and load units, respectively. The set of operation modes are denoted as $\mathcal{V}_0$, $\mathcal{S}_0$, $\mathcal{M}_0$, and $\mathcal{N}_0$. $\mathcal{T}$ denotes the set of all time instants considered in the operation horizon.

The binary coefficients δ_{vt}, μ_{st}, α_{mt} and β_{nt} defined in constraints (12.36) are determined by the CSC. For example, they can be defined as follows during time t, respectively:

$$\delta_{vt} = \begin{cases} 1, & v = P_{10} \\ 0, & v = P_{11} \end{cases} \qquad \mu_{st} = \begin{cases} 1, & s = P_8,\ P_6 \\ 0, & s = P_7,\ P_9 \end{cases}$$

$$\alpha_{mt} = \begin{cases} 1, & m = P_1 \\ 0, & m = P_2,\ P_3 \end{cases} \qquad \beta_{nt} = \begin{cases} 1, & n = P_5 \\ 0, & n = P_4 \end{cases}$$

By solving Eq. (12.21), the optimal power regulation is implemented to the lower-level unit agents. Therefore, to make the best utility of each generator and realize a system balance, the reference power is assigned proportionally to its capacity through Eq. (12.37) among the homogeneous lower-level unit agent. The distributed optimal control is implemented in the lower-level agents based on the DMPC method with much reduced computation and communication free manner. The dynamic ED and optimal control structure of the IMS is shown in Fig. 12.4.

$$P_{ref}^{r} = P_g \frac{P_{rate}^{r}}{\sum P_{rate}^{r}},\ \forall r \in \mathcal{R},\ \forall g \in \mathcal{G} \tag{12.37}$$

where $\mathcal{R} = \mathcal{R}_v \cup \mathcal{R}_s \cup \mathcal{R}_m$. P_g is obtained by solving (12.21) which indicates the sum of generations of homogeneous lower-level unit agents. $\mathcal{G} = \mathcal{V} \cup \mathcal{S} \cup \mathcal{M}$ and P_{rate}^{r} denotes the rated power of rth units.

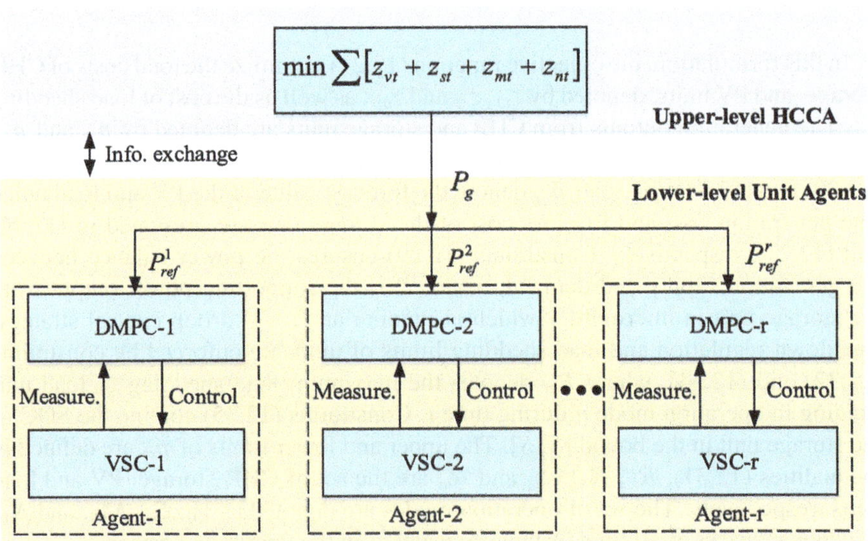

Fig. 12.4 The dynamic ED and control structure of the IMS

12.3.2 DMPC-Based Distributed Optimal Control

A typical inverter-based DER and its PQ control-based controller are shown in Fig. 12.5. ρ and ω provided by the PLL represent the phase angle of the PCC voltage and the frequency of power system, respectively. u_{oq}, decoupling by the synchronous dq reference frame, is set as zero [5]. Therefore, the current reference values can be calculated by:

$$i_{odref}^{r} = \frac{P_{ref}^{r}}{u_{od}} \tag{12.38}$$

$$i_{oqref}^{r} = -\frac{Q_{ref}^{r}}{u_{od}} \tag{12.39}$$

where $Q_{ref}^{r} = P_{ref}^{r} \cdot \tan \varphi_r, \ \ 0.85 \leq \cos \varphi_r \leq 1, \ \ \forall r \in \mathcal{R}$. For convenience, the index r is omitted in the following analysis.

As shown in Fig. 12.6, the PQ control is realized by regulating i_{od} and i_{oq} to meet the reference value i_{odref} and i_{oqref}, and ω is used to decouple the control of i_{od} and i_{oq}. Based on the control scheme shown in Fig. 12.6 and with the reference to Fig. 12.5, the state-space equations are given as below:

$$\dot{\boldsymbol{x}}(t) = \boldsymbol{A}_c \boldsymbol{x}(t) + \boldsymbol{B}_c \boldsymbol{u}(t) \tag{12.40}$$

where

$$\boldsymbol{x} = \left[i_{od} \ i_{oq} \right]^T, \boldsymbol{u} = \left[m_d \ m_q \ u_{od} \ u_{oq} \right]^T$$

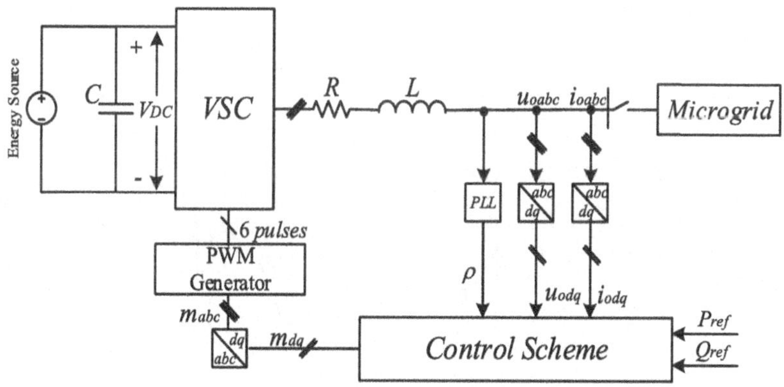

Fig. 12.5 The control schematic diagram of an inverter-based DER

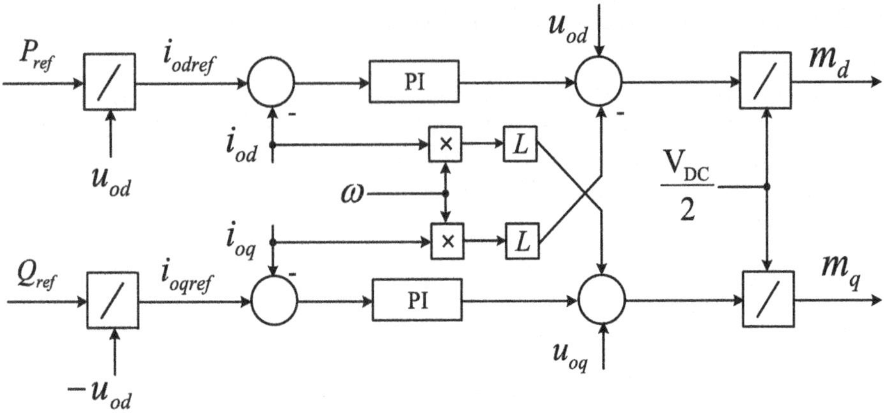

Fig. 12.6 Block diagram of the PQ control scheme

$$A_c = \begin{bmatrix} -R/L & \omega \\ -\omega & -R/L \end{bmatrix} B_c = \begin{bmatrix} V_{DC}/2L & 0 & -1/L & 0 \\ 0 & V_{DC}/2L & 0 & -1/L \end{bmatrix}$$

The discrete-time state-space equations are give as:

$$x(k+1) = A_d x(k) + B_d u(k) \tag{12.41}$$

where T_s is the sampling time and

$$A_d = e^{A_c T_s},\ B_d = A^{-1}(e^{A_c T_s} - I)B_c$$

where I is a unit matrix with appropriate dimensions. Based on (12.41), we can get the discretization equation:

$$\begin{aligned} X(k+1) &= AX(k) + B\Delta u(k) \\ y(k) &= CX(k) \end{aligned} \tag{12.42}$$

where

$$X(k) = \begin{bmatrix} x(k) \\ u(k-1) \end{bmatrix},\ A = \begin{bmatrix} A_d & B_d \\ 0 & I \end{bmatrix}$$
$$B = \begin{bmatrix} B_d \\ I \end{bmatrix}, \quad C = \begin{bmatrix} I_{2\times 2} & 0 \end{bmatrix},$$

$$\Delta u(k) = u(k) - u(k-1)$$

The objective function that is optimized for the DMPC is defined as:

$$\begin{aligned} J(k) &= \boldsymbol{\Delta Y}^T \boldsymbol{Q} \boldsymbol{\Delta Y} + \boldsymbol{\Delta U}^T \boldsymbol{R} \boldsymbol{\Delta U} \\ &= (\boldsymbol{R}_s - \boldsymbol{Y})^T \boldsymbol{Q} (\boldsymbol{R}_s - \boldsymbol{Y}) + \boldsymbol{\Delta U}^T \boldsymbol{R} \boldsymbol{\Delta U} \end{aligned} \tag{12.43}$$

s.t.

$$\begin{aligned} \boldsymbol{u}_{min}(k+i|k) \le \boldsymbol{u}(k+i|k) \le \boldsymbol{u}_{max}(k+i|k), \\ i = 0, 1, \ldots, N_c - 1 \end{aligned} \tag{12.44}$$

$$\begin{aligned} \boldsymbol{\Delta u}_{min}(k+i|k) \le \boldsymbol{\Delta u}(k+i|k) \le \boldsymbol{\Delta u}_{max}(k+i|k), \\ i = 0, 1, \ldots, N_c - 1 \end{aligned} \tag{12.45}$$

$$\begin{aligned} \boldsymbol{y}_{min}(k+i|k) \le \boldsymbol{y}(k+i|k) \le \boldsymbol{y}_{max}(k+i|k), \\ i = 1, 2, \ldots, N_p \end{aligned} \tag{12.46}$$

where

$$\begin{aligned} \boldsymbol{Y} &= [\boldsymbol{y}(k+1|k) \ \cdots \ \boldsymbol{y}(k+N_p|k)]^T \\ &= \boldsymbol{\Psi} \boldsymbol{X}(k|k) + \boldsymbol{\Theta} \boldsymbol{\Delta U} \end{aligned}$$

$$\boldsymbol{\Psi} = [\boldsymbol{CA} \ \boldsymbol{CA}^2 \ \cdots \ \boldsymbol{CA}^{N_p}]^T$$

$$\boldsymbol{\Delta U} = [\boldsymbol{\Delta u}(k|k) \ \boldsymbol{\Delta u}(k+1|k) \ \cdots \ \boldsymbol{\Delta u}(k+N_c-1|k)]^T$$

$$\boldsymbol{\Theta} = \begin{bmatrix} \boldsymbol{CB} & \boldsymbol{0} & & \boldsymbol{0} \\ \boldsymbol{CAB} & \boldsymbol{CB} & & \boldsymbol{0} \\ \vdots & \vdots & \cdots & \vdots \\ \boldsymbol{CA}^{N_p-1}\boldsymbol{B} & \boldsymbol{CA}^{N_p-2}\boldsymbol{B} & & \boldsymbol{CA}^{N_p-N_c}\boldsymbol{B} \end{bmatrix}$$

$$\boldsymbol{R}_s = \underbrace{[\boldsymbol{I}_{2\times2} \ \boldsymbol{I}_{2\times2} \ \cdots \ \boldsymbol{I}_{2\times2}]}_{N_p} \boldsymbol{r}(k)$$

$$\boldsymbol{r}(k) = [i_{odref}(k), i_{oqref}(k)]^T$$

with $\boldsymbol{\Delta Y}$ and $\boldsymbol{\Delta U}$ denote the output variable differences and input variable differences, respectively. $k+i|k$ is the instant of which the value is evaluated based on instant k. N_p and N_c are the prediction and control horizons, respectively. $\boldsymbol{Q}$ and $\boldsymbol{R}$ are the weight diagonal matrices, with $\boldsymbol{Q} \ge 0$ and $\boldsymbol{R} > 0$.

The issue of minimizing the objective function (12.43) subjected to constraints (12.44)–(12.46) at each time step k yields the optimal sequence of control variable differences:

$$\boldsymbol{\Delta U}_k^* = [\boldsymbol{\Delta u}^*(k|k) \ \boldsymbol{\Delta u}^*(k+1|k) \ \cdots \ \boldsymbol{\Delta u}^*(k+N_c-1|k)]^T \tag{12.47}$$

It's worth noting that only the first element of the optimal sequence, i.e., $\boldsymbol{\Delta u}^*(k|k)$ is applied for the predictive control of each DER unit. By formulating the inverter-based model and implementing the optimal sequence to deal with the predictive control of the DER units, the local dynamic control can be realized in an distributed manner which coordinates different types of DERs to achieve the equality constraints (12.29).

Remark 12.5 For stability, since the objective function and feasible region is convex, the proposed DMPC satisfies Lemmas 1 and 2 in literature [6], which can establish the exponential closed-loop stability property.

12.4 Simulation Results

In this section, 2 case studies are introduced to verify the validity of the proposed approach, which consist of different numbers of the DER and load units. Case study 1 is designed without the limitation of the thermal load, which means that the CHP units are formulated in ED. On the contrary, case study 2 is designed based on the limitation of the thermal load, which indicates that the CHP units are saturated. For convenience, the dynamic ED proposed in this chapter is simulated in a 24 h time period. The simulations in our numerical studies are implemented on the MATLAB platform.

12.4.1 Case Study 1

12.4.1.1 Dynamic Optimization

In this case study, the system is simulated with 3 CHPs, 1 storage and 4 PV units, as well as a noncritical load demand. The evolution of the bus voltage at the monitoring point is shown in Fig. 12.7, in which the areas shadowed in golden and grey color represent the insecurity events $E_1(t)$ and $E_2(t)$, respectively. Furthermore, the event $E_2(t)$ is triggered when the bus voltage beyond the upper bound: $(1 + 5\%)$ p.u., the event $E_1(t)$ is triggered when the bus voltage drops down the lower bound: (1–5%) p.u.

From Fig. 12.7, it can be seen that $E_1(t)$ is triggered in the golden-color-shadowed area (e.g. t = 5 h, 7 h, 8 h, ...), and $E_2(t)$ is triggered in the grey-color-shadowed area (e.g. t = 3 h, 10 h, 19 h,...). By using the proposed event-triggered hybrid switching control, the operation modes of the DER and load units are switched appropriately at each trigger time. For example, at time 5 h ($E_1(t)$ is triggered), the transition T_9 is triggered so that the storage unit is switched from charging mode to discharging mode. After one switching interval time, T_1 is triggered result in the PV units are switched from PQ mode to MPPT mode. After another one switching interval time, T_7 is triggered and the load unit is switched from normal operation mode to load shedding mode. More details about the operation mode switching are shown in Table 12.3,

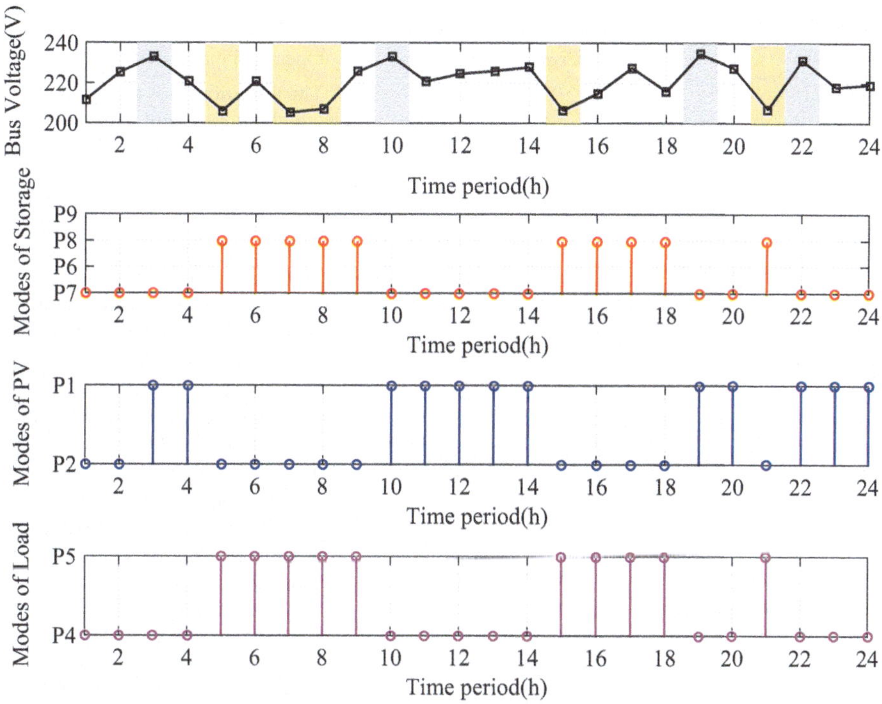

Fig. 12.7 Bus voltage profile and operation modes of three DER units

which can prove that the proposed switching control method performed well while handling the insecurity events.

The curves of the load and the total cost calculated by the proposed ED can be seen in Fig. 12.8 and the optimal solutions of each DER and load unit are shown in Fig. 12.9. It proves that the proposed dynamic ED model can regulate the generations well while satisfying all the constraints during the whole time periods. Besides, it can be seen that the CHP unit and the PV unit are prone to regulate the power output in the supply side, furthermore, the storage unit is basically in a balanced state, which means that it has no output power unless the events of insufficient power supply(e.g. $E_1(t)$) are triggered. The reason is that the cost of storage life loss is considered in constraints (12.27). Therefore, this result helps to prolong the service life of the storage unit in the IMS.

The convergence of cost in different scenarios are shown in Fig. 12.10, it can be seen that the convergence speed is assured in normal conditions(e.g. at time 1 h) as well as when the insecurity events triggered (e.g. at time 8 h or 10 h). This indicates that the proposed integrated design of ED and security control possess a robust property when confronting the challenge of insecurity events for the IMS.

From Fig. 12.11, it can be seen that the deviation of the power output between supply-demand is under effectively control. The average deviation is 0.047%p.u., a

Table 12.3 Operation mode switching in the CSCs in case study 1

Times (h)	Insecurity events	Operation mode switching
5, 15, 21	E_1	The transition T_9 is triggered so that the storage unit is switched from charging mode to discharging mode. After one switching interval time, T_1 is triggered so that the PV unit is switched from PQ mode to MPPT mode. After another one switching interval time, T_7 is triggered so that the load unit is switched from normal operation mode to load shedding mode
3	E_2	The transition T_2 is triggered so that the PV unit is switched from MPPT mode to PQ mode
10, 19, 22	E_2	The transition T_{13} is triggered so that the storage unit is switched from discharging mode to charging mode. After one switching interval time, T_2 is triggered so that the PV unit is switched from MPPT mode to PQ mode. After another one switching interval time, T_6 is triggered so that the load unit is switched from load shedding mode to normal operation mode

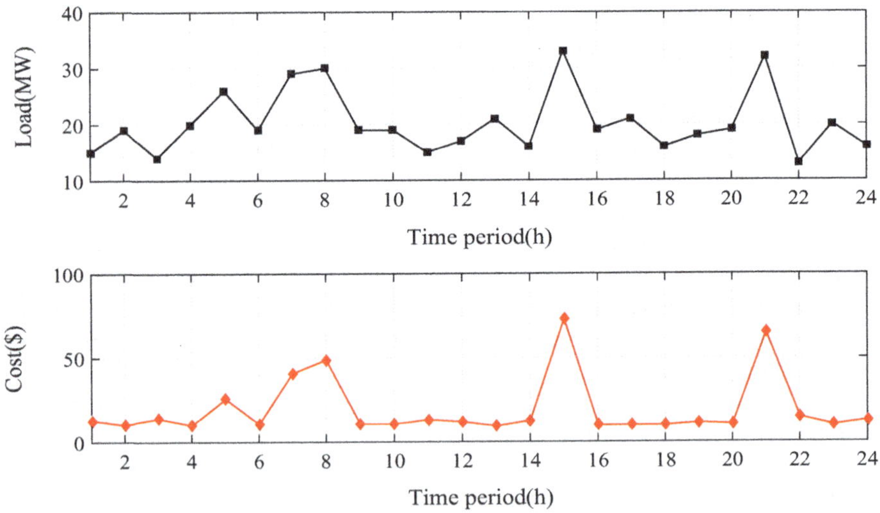

Fig. 12.8 Evolution of loads and the total cost of DERs

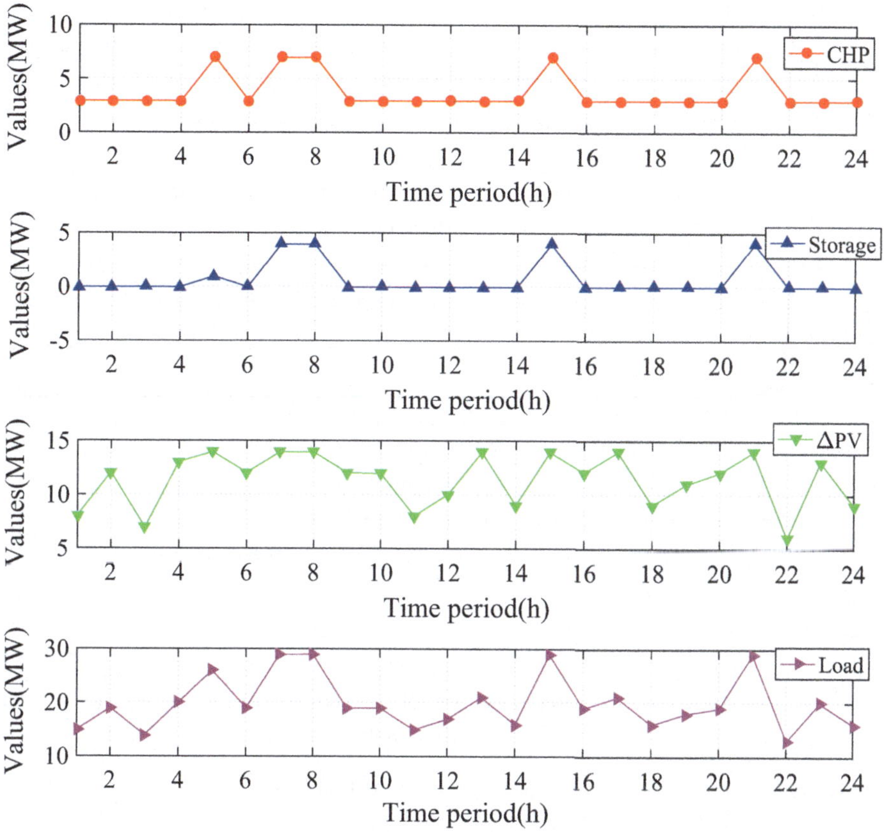

Fig. 12.9 Optimal solution of dynamic ED during 24 h time period

rather good performance, which means that the proposed method assures the security operation of the IMS.

12.4.1.2 Comparative Experiments

In this subsection, comparative experiments between the proposed ED and the central PSO(CPSO) method without a multi-mode switching are introduced, which indicate the effectiveness of the proposed method. The details of the experiments are as follows: Fig. 12.12 give the power output of 4 PV units using different methods in each time period, respectively. According to the comparison results, it can be seen that the power output of the 4 PV units are well-proportioned by using the proposed method, which assure a balance of the system state.

The comparative experiment about the convergence speed is simulated according to Fig. 12.13. Here, the biased evolution of the total cost associated with iterations

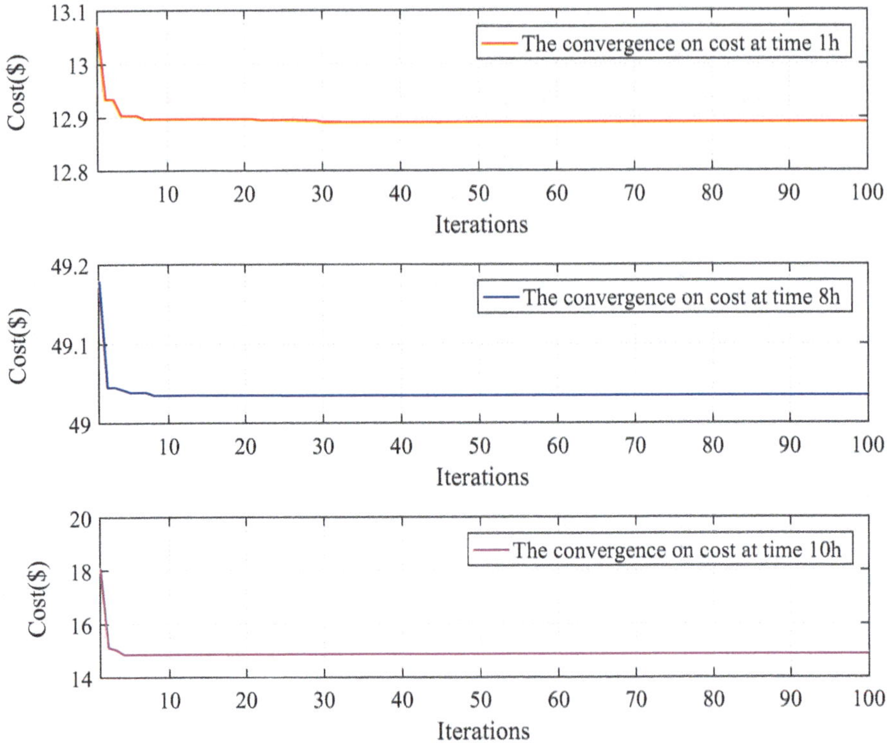

Fig. 12.10 The convergence performance on cost at time 1, 8, 10 h

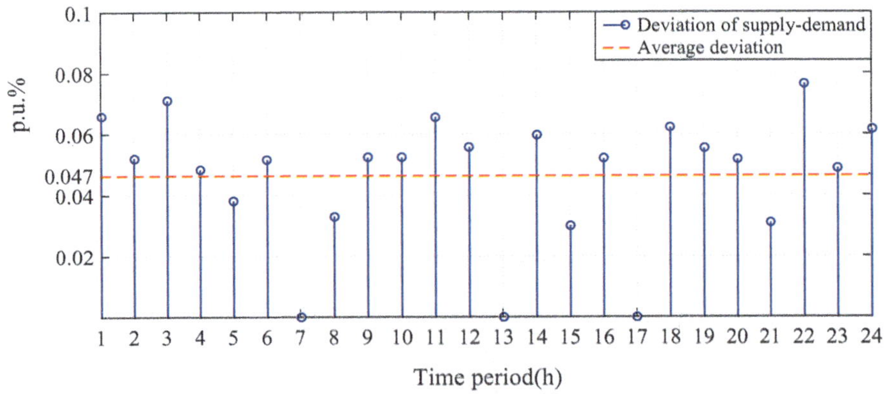

Fig. 12.11 The deviation of power output between supply-demand

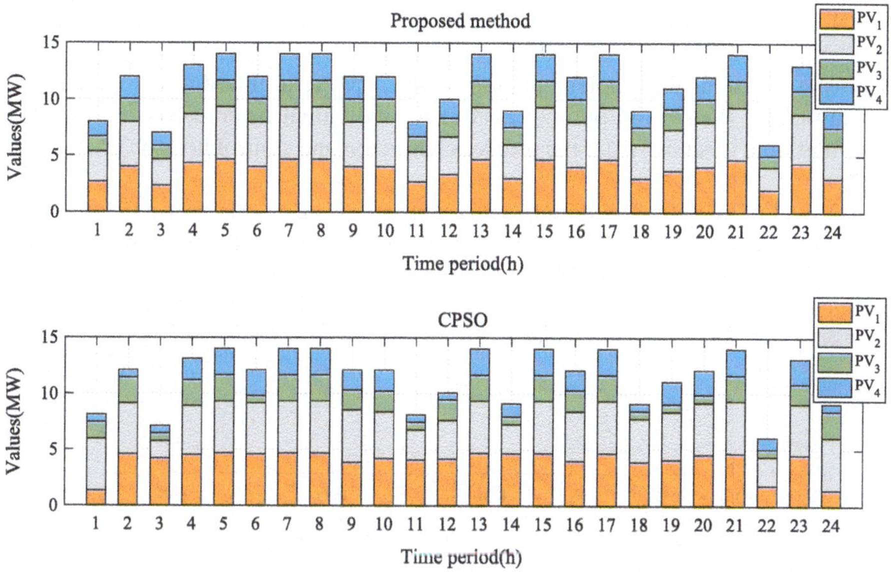

Fig. 12.12 The output of PV_1–PV_4 by the proposed method and CPSO

Fig. 12.13 A comparison of $e_{conv}(k)$ by different methods at time 1 h

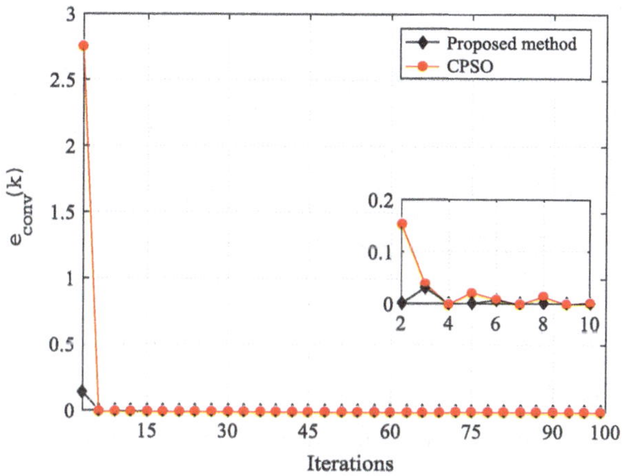

k is defined as $e_{conv}(k) = \|f(k) - f(k-1)\|_2$, where $\| \cdot \|$ denotes the l_2 norm of adjacent iterations. From Fig. 12.13, it can be shown that the proposed ED in this chapter enjoys a better convergence performance in which the corresponding curve converges to 0 faster than the CPSO method. The reason is that the computational complexity is reduced after the DHPN modeling of the DER and load units, therefore, it does not need to dispatch all the generations after the switching control which coordinates the lower-level agents in multi operation modes.

12.4.1.3 The Implementation of the DMPC

In this subsection, the implementation of the DMPC is introduced to verify the validity of the distributed optimal control for individual inverter-based DER. The sampling time is set as 10 μs, the prediction horizon and control horizon are both set as 20.

The evolution of the decoupling component of current i_{oabc} is shown in Fig. 12.14, it can be seen that both i_{od} and i_{oq} track the set point well by the moving optimization

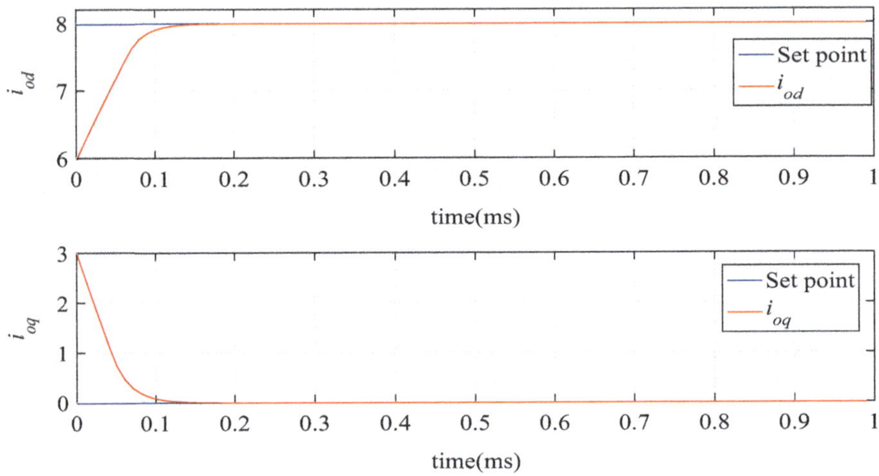

Fig. 12.14 The evolution of i_{od} and i_{oq} by giving a set point

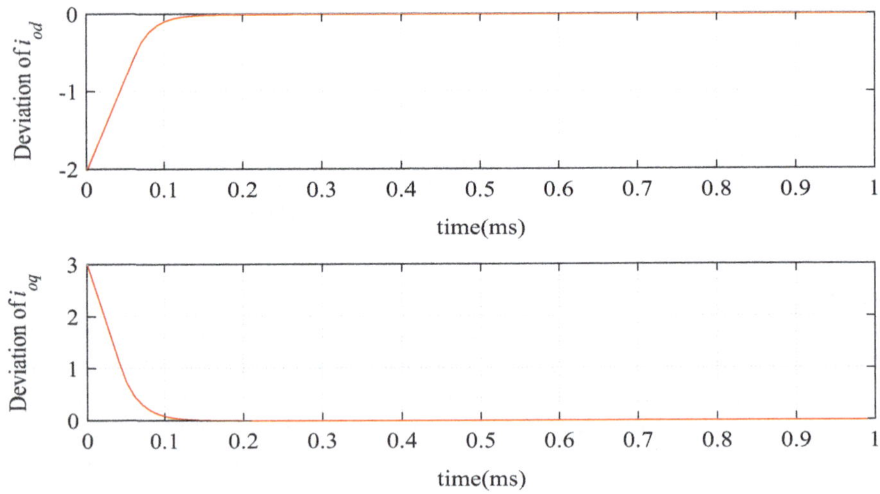

Fig. 12.15 Evolution of the deviation of i_{od} and i_{oq}

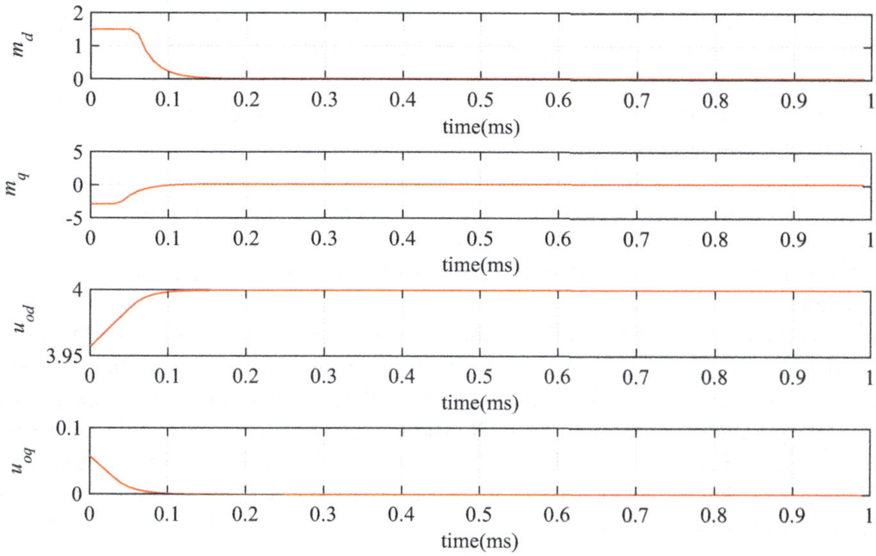

Fig. 12.16 The evolution of control variables m_d, m_q, u_{od} and u_{oq}

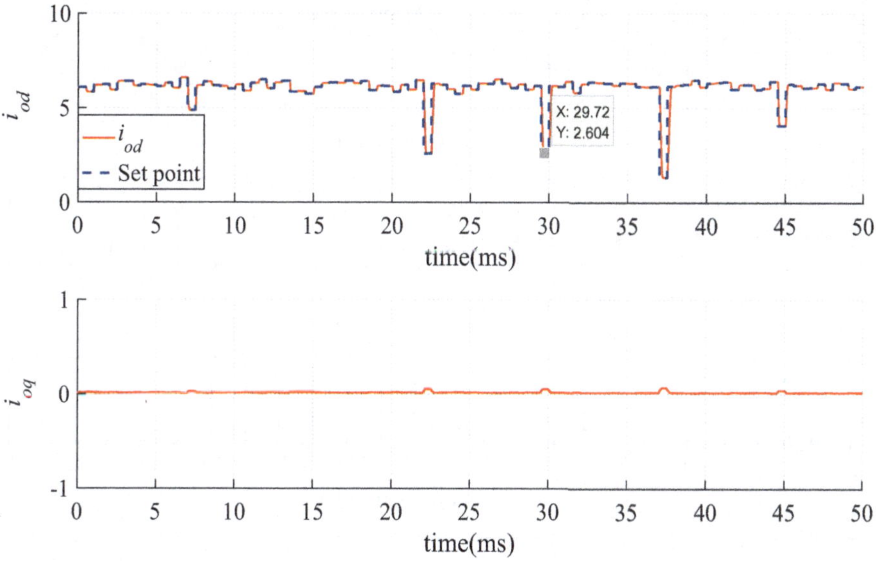

Fig. 12.17 The evolution of i_{od} and i_{oq} in a 50 ms time period considering disturbance

of the objective function (12.43). The set point can be obtained by Eqs. (12.38) and (12.39) which are related to the power output of the DERs. From Fig. 12.15, it can be shown that the deviation of i_{od} and i_{oq} from the set point converge to 0 in a short time, the well-performed convergence speed assures the real-time control of the proposed ED in the IMS.

From Fig. 12.16, it can be shown that the convergence evolution of control variables m_d, m_q, u_{od} and u_{oq} in 1ms time period. It proves that all the control variables converge in a rather fast speed with the limitation of all the constraints (12.44)–(12.46). These results indicated that the implementation of DMPC can be properly handled.

As shown in Fig. 12.17, the dynamic control property of the proposed DMPC is simulated in a 50 ms' time period. It can be seen that there's some fluctuation in the simulation progress. For example, at time 29.72 ms, the setpoint dropped down to 2.604 p.u., and the proposed DMPC showed a robust property while facing the disturbance. Therefore, i_{oq} is basically staying at 0 in the whole time period. This result proves that the proposed DMPC is of remarkable strengths in the distributed dynamic control of the IMS.

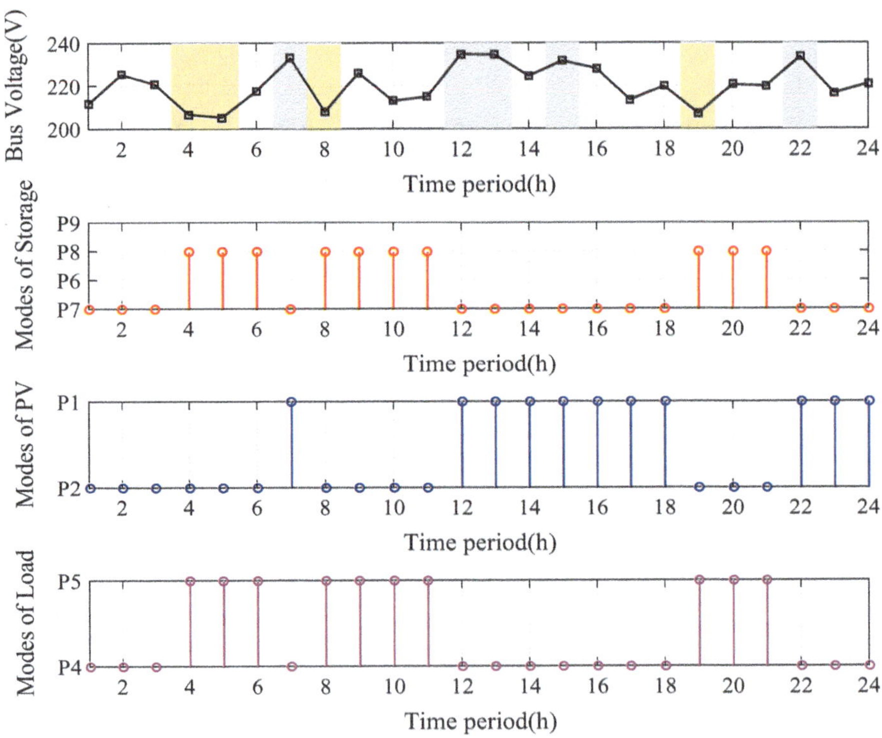

Fig. 12.18 Bus voltage profile and operation modes of three DER units

12.4.2 Case Study 2

In this case study, the thermal loads of the CHP units are under limitation, and the number of PV units extends to double scales of case study 1. Besides, the load demands are divided into noncritical load and critical load as shown in Fig. 12.20.

The evolution of bus voltage is showed in Fig. 12.18, which also contains insecurity events $E_1(t)$ and $E_2(t)$ during different time instants. The switching control of operation modes is similar to the results in Fig. 12.7, it can be seen that the operation modes of each DER unit and load unit switched properly according to the implementations of the CSCs. From Fig. 12.19, it can be shown that the PV units are fully used in each period so that the times of charging and discharging of the storage unit can be minimized as much as possible, and the noncritical load is more likely used to regulate the balance of the supply-demand side to guarantee the safe and stable running of the critical load. The deviation of power between the supply-demand side is

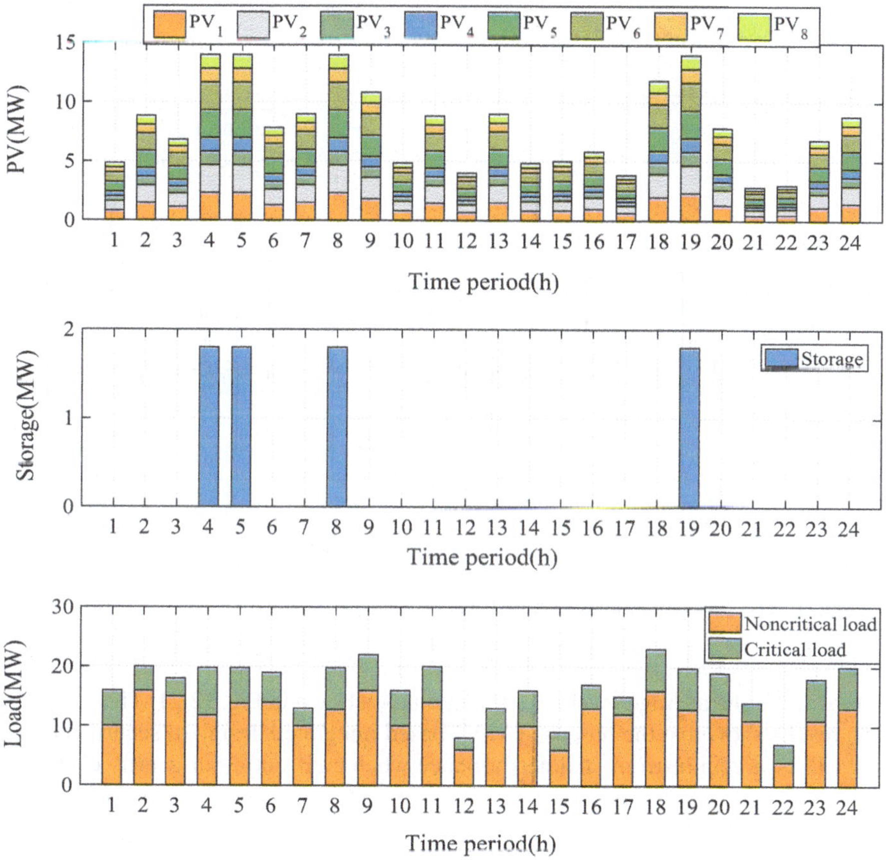

Fig. 12.19 Optimal solutions during 24 h time period

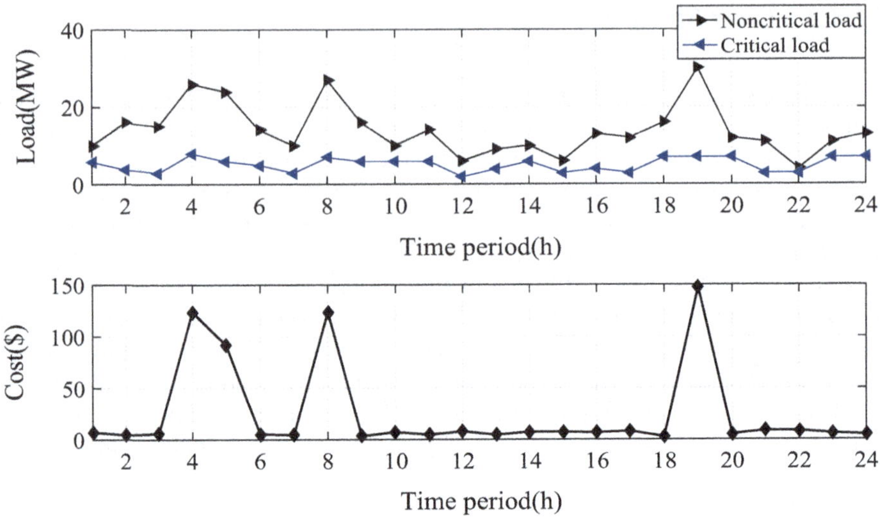

Fig. 12.20 Evolution of all loads and the total costs

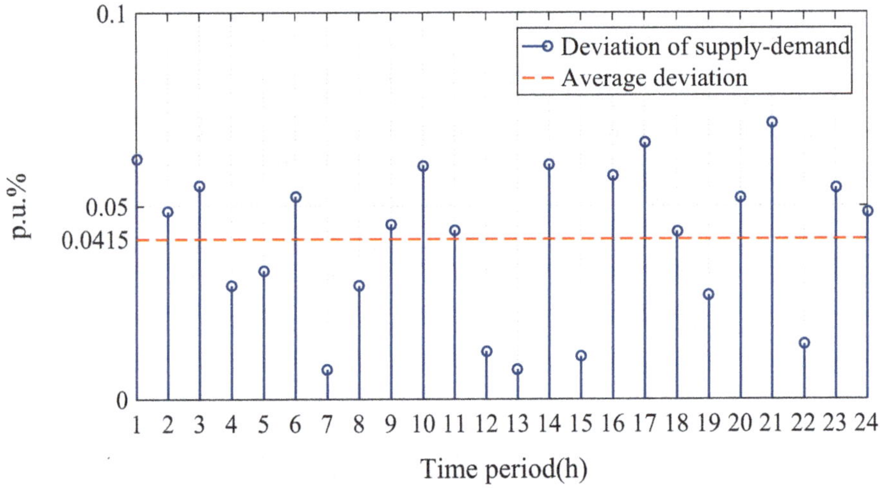

Fig. 12.21 The deviation of power output between supply-demand

controlled at the average level of 0.0415%p.u. as shown in Fig. 12.21. These simulation results show that the proposed MAS-based integrated design of security control and ED assure an instrumental performance when confronting insecurity events.

12.5 Conclusion

This chapter proposed an integrated design for the security control and economic dispatch from a new scope, by which a tentative response to the challenges of energy supplying while confronting big disturbance in microgrids is given. The MAS-based hierarchical control scheme is utilized to deal with the hybrid characteristics of the IMS. Furthermore, the switching control and dynamic regulation are naturally embedded in the two-level unit agents by means of the DHPN model. The ED is implemented by dividing into several subsystems, in which the subsystems are linked by the equality constraints so that the communication free manner is realized by means of the DMPC method. The simulation results verify the effectiveness of the proposed method in comparison with the traditional approach without a multi-mode switching.

In addition, the proposed method is not limited in the scope of this chapter. For example, it can be extended into other systems e.g. smart grids, and the plug-and-play property of the MAS-based control scheme owns the ability of extending the system scales by adding new DERs through developing additional logic functions of agents. These results will bring some interest in the future work for the IMS.

References

1. C. Dou, D. Yue, X. Li, Y. Xue, Mas-based management and control strategies for integrated hybrid energy system. IEEE Trans. Industr. Inf. **12**(4), 1332–1349 (2016)
2. C. Dou, B. Liu, D.J. Hill, Hybrid control for high-penetration distribution grid based on operational mode conversion. IET Gener. Transm. Distrib. **7**(7), 700–708 (2013)
3. C. Dou, D. Yue, J.M. Guerrero, Multiagent system-based event-triggered hybrid controls for high-security hybrid energy generation systems. IEEE Trans. Industr. Inf. **13**(2), 584–594 (2017)
4. L. Zhang, G. Wang, G.B. Giannakis, Real-time power system state estimation and forecasting via deep neural networks (2018). arXiv:1811.06146
5. A. Saleh, A. Deihimi, R. Iravani, Model predictive control of distributed generations with feed-forward output currents. IEEE Trans. Smart Grid **10**(2), 1488–1500 (2017)
6. A.N. Venkat, I.A. Hiskens, J.B. Rawlings, S.J. Wright, Distributed mpc strategies with application to power system automatic generation control. IEEE Trans. Control Syst. Technol. **16**(6), 1192–1206 (2008)

Part V
Distributed Cooperative Control of Islanded AC Microgrids

Chapter 13
Distributed Event-Triggered Cooperative Control for Frequency and Voltage Stability and Power Sharing in Isolated Inverter-Based Microgrid

The distributed cooperative control for frequency and voltage stability and power sharing in microgrid considering the limitation of communication network is concerned in this chapter. Two types of novel event-triggered mechanism with distributed architecture are firstly proposed, which can greatly reduce the communication burdens among power source inverters. Based on the event-triggered schemes, distributed restoration mechanism is constructed, which can restore the frequency and voltage magnitude of microgrid and realize the fair utilization of all power sources with comparative less requirements for the transmission data. Simulation is carried out to verify the effectiveness of the proposed method.

13.1 Problem Formulation and Preliminaries

13.1.1 Inverter and Load Dodels

For the isolated microgrid control problem, the microgrid with n buses is considered, where the buses labeled from 1 to $m(m < n)$ are power sources embedded with inverters, and the buses labeled from $m + 1$ to n are loads. The susceptance of the electric line between bus i and j is denoted as B_{ij}, where $B_{ij} \neq 0$ if buses i and j are connected directly via the electric line, otherwise $B_{ij} = 0$. It is assumed that each electric line is lossless in this chapter. The active power P_i and reactive power Q_i injected into the network at bus $i \in \{1, ..., n\}$ at time t are given by [1] as follows

$$P_i(t) = \sum_{j=1}^{n} V_i(t)V_j(t)B_{ij}sin(\theta_i(t) - \theta_j(t)) \tag{13.1}$$

D. Yue et al., *Cooperative Optimal Control of Hybrid Energy Systems*,
https://doi.org/10.1007/978-981-33-6722-7_13

$$Q_i(t) = -\sum_{j=1}^{n} V_i(t)V_j(t)B_{ij}cos(\theta_i(t) - \theta_j(t)) \tag{13.2}$$

where $V_i(t)$ and $\theta_i(t)$ are the voltage magnitude and voltage phase angle of bus i at time t respectively.

Conventionally, the real power-frequency $P - f$ droop control and reactive power-voltage magnitude $Q - V$ droop control are applied to achieve the power sharing in microgrid. However, the $Q - V$ droop control method may bring adverse effect on the reactive power sharing [2, 3]. In order to resolve this problem, the $P - f/Q - \dot{V}$ droop control method, which has proven to be effective in the reactive power sharing in [4], is adopted in this chapter. The $P - f/Q - \dot{V}$ droop control model of inverter $i \in \{1, ..., m\}$ is represented as follows according to [5]

$$D_{p,i}\frac{d}{dt}\theta_i(t) = P_i^* - P_i(t) - p_i(t) \tag{13.3}$$

$$D_{q,i}\frac{d}{dt}V_i(t) = Q_i^* - Q_i(t) - q_i(t) \tag{13.4}$$

where the positive constants $D_{p,i}$ and $D_{q,i}$ are the droop coefficients, P_i^* and Q_i^* are the nominal active and reactive power ratings of inverter i respectively, $P_i(t)$ and $Q_i(t)$ are the active and reactive power injections of inverter i at time t as described in (13.1) and (13.2), $p_i(t)$ and $q_i(t)$ indicate the restoration mechanism designed for regulating the frequency and voltage magnitude. For simplicity, it is assumed that the power source inverters do not reach their power output limitations.

The structure-preserving model is adopted in this chapter, and thus we can consider the constant power load which is realistic to capture the load behaviors [6]. The following algebraic equations describe the active and reactive power flow at load $k \in \{m + 1, ..., n\}$

$$P_{Lk} + P_k(t) = 0 \tag{13.5}$$

$$Q_{Lk} + Q_k(t) = 0 \tag{13.6}$$

where the constants P_{Lk} and Q_{Lk} denote the load demands of active and reactive power of load k. Since the active power load is relatively insensitive to change in the voltage magnitude while the reactive power load is relatively insensitive to change in the phase angle as known in [7, 8], the load model (13.5)–(13.6) can be transformed into the following form

$$D_{Lp,k}\frac{d}{dt}\theta_k(t) = -P_{Lk} - P_k(t) \tag{13.7}$$

$$D_{Lq,k}\frac{d}{dt}V_k(t) = -Q_{Lk} - Q_k(t) \tag{13.8}$$

where $D_{Lp,k}$ and $D_{Lq,k}$ are sufficiently small positive constants. According to the singular perturbation approach [9], the equilibrium of the differential equations (13.3)–(13.4) and (13.7)–(13.8) shares the same stability properties as the same equilibrium of the differential algebraic equations (13.3)–(13.6). The inverter and load models (13.3)–(13.4) and (13.7)–(13.8) combine to make the frequency and voltage control model of isolated microgrid in this chapter.

13.1.2 Communication Network

The proposed frequency and voltage control law is performed on each inverter in a distributed way to avoid the drawbacks of the centralized control architecture as mentioned in the introduction. This implies that the neighboring inverters in the microgrid need to be able to communicate with each other. The communication technology for microgrid, which includes the performance of web-based communication and several industrial communication standards, is discussed in [10, 11]. The implementation over wireless networks is feasible since it increases the stability of secondary control and ensre plug-and-play ability. The communication network is represented by an undirected connected graph $\mathcal{G} \triangleq (\mathcal{V}, \mathcal{E}, C)$. $\mathcal{V} \triangleq \{v_1, v_2, ..., v_m\}$ is the node set where v_i signifies inverter i for $i \in \{1, ..., m\}$. $\mathcal{E} \triangleq \{(v_j, v_i), if\ j \to i\}$ is the edge set where $j \to i$ denotes that inverter j can transmit its own information to inverter i. Define the neighbouring set of inverter i as $N_i \triangleq \{v_j \in \mathcal{V} | (v_j, v_i) \in \mathcal{E}\}$, whose cardinal number is $|N_i|$. Define the Laplacian matrix of graph $\mathcal{G}$ as $L \triangleq \{l_{ij}\} \in R^{m\times m}$, where $l_{ij} = -1$ if and only if $(i, j) \in \mathcal{E}$, otherwise $l_{ij} = 0$ and $l_{ii} = -\sum_{j\neq i} l_{ij}$. It is known that the matrix L is semi-positive definite according to [12] since the graph $\mathcal{G}$ is undirected connected. Note that the communication network does not necessarily have to coincide with the electrical network between the inverters in the microgrid.

13.1.3 Control Purposes

When the power source or load patterns of the isolated microgrid are changed subject to certain disturbances, the frequencies and voltage magnitudes of the buses will deviate from their specific values. Under this situation, the restoration mechanisms $p_i(t)$ and $q_i(t)$ of each inverter as shown in (13.3)–(13.4) should be designed in order to accomplish the following two purposes.

Purpose 13.1 Restore the frequency and voltage magnitude. That is, the frequency and voltage magnitude of each bus in the microgrid are adjusted at the specific values respectively, i.e. $P_i^* - P_i(t) - p_i(t) \equiv 0$ and $Q_i^* - Q_i(t) - q_i(t) \equiv 0$ for inverter $i \in \{1, ..., m\}$, $P_{Lk} + P_k(t) \equiv 0$ and $Q_{Lk} + Q_k(t) \equiv 0$ for load $k \in \{m+1, ..., n\}$ at the steady state. This purpose is the basis of the isolated microgrid safe and stable operation.

Purpose 13.2 Realize the fair utilization of all power sources. This purpose is mathematically stated as $\frac{P_i(t)}{P_i^*} = \frac{P_j(t)}{P_j^*}$ and $\frac{Q_i(t)}{Q_i^*} = \frac{Q_j(t)}{Q_j^*}$ for $i, j \in \{1, 2, ..., m\}$ at the steady state, where the functions $\frac{P_i(t)}{P_i^*}$ and $\frac{Q_i(t)}{Q_i^*}$ denote the active and reactive power utilization ratio of power source inverter i respectively. Purpose 13.2 indicates that all the m power sources fairly share the entire loads in the microgrid.

The following restoration mechanism was designed in [5] to achieve Purpose 13.1 and 13.2

$$D_{p,i}\frac{d}{dt}p_i(t) = P_i^* - P_i(t) - p_i(t) + \sum_{j \in N_i}\left(\frac{p_j(t)}{D_{p,j}} - \frac{p_i(t)}{D_{p,i}}\right) \tag{13.9}$$

$$D_{q,i}\frac{d}{dt}q_i(t) = Q_i^* - Q_i(t) - q_i(t) + \sum_{j \in N_i}\left(\frac{q_j(t)}{D_{q,j}} - \frac{q_i(t)}{D_{q,i}}\right) \tag{13.10}$$

Considering the close-loop system constructed by (13.3)–(13.4), (13.7)–(13.8) and (13.9)–(13.10), the following assumption is mild as shown in [5, 13] for practical microgrid and it will be used in our proofs to given the existence of close-loop system equilibrium.

Assumption 13.1 The equilibrium of the following closed-loop system exist

$$\begin{aligned}
D_{p,i}\frac{d}{dt}\theta_i(t) &= P_i^* - P_i(t) - p_i(t)\\
D_{p,i}\frac{d}{dt}p_i(t) &= P_i^* - P_i(t) - p_i(t) + \sum_{j \in N_i}\left(\frac{p_j(t)}{D_{p,j}} - \frac{p_i(t)}{D_{p,i}}\right)\\
D_{q,i}\frac{d}{dt}V_i(t) &= Q_i^* - Q_i(t) - q_i(t)\\
D_{q,i}\frac{d}{dt}q_i(t) &= Q_i^* - Q_i(t) - q_i(t) + \sum_{j \in N_i}\left(\frac{q_j(t)}{D_{q,j}} - \frac{q_i(t)}{D_{q,i}}\right)\\
D_{Lp,k}\frac{d}{dt}\theta_k(t) &= -P_{Lk} - P_k(t)\\
D_{Lq,k}\frac{d}{dt}V_k(t) &= -Q_{Lk} - Q_k(t)
\end{aligned}$$

where inverter $i \in \{1, ..., m\}$ and load $k \in \{m + 1, ..., n\}$.

As mentioned before, the restoration mechanisms $p_i(t)$ and $q_i(t)$ should be designed in a distributed way, which implies that the implementation of the restoration mechanisms of inverter i only requires the information of itself and its neighbors in N_i. Although the conventional restoration mechanisms (13.9)–(13.10) satisfy the distributed architecture, the implementation of which is based on continuous information transmission. The restoration mechanism based on event-triggered strategy will be constructed in this chapter to reduce the communication burdens. The event-triggered mechanism is configured at each inverter which is used to determine when the local information should be transmitted to its neighbors. Similar to the restoration mechanisms, the event-triggered mechanism should also adopt the distributed implementation.

13.2 Main Results

13.2.1 Distributed Event-Triggered Restoration Mechanisms

Since the event-triggered mechanism is introduced, each inverter in the microgrid transmits its own information to its neighbors at some discrete triggering times. Denote $\{t_0^i, t_1^i, ..., t_g^i, ...\}$ and $\{\tau_0^i, \tau_1^i, ..., \tau_h^i ...\}$, where $t_0^i < t_1^i < t_2^i < ...$ and $\tau_0^i < \tau_1^i < \tau_2^i < ...$, as the triggering time sequences corresponding to the restoration mechanisms $p_i(t)$ and $q_i(t)$ of inverter i respectively. It should be mentioned that, $[t_0^i, t_1^i) \cup [t_1^i, t_2^i) \cup [t_2^i, t_3^i) \cup ... = [0, +\infty)$ and $[\tau_0^i, \tau_1^i) \cup [\tau_1^i, \tau_2^i) \cup [\tau_2^i, \tau_3^i) \cup ... = [0, +\infty)$. For any $t \in [0, +\infty)$, there must be some constants $g \in Z^+$ and $h \in Z^+$ such that $t \in [t_g^i, t_{g+1}^i)$ and $t \in [\tau_h^i, \tau_{h+1}^i)$. Based on the event-triggered mechanism, the distributed restoration mechanisms $p_i(t)$ and $q_i(t)$ is designed as follows:

$$D_{p,i}\frac{d}{dt}p_i(t) = P_i^* - P_i(t) - p_i(t) + \sum_{j\in N_i}\left(\frac{p_j(t_{g'(t)}^j)}{D_{p,j}} - \frac{p_i(t_g^i)}{D_{p,i}}\right), \quad t \in [t_g^i, t_{g+1}^i) \tag{13.11}$$

$$D_{q,i}\frac{d}{dt}q_i(t) = Q_i^* - Q_i(t) - q_i(t) + \sum_{j\in N_i}\left(\frac{q_j(\tau_{h'(t)}^j)}{D_{q,j}} - \frac{q_i(\tau_h^i)}{D_{q,i}}\right), \quad t \in [\tau_h^i, \tau_{h+1}^i) \tag{13.12}$$

where $t_{g'(t)}^j$ and $\tau_{h'(t)}^j$ are the latest triggering times corresponding to the restoration mechanisms $p_j(t)$ and $q_j(t)$ of inverter j respectively.

Remark 13.1 Note that only the information of inverter i and its neighbors in N_i is required to execute the restoration mechanism (13.11) and (13.12), which is compliant with the distributed control architecture. Moreover, inverter i receives the triggered information $p_j(t_{g'(t)}^j)$ and $q_j(\tau_{h'(t)}^j)$ and remains it unchanged until the next corresponding triggered information is transmitted from inverter j when the triggering event occurs at inverter j for $j \in N_i$. In other word, in each time interval $t \in [t_g^i, t_{g+1}^i) \cap [\tau_h^i, \tau_{h+1}^i)$, inverter i only uses its neighbors' latest triggered information, which is $p_j(t_{g'(t)}^j)$ and $q_j(\tau_{h'(t)}^j)$ for $j \in N_i$, instead of their current information $p_j(t)$ and $q_j(t)$. This implies that less information transmission is required under the event-triggered restoration mechanisms compared with the continuous information transmission restoration mechanisms in [5].

The next step is to design the event-triggered mechanism of inverter $i \in \{1, ..., m\}$, which is used to determine the triggering times $\{t_g^i\}_{g=0}^{\infty}$ and $\{\tau_h^i\}_{h=0}^{\infty}$.

13.2.2 Distributed Event-Triggered Mechanism

Define the measurement errors with respect to the restoration mechanisms $p_i(t)$ and $q_i(t)$ for inverter i as

$$e_{p,i}(t) \triangleq p_i(t) - p_i(t_g^i), \qquad t \in [t_g^i, t_{g+1}^i) \tag{13.13}$$

$$e_{q,i}(t) \triangleq q_i(t) - q_i(\tau_h^i), \qquad t \in [\tau_h^i, \tau_{h+1}^i) \tag{13.14}$$

The event-triggered mechanism of inverter $i \in \{1, ..., m\}$ is designed as follows.

Event-triggered mechanism 1: Construct the following two triggering conditions for p_i and q_i respectively as

$$e_{p,i}^2(t) > \frac{D_{p,i}^2}{4|N_i|} \sum_{j \in N_i} \left(\frac{p_j(t_{g'(t)}^j)}{D_{p,j}} - \frac{p_i(t_g^i)}{D_{p,i}} \right)^2 + \frac{D_{p,i}}{2|N_i|} \left(P_i^* - P_i(t) - p_i(t)\right)^2 \tag{13.15}$$

$$e_{q,i}^2(t) > \frac{D_{q,i}^2}{4|N_i|} \sum_{j \in N_i} \left(\frac{q_j(\tau_{h'(t)}^j)}{D_{q,j}} - \frac{q_i(\tau_h^i)}{D_{q,i}} \right)^2 + \frac{D_{q,i}}{2|N_i|} \left(Q_i^* - Q_i(t) - q_i(t)\right)^2 \tag{13.16}$$

Denoting the last triggering times for $p_i(t)$ and $q_i(t)$ as t_g^i and τ_h^i, the next triggering about $p_i(t)$ (resp. q_i) occurs at time t if the triggering condition (13.15) (resp. (13.16)) is satisfied. Denoting the new triggering time $t_{g+1}^i = t$ (resp. $\tau_{h+1}^i = t$), the value of $p_i(t_{g+1}^i)$ (resp. $q_i(\tau_{h+1}^i)$) will be sampled and transmitted to the inverters in N_i.

Remark 13.2 The triggering condition (13.15) determines the triggering times $\{t_g^i\}_{g=0}^{\infty}$ about the restoration mechanism p_i while the triggering condition (13.16) determines the triggering times $\{\tau_h^i\}_{h=0}^{\infty}$ about the restoration mechanism q_i. If the triggering condition is satisfied, inverter i will transmit its own information to its neighbors; otherwise, no information needs to be transmitted. Since the information transmission occurs only at the discrete triggering times, the communication burdens can be sharply reduced. Only the information of the neighbors of inverter i is required for the execution of Event-triggered mechanism 1, which is in harmony with distributed architecture. Moreover, similar to the restoration mechanisms (13.11) and (13.12), Event-triggered mechanism 1 only uses the neighbors' latest triggered information instead of their real-time information.

Theorem 13.1 *The distributed event-triggered restoration mechanisms (13.11) and (13.12) with Event-triggered mechanism 1 restore the frequency and voltage magnitude in the microgrid. Moreover, if the droop coefficients of inverters satisfy* $\frac{D_{p,i}}{P_i^*} = \frac{D_{p,j}}{P_j^*}$ *and* $\frac{D_{q,i}}{Q_i^*} = \frac{D_{q,j}}{Q_j^*}$ *for* $i, j \in \{1, ..., m\}$, *the fair utilization of all power sources is realized.*

Proof 13.1 *Construct the function*

$$U = -\frac{1}{2}\left(\sum_{i=1}^{n}\sum_{j=1}^{n} V_i V_j B_{ij} cos(\theta_i - \theta_j)\right) - \sum_{i=1}^{m} P_i^* \theta_i$$
$$+ \sum_{i=m+1}^{n} P_{Li}^* \theta_i - \sum_{i=1}^{m} Q_i^* \ln V_i + \sum_{i=m+1}^{n} Q_{Li} \ln V_i$$
$$+ \frac{1}{2}\sum_{i=1}^{m} p_i^2 + \frac{1}{2}\sum_{i=1}^{m} q_i^2 \tag{13.17}$$

Define the variable $v_i \triangleq \ln V_i$ *for* $i \in \{1, ..., n\}$. *According to (13.1) and (13.2), we have*

$$\frac{\partial U}{\partial \theta_i} = \begin{cases} P_i(t) - P_i^* & i \in \{1, ..., m\} \\ P_i(t) + P_{Li} & i \in \{m+1, ..., n\} \end{cases}$$
$$\frac{\partial U}{\partial v_i} = \begin{cases} Q_i(t) - Q_i^* & i \in \{1, ..., m\} \\ Q_i(t) + Q_{Li} & i \in \{m+1, ..., n\} \end{cases}$$
$$\frac{\partial U}{\partial p_i} = p_i(t), \qquad \frac{\partial U}{\partial q_i} = q_i(t)$$

Then, from (13.3)–(13.4) and (13.7)–(13.14), the closed-loop system can be written as follows.
For inverter $i \in \{1, ..., m\}$,

$$D_{p,i}\frac{d}{dt}\theta_i(t) = -\frac{\partial U}{\partial \theta_i} - \frac{\partial U}{\partial p_i} \tag{13.18}$$
$$D_{p,i}\frac{d}{dt}p_i(t) = -\frac{\partial U}{\partial \theta_i} - \frac{\partial U}{\partial p_i} - \sum_{j=1}^{m}\frac{l_{ij}}{D_{p,j}}(p_j(t) - e_{p,j}(t)) \tag{13.19}$$
$$e^{v_i(t)} D_{q,i}\frac{d}{dt}v_i(t) = -\frac{\partial U}{\partial v_i} - \frac{\partial U}{\partial q_i} \tag{13.20}$$
$$D_{q,i}\frac{d}{dt}q_i(t) = -\frac{\partial U}{\partial v_i} - \frac{\partial U}{\partial q_i} - \sum_{j=1}^{m}\frac{l_{ij}}{D_{q,j}}(q_j(t) - e_{q,j}(t)) \tag{13.21}$$

where l_{ij} is the the element in the Laplacian matrix L.
For load $i \in \{m+1, ..., n\}$,

$$D_{p,i}\frac{d}{dt}\theta_i(t) = -\frac{\partial U}{\partial \theta_i} \tag{13.22}$$

$$e^{v_i(t)}D_{q,i}\frac{d}{dt}v_i(t) = -\frac{\partial U}{\partial v_i} \tag{13.23}$$

Denote the $m \times n$ dimensional zero and identity matrix as $0_{m\times n}$ and $I_{m\times n}$ respectively. Define the vectors $\theta(t) \triangleq (\theta_1(t), ..., \theta_n(t))^T$, $p(t) \triangleq (p_1(t), ..., p_m(t))^T$, $v(t) \triangleq (v_1(t), ..., v_n(t))^T$, $q(t) \triangleq (q_1(t), ..., q_m(t))^T$, $e_p(t) \triangleq (e_{p,1}(t), ..., e_{p,m}(t))^T$, $e_q(t) \triangleq (e_{q,1}(t), ..., e_{q,m}(t))^T$, $x(t) \triangleq (\theta^T(t), p^T(t), v^T(t), q^T(t))^T$ and $e(t) \triangleq (0_{1\times n}, e_p^T(t), 0_{1\times n}, e_q^T(t))^T$. Moreover, set the matrix $D_p \triangleq diag(D_{p,1}, ..., D_{p,m})$, $D_{Lp} \triangleq diag(D_{Lp,m+1}, ..., D_{Lp,n})$, $\mathbf{e}^{v(t)} \triangleq diag(e^{v_1(t)}, ..., e^{v_m(t)})$, $\mathbf{e}^{v_L(t)} \triangleq diag(e^{v_{m+1}(t)}, ..., e^{v_n(t)})$, $D_q \triangleq diag(D_{q,1}, ..., D_{q,m})$, $D_{Lq} \triangleq diag(D_{Lq,m+1}, ..., D_{Lq,n})$. Construct the following $2(m+n) \times 2(m+n)$ dimensional block matrix $D(t) \triangleq diag(D_p, D_{Lp}, D_p, \mathbf{e}^{v(t)}D_q, \mathbf{e}^{v_L(t)}D_{Lq}, D_q)$ and $A \triangleq A_1 + A_2$, where A_1 is shown as (*) and $A_2 \triangleq diag(0_{n\times n}, L\cdot D_p^{-1}, 0_{n\times n}, L\cdot D_q^{-1})$.

$$A_1 \triangleq \begin{pmatrix} I_{n\times n} & \begin{pmatrix} I_{m\times m} \\ 0_{(n-m)\times m} \end{pmatrix} & 0_{n\times n} & 0_{n\times m} \\ \left(I_{m\times m}, 0_{m\times(n-m)}\right) & I_{m\times m} & 0_{m\times n} & 0_{m\times m} \\ 0_{n\times n} & 0_{n\times m} & I_{n\times n} & \begin{pmatrix} I_{m\times m} \\ 0_{(n-m)\times m} \end{pmatrix} \\ 0_{m\times n} & 0_{m\times m} & \left(I_{m\times m}, 0_{m\times(n-m)}\right) & I_{m\times m} \end{pmatrix} \tag{*}$$

Based on the above vectors and matrix, the closed-loop system (13.18)–(13.23) can be rewritten in a compact form as

$$D\frac{d}{dt}x = -A\frac{\partial U}{\partial x} + A_2 e(t) \tag{13.24}$$

The derivative of U with respect to time t along with the solution of system (13.24) is

$$\begin{aligned}
\frac{d}{dt}U &= \left(\frac{\partial U}{\partial x}\right)^T \frac{dx}{dt} \\
&= -\left(\frac{\partial U}{\partial x}\right)^T D^{-1}A\frac{\partial U}{\partial x} + \left(\frac{\partial U}{\partial x}\right)^T D^{-1}A_2 e(t) \\
&= -\frac{1}{2}\left(\frac{\partial U}{\partial x}\right)^T D^{-1}A\frac{\partial U}{\partial x} - \frac{1}{2}\left(\frac{\partial U}{\partial x}\right)^T D^{-1}A_1\frac{\partial U}{\partial x} \\
&\quad -\frac{1}{2}\left(\frac{\partial U}{\partial x}\right)^T D^{-1}A_2\frac{\partial U}{\partial x} + \left(\frac{\partial U}{\partial x}\right)^T D^{-1}A_2 e(t)
\end{aligned} \tag{13.25}$$

By simple calculation, we can get

$$\left(\frac{\partial U}{\partial x}\right)^T D^{-1} A_2 \frac{\partial U}{\partial x} = p^T(t) D_p^{-1} L D_p^{-1} p(t) + q^T(t) D_q^{-1} L D_q^{-1} q(t) \quad (13.26)$$

and

$$\left(\frac{\partial U}{\partial x}\right)^T D^{-1} A_2 e(t) = p^T(t) D_p^{-1} L D_p^{-1} e_p(t) + q^T(t) D_q^{-1} L D_q^{-1} e_q(t) \quad (13.27)$$

Defining the vectors $\widehat{p}(t) \triangleq (p_1(t^1_{g'(t)}), \ldots, p_m(t^m_{g'(t)}))^T$ *and* $\widehat{q}(t) \triangleq (q_1(\tau^1_{h'(t)}), \ldots, q_m(\tau^m_{h'(t)}))^T$, *it can be derived from (13.13) and (13.14) that*

$$\begin{aligned}
&-\frac{1}{2} p^T(t) D_p^{-1} L D_p^{-1} p(t) + p^T(t) D_p^{-1} L D_p^{-1} e_p(t) \\
=& p^T(t) D_p^{-1} L D_p^{-1} \left(e_p(t) - \frac{1}{2} p(t)\right) \\
=& \frac{1}{2}\left(e_p(t) + \widehat{p}(t)\right) D_p^{-1} L D_p^{-1} \left(e_p(t) - \widehat{p}(t)\right) \\
=& \frac{1}{2}\left(e_p(t) D_p^{-1} L D_p^{-1} e_p(t) - \widehat{p}(t) D_p^{-1} L D_p^{-1} \widehat{p}(t)\right)
\end{aligned} \quad (13.28)$$

Since the communication topology G is undirected and connected, we have

$$\begin{aligned}
&\widehat{p}(t) D_p^{-1} L D_p^{-1} \widehat{p}(t) \\
=& \sum_{i=1}^m \frac{p_i(t^i_{g'(t)})}{D_{p,i}} \sum_{j \in N_i} \left(\frac{p_i(t^i_{g'(t)})}{D_{p,i}} - \frac{p_j(t^j_{g'(t)})}{D_{p,j}}\right) \\
=& \sum_{i=1}^m \sum_{j \in N_i} \left(\frac{p_i^2(t^i_{g'(t)})}{D_{p,i}^2} - \frac{p_i(t^i_{g'(t)}) p_j(t^j_{g'(t)})}{D_{p,i} D_{p,j}}\right) \\
=& \sum_{i=1}^m \sum_{j \in N_i} \left(\frac{p_i^2(t^i_{g'(t)})}{2D_{p,i}^2} + \frac{p_j^2(t^j_{g'(t)})}{2D_{p,j}^2} - \frac{p_i(t^i_{g'(t)}) p_j(t^j_{g'(t)})}{D_{p,i} D_{p,j}}\right) \\
=& \sum_{i=1}^m \sum_{j \in N_i} \frac{1}{2} \left(\frac{p_i(t^i_{g'(t)})}{D_{p,i}} - \frac{p_j(t^j_{g'(t)})}{D_{p,j}}\right)^2
\end{aligned} \quad (13.29)$$

and

$$
\begin{aligned}
& e_p(t) D_p^{-1} L D_p^{-1} e_p(t) \\
&= \sum_{i=1}^{m} \frac{e_{p,i}(t)}{D_{p,i}} \sum_{j\in N_i} \left(\frac{e_{p,i}(t)}{D_{p,i}} - \frac{e_{p,j}(t)}{D_{p,j}} \right) \\
&\leq \sum_{i=1}^{m} |N_i| \frac{e_{p,i}^2(t)}{D_{p,i}^2} + \sum_{i=1}^{m} \sum_{j\in N_i} |\frac{e_{p,i}(t)}{D_{p,i}}| \cdot |\frac{e_{p,j}(t)}{D_{p,j}}| \\
&\leq \sum_{i=1}^{m} |N_i| \frac{e_{p,i}^2(t)}{D_{p,i}^2} + \sum_{i=1}^{m} \sum_{j\in N_i} \left(\frac{e_{p,i}^2(t)}{2D_{p,i}^2} + \frac{e_{p,j}^2(t)}{2D_{p,j}^2} \right) \\
&= \sum_{i=1}^{m} |N_i| \frac{e_{p,i}^2(t)}{D_{p,i}^2} + \frac{1}{2} \sum_{i=1}^{m} |N_i| \frac{e_{p,i}^2(t)}{D_{p,i}^2} + \frac{1}{2} \sum_{i=1}^{m} \sum_{j\in N_i} \frac{e_{p,j}^2(t)}{D_{p,j}^2} \\
&= \sum_{i=1}^{m} |N_i| \frac{e_{p,i}^2(t)}{D_{p,i}^2} + \frac{1}{2} \sum_{i=1}^{m} |N_i| \frac{e_{p,i}^2(t)}{D_{p,i}^2} + \frac{1}{2} \sum_{i=1}^{m} |N_i| \frac{e_{p,i}^2(t)}{D_{p,i}^2} \\
&= \sum_{i=1}^{m} \frac{2|N_i| e_{p,i}^2(t)}{D_{p,i}^2}
\end{aligned}
\tag{13.30}
$$

Substituting (13.29) and (13.30) into (13.28) yields

$$
\begin{aligned}
& -\frac{1}{2} p^T(t) D_p^{-1} L D_p^{-1} p(t) + p^T(t) D_p^{-1} L D_p^{-1} e_p(t) \\
\leq & \sum_{i=1}^{m} \left(\frac{|N_i| e_{p,i}^2(t)}{D_{p,i}^2} - \sum_{j\in N_i} \frac{1}{4} \left(\frac{p_i(t_{g'(t)}^i)}{D_{p,i}} - \frac{p_j(t_{g'(t)}^j)}{D_{p,j}} \right)^2 \right)
\end{aligned}
\tag{13.31}
$$

The following inequation can be obtained by a similar analysis

$$
\begin{aligned}
& -\frac{1}{2} q^T(t) D_q^{-1} L D_q^{-1} q(t) + q^T(t) D_q^{-1} L D_q^{-1} e_q(t) \\
\leq & \sum_{i=1}^{m} \left(\frac{|N_i| e_{q,i}^2(t)}{D_{q,i}^2} - \sum_{j\in N_i} \frac{1}{4} \left(\frac{q_i(\tau_{h'(t)}^i)}{D_{q,i}} - \frac{q_j(\tau_{h'(t)}^j)}{D_{q,j}} \right)^2 \right)
\end{aligned}
\tag{13.32}
$$

By the definition of the matrix D and A_1, we can obtain

$$
\begin{aligned}
&\left(\frac{\partial U}{\partial x}\right)^T D^{-1} A_1 \frac{\partial U}{\partial x} \\
&\geq \sum_{i=1}^{m} \frac{1}{D_{p,i}} (P_i(t) - P_i^*)^2 + \sum_{i=1}^{m} \frac{1}{D_{p,i}} (P_i(t) - P_i^*) p_i(t) \\
&\quad + \sum_{i=1}^{m} \frac{1}{D_{p,i}} (P_i(t) - P_i^*) p_i(t) + \sum_{i=1}^{m} \frac{1}{D_{p,i}} p_i^2(t) \\
&\quad + \sum_{i=1}^{m} \frac{1}{D_{q,i}} (Q_i(t) - Q_i^*)^2 + \sum_{i=1}^{m} \frac{1}{D_{q,i}} (Q_i(t) - Q_i^*) q_i(t) \\
&\quad + \sum_{i=1}^{m} \frac{1}{D_{q,i}} (Q_i(t) - Q_i^*) q_i(t) + \sum_{i=1}^{m} \frac{1}{D_{q,i}} q_i^2(t) \\
&\quad + \sum_{i=m+1}^{n} \frac{1}{D_{p,Li}} (P_i(t) + P_{Li})^2 \\
&\quad + \sum_{i=m+1}^{n} \frac{1}{D_{q,Li}} (Q_i(t) + Q_{Li})^2 \\
&\geq \sum_{i=1}^{m} \frac{1}{D_{p,i}} \left(P_i(t) - P_i^* + p_i(t)\right)^2 \\
&\quad + \sum_{i=1}^{m} \frac{1}{D_{q,i}} \left(Q_i(t) - Q_i^* + q_i(t)\right)^2
\end{aligned} \tag{13.33}
$$

Substituting (13.26)–(13.27) and (13.31)–(13.33) into (13.25) yields

$$
\begin{aligned}
\frac{d}{dt} U \leq & -\frac{1}{2} \left(\frac{\partial U}{\partial x}\right)^T D^{-1} A \frac{\partial U}{\partial x} + \sum_{i=1}^{m} \left(\frac{|N_i| e_{p,i}^2(t)}{D_{p,i}^2} \right. \\
& - \sum_{j \in N_i} \frac{1}{4} \left(\frac{p_i(t_{g'(t)}^i)}{D_{p,i}} - \frac{p_j(t_{g'(t)}^j)}{D_{p,j}} \right)^2 \\
& \left. - \frac{1}{2D_{p,i}} \left(P_i(t) - P_i^* + p_i(t)\right)^2 \right) \\
& + \sum_{i=1}^{m} \left(\frac{|N_i| e_{q,i}^2(t)}{D_{q,i}^2} - \sum_{j \in N_i} \frac{1}{4} \left(\frac{q_i(\tau_{h'(t)}^i)}{D_{q,i}} - \frac{q_j(\tau_{h'(t)}^j)}{D_{q,j}} \right)^2 \right. \\
& \left. - \frac{1}{2D_{q,i}} \left(Q_i(t) - Q_i^* + q_i(t)\right)^2 \right)
\end{aligned} \tag{13.34}
$$

The triggering conditions (13.15) and (13.16) derive that

$$e_{p,i}^2(t) \leq \sum_{j \in N_i} \left(\frac{D_{p,i}^2}{4|N_i|} \left(\frac{p_j(t_{g'(t)}^j)}{D_{p,j}} - \frac{p_i(t_g^i)}{D_{p,i}} \right)^2 \right.$$
$$\left. + \frac{D_{p,i}}{2|N_i|} \left(P_i^* - P_i(t) - p_i(t)\right)^2 \right), \quad t \in [t_g^i, t_{g+1}^i) \tag{13.35}$$

and

$$e_{q,i}^2(t) \leq \sum_{j \in N_i} \left(\frac{D_{q,i}^2}{4|N_i|} \left(\frac{q_j(\tau_{h'(t)}^j)}{D_{q,j}} - \frac{q_i(\tau_h^i)}{D_{q,i}} \right)^2 \right.$$
$$\left. + \frac{D_{q,i}}{2|N_i|} \left(Q_i^* - Q_i(t) - q_i(t)\right)^2 \right), \quad t \in [\tau_h^i, \tau_{h+1}^i) \tag{13.36}$$

Combining (13.34), (13.35) and (13.36) leads to

$$\frac{d}{dt}U \leq -\frac{1}{2} \left(\frac{\partial U}{\partial x} \right)^T D^{-1} A \frac{\partial U}{\partial x} \leq 0 \tag{13.37}$$

where the last inequation is true since the matrix $D^{-1}A$ is positive semi-definite.

Since the fact that the boundedness of x can be deduced according to the boundedness of $U(x)$ though a similar analysis in [14], (13.37) implies that the solution of the closed-loop system (13.24) converges into the set $S \triangleq \{x | D^{-1} A \frac{\partial U}{\partial x} = 0\}$ by applying LaSalle's invariance principle. It should be mentioned that, $D^{-1} A \frac{\partial U}{\partial x} = 0$ indicates $A \frac{\partial U}{\partial x} = 0$, which further implies $P_i^* - P_i(t) - p_i(t) = 0$, $P_i^* - P_i(t) - p_i(t) + \sum_{j \in N_i} \left(\frac{p_j(t)}{D_{p,j}} - \frac{p_i(t)}{D_{p,i}} \right) = 0$, $Q_i^* - Q_i(t) - q_i(t) = 0$, $Q_i^* - Q_i(t) - q_i(t) + \sum_{j \in N_i} \left(\frac{q_j(t)}{D_{q,j}} - \frac{q_i(t)}{D_{q,i}} \right) = 0$ for inverter $i \in \{1, ..., m\}$, and $P_{Lk} + P_k(t) = 0$, $Q_{Lk} + Q_k(t) = 0$ for load $k \in \{m+1, ..., n\}$ by algebraic calculating and Assumption 13.1. This shows that S is the set of all equilibriums of the closed-loop system (13.18)–(13.23). Therefore, the solution of (13.18)–(13.23) starting from a neighborhood of the initial equilibrium asymptotically converges to an equilibrium in the neighborhood, which means that the frequency and voltage magnitude in the microgrid can be restored.

By the above analysis, it can be obtained that the equations $\sum_{j \in N_i} \left(\frac{p_j(t)}{D_{p,j}} - \frac{p_i(t)}{D_{p,i}} \right) = 0$ and $\sum_{j \in N_i} \left(\frac{q_j(t)}{D_{q,j}} - \frac{q_i(t)}{D_{q,i}} \right) = 0$ hold at the equilibrium for all $i \in \{1, ..., m\}$. This indicates that $\frac{p_i(t)}{D_{p,i}} = \frac{p_j(t)}{D_{p,j}}$ and $\frac{q_i(t)}{D_{q,i}} = \frac{q_j(t)}{D_{q,j}}$ for all $i, j \in \{1, ..., m\}$. Since the droop coefficients of inverters are selected as $\frac{D_{p,i}}{P_i^*} = \frac{D_{p,j}}{P_j^*}$ and $\frac{D_{q,i}}{Q_i^*} = \frac{D_{q,j}}{Q_j^*}$, and the fact that $P_i^* - P_i(t) - p_i(t) = 0$ and $Q_i^* - Q_i(t) - q_i(t) = 0$ at the equilibrium, we can obtain $\frac{P_i(t)}{P_i^*} = \frac{P_j(t)}{P_j^*}$ and $\frac{Q_i(t)}{Q_i^*} = \frac{Q_j(t)}{Q_j^*}$ in set S. Therefore, the fair utilization of all power sources can be realized. This concludes the proof.

Theorem 13.1 shows that Purpose 13.1 and 13.2 can be achieved by the distributed restoration mechanisms (13.11)–(13.12) under Event-triggered mechanism

1. One key issue for event-triggered mechanism is to estimate the upper bound of the communication frequency of each inverter. The subsequent development analyze the inter-event time intervals $t^i_{g+1} - t^i_g$ and $\tau^i_{h+1} - \tau^i_h$ of the triggering conditions (13.15) and (13.16) respectively to establish the maximum communication rate that is required to achieve the control purpose.

Theorem 13.2 *The lower bounds of the inter-event time intervals for the triggering conditions (13.15) and (13.16) are described as*

$$t^i_{g+1} - t^i_g > \frac{1}{M_{p,i}}\left(\frac{D^2_{p,i}}{4|N_i|}\sum_{j\in N_i}\left(\frac{p_j(t^j_{g'(t^i_{g+1})})}{D_{p,j}} - \frac{p_i(t^i_g)}{D_{p,i}}\right)^2 + \frac{D_{p,i}}{2|N_i|}\left(P^*_i - P_i(t^i_{g+1}) - p_i(t^i_{g+1})\right)^2\right)^{\frac{1}{2}} \tag{13.38}$$

and

$$\tau^i_{h+1} - \tau^i_h > \frac{1}{M_{q,i}}\left(\frac{D^2_{q,i}}{4|N_i|}\sum_{j\in N_i}\left(\frac{q_j(\tau^j_{h'(\tau^i_{h+1})})}{D_{q,j}} - \frac{q_i(\tau^i_h)}{D_{q,i}}\right)^2 + \frac{D_{q,i}}{2|N_i|}\left(Q^*_i - Q_i(\tau^i_{h+1}) - q_i(\tau^i_{h+1})\right)^2\right)^{\frac{1}{2}} \tag{13.39}$$

with the positive constants $M_{p,i}$ and $M_{q,i}$, and the Zeno behavior can be excluded.

Proof 13.2 *For $t \in [t^i_g, t^i_{g+1})$, the following inequation can be obtained according to (13.11) and (13.13)*

$$\frac{d}{dt}|e_{p,i}(t)| \le |\frac{d}{dt}e_{p,i}(t)| = |\frac{d}{dt}p_i(t)| = \frac{1}{D_{p,i}}|P^*_i - P_i(t) - p_i(t) + \sum_{j\in N_i}\left(\frac{p_j(t^j_{g'(t)})}{D_{p,j}} - \frac{p_i(t^i_g)}{D_{p,i}}\right)| \le M_{p,i}$$

where the last inequation holds since the continuously differentiable functions $P^*_i - P_i(t) - p_i(t) + \sum_{j\in N_i}\left(\frac{p_j(t)}{D_{p,j}} - \frac{p_i(t)}{D_{p,i}}\right)$ converges to 0 as shown in the proof of Theorem 13.1. This implies $|e_{p,i}(t)| \le M_{p,i}(t - t^i_g)$ for $t \in [t^i_g, t^i_{g+1})$ since $e_{p,i}(t^i_g) = 0$. Based on the triggering condition (13.15), the next triggering time $t = t^i_{g+1}$ for inverter i occurs when $|e_{p,i}(t)| > \left(\frac{D^2_{p,i}}{4|N_i|}\sum_{j\in N_i}\left(\frac{p_j(t^j_{g'(t)})}{D_{p,j}} - \frac{p_i(t^i_g)}{D_{p,i}}\right)^2 + \frac{D_{p,i}}{2|N_i|}\left(P^*_i - P_i(t) - p_i(t)\right)^2\right)^{\frac{1}{2}}$ holds, which means that $t^i_{g+1} - t^i_g > \frac{1}{M_{p,1}}\left(\frac{D^2_{p,i}}{4|N_i|}\sum_{j\in N_i}\left(\frac{p_j(t^j_{g'(t^i_{g+1})})}{D_{p,j}} - \frac{p_i(t^i_g)}{D_{p,i}}\right)^2 + \frac{D_{p,i}}{2|N_i|}\left(P^*_i - P_i(t^i_{g+1}) - p_i(t^i_{g+1})\right)^2\right)^{\frac{1}{2}}$, i.e. inequation (13.38).

Two cases are considered here. Note that $\frac{D_{p,i}^2}{4|N_i|}\sum_{j\in N_i}\left(\frac{p_j(t^j_{g'(t^i_{g+1})})}{D_{p,j}}-\frac{p_i(t^i_g)}{D_{p,i}}\right)^2+\frac{D_{p,i}}{2|N_i|}\left(P_i^*-P_i(t^i_{g+1})-p_i(t^i_{g+1})\right)^2\neq 0$ *in the first case. Obviously, the inter-event time interval* $t^i_{g+1}-t^i_g$ *is strictly positive, which means that the execution of the* $(g+1)$*th event triggering occurs after a finite time interval since the gth one. The other case is when* $\frac{D_{p,i}^2}{4|N_i|}\sum_{j\in N_i}\left(\frac{p_j(t^j_{g'(t^i_{g+1})})}{D_{p,j}}-\frac{p_i(t^i_g)}{D_{p,i}}\right)^2+\frac{D_{p,i}}{2|N_i|}\left(P_i^*-P_i(t^i_{g+1})-p_i(t^i_{g+1})\right)^2=0.$ *Note that* $\sum_{j\in N_i}\left(\frac{p_j(t^j_{g'(t^i_{g+1})})}{D_{p,j}}-\frac{p_i(t^i_g)}{D_{p,i}}\right)=0$ *and* $\left(P_i^*-P_i(t^i_{g+1})-p_i(t^i_{g+1})\right)=0$ *in this case, which shows that* $\theta_i(t^i_{g+1})$ *and* $p_i(t^i_{g+1})$ *are the equilibrium points according to (13.3) and (13.11). By a similar analysis in [15], the event triggering is unnecessary to occur at* $t=t^i_{g+1}$*. Therefore, the Zeno behavior about triggering condition (13.15) can be excluded according to the above analysis. The result about the triggering condition (13.16) can be proved similarly.*

Theorem 13.2 describes that the Zeno behavior can be avoided under Event-triggered mechanism 1. However, it can be seen from (13.38) and (13.39) that, the lower bounds of inter-event time intervals converge to zero when the Purpose 13.1 and 13.2 trend to be accomplished. It indicates a probable high communication frequency when the control purposes are almost completed, which would result serious pressure in communication network.

13.2.3 Modified Distributed Event-Triggered Mechanism

In this section, the event-triggered mechanism with a modified triggering threshold, which introduces some positive parameters to guarantee the uniform positive lower bounds of inter-event time intervals, is established to avoid the issue of the potential high communication frequency caused by Event-triggered mechanism 1. *Event-triggered mechanism* 2: Construct the following two triggering conditions for p_i and q_i respectively as

$$e_{p,i}^2(t)>\frac{D_{p,i}^2}{4|N_i|}\sum_{j\in N_i}\left(\frac{p_j(t^j_{g'(t)})}{D_{p,j}}-\frac{p_i(t^i_g)}{D_{p,i}}\right)^2+\frac{D_{p,i}}{2|N_i|}(P_i^*-P_i(t)-p_i(t))^2+\frac{D_{p,i}^2}{|N_i|}\eta_{p,i} \tag{13.40}$$

$$e_{q,i}^2(t)>\frac{D_{q,i}^2}{4|N_i|}\sum_{j\in N_i}\left(\frac{q_j(\tau^j_{h'(t)})}{D_{q,j}}-\frac{q_i(\tau^i_h)}{D_{q,i}}\right)^2+\frac{D_{q,i}}{2|N_i|}\left(Q_i^*-Q_i(t)-q_i(t)\right)^2+\frac{D_{q,i}^2}{|N_i|}\eta_{q,i} \tag{13.41}$$

where $\eta_{p,i}$ and $\eta_{q,i}$ are positive constants. Denoting the last triggering times for $p_i(t)$ and $q_i(t)$ as t_g^i and τ_h^i, the next triggering about $p_i(t)$ (resp. q_i) occurs at time t if the triggering condition (13.40) (resp. (13.41)) is satisfied. Denoting the new triggering time $t_{g+1}^i = t$ (resp. $\tau_{h+1}^i = t$), the value of $p_i(t_{g+1}^i)$ (resp. $q_i(\tau_{h+1}^i)$) will be sampled and transmitted to the inverters in N_i.

Similar to Event-triggered mechanism 1, the triggering times $\{t_g^i\}_{g=0}^{\infty}$ and $\{\tau_h^i\}_{h=0}^{\infty}$ are determined by the triggering conditions (13.40) and (13.41) of Event-triggered mechanism 2, which also satisfies the distributed implementation. The difference between the two types of event-triggered mechanism is that the positive constant parameters $\eta_{p,i}$ and $\eta_{q,i}$ are used here to avoid the potential frequent information transmission which is possible in Event-triggered mechanism 1.

Theorem 13.3 *The restoration mechanisms (13.11) and (13.12) with Event-triggered mechanism 2 restore the frequency and voltage magnitude in the microgrid into an arbitrary small neighborhood of the corresponding specific values with the fluctuation range* $\sum_{i=1}^{m}(\eta_{p,i} + \eta_{q,i})$. *Moreover, if the droop coefficients of inverters satisfy* $\frac{D_{p,i}}{P_i^*} = \frac{D_{p,j}}{P_j^*}$ *and* $\frac{D_{q,i}}{Q_i^*} = \frac{D_{q,j}}{Q_j^*}$ *for* $i, j \in \{1, ..., m\}$, *the utilization ratios of all power sources achieve bounded consensus.*

Proof 13.3 *By defining the same functions and variables as those in the proof of Theorem 13.1, we can obtain the same closed-loop system (13.24) here since the restoration mechanisms are not changed, and the derivative of U satisfies (13.34). According to the triggering conditions (13.40) and (13.41), we have*

$$e_{p,i}^2(t) \le \frac{D_{p,i}^2}{4|N_i|}\sum_{j\in N_i}\left(\frac{p_j(t_{g'(t)}^j)}{D_{p,j}} - \frac{p_i(t_g^i)}{D_{p,i}}\right)^2 + \frac{D_{p,i}}{2|N_i|}(P_i^* - P_i(t) - p_i(t))^2 + \frac{D_{p,i}^2}{|N_i|}\eta_{p,i} \tag{13.42}$$

$$e_{q,i}^2(t) \le \frac{D_{q,i}^2}{4|N_i|}\sum_{j\in N_i}\left(\frac{q_j(\tau_{h'(t)}^j)}{D_{q,j}} - \frac{q_i(\tau_h^i)}{D_{q,i}}\right)^2 + \frac{D_{q,i}}{2|N_i|}(Q_i^* - Q_i(t) - q_i(t))^2 + \frac{D_{q,i}^2}{|N_i|}\eta_{p,i} \tag{13.43}$$

Substituting (13.42) and (13.43) into (13.34) yields

$$\begin{aligned}\frac{d}{dt}U &\le -\frac{1}{2}\left(\frac{\partial U}{\partial x}\right)^T D^{-1}A\frac{\partial U}{\partial x} + \sum_{i=1}^{m}(\eta_{p,i} + \eta_{q,i})\\ &= -\sum_{i=1}^{m}\frac{1}{2D_{p,i}}(P_i(t) - P_i^* + p_i(t))^2\\ &\quad -\sum_{i=1}^{m}\frac{1}{2D_{q,i}}(Q_i(t) - Q_i^* + q_i(t))^2\end{aligned}$$

$$
\begin{aligned}
&-\sum_{i=m+1}^{n}\frac{1}{2D_{p,Li}}(P_i(t)+P_{Li})^2\\
&-\sum_{i=m+1}^{n}\frac{1}{2D_{q,Li}}(Q_i(t)+Q_{Li})^2\\
&-\sum_{i=1}^{m}\sum_{j\in N_i}\frac{1}{4}\left(\frac{p_i(t)}{D_{p,i}}-\frac{p_j(t)}{D_{p,j}}\right)^2\\
&-\sum_{i=1}^{m}\sum_{j\in N_i}\frac{1}{4}\left(\frac{q_i(t)}{D_{q,i}}-\frac{q_j(t)}{D_{q,j}}\right)^2\\
&+\sum_{i=1}^{m}(\eta_{p,i}+\eta_{q,i})
\end{aligned}
\tag{13.44}
$$

where the last equation holds based on (13.26) and (13.33). Equation (13.44 indicates that the solution of the closed-loop system converges into the set $\overline{S}\triangleq\{(\theta_1,...\theta_n,p_1,...,p_m,V_1,...,V_n,q_1,...,q_m)|\sum_{i=1}^{m}\frac{1}{2D_{p,i}}\left(P_i(t)-P_i^*+p_i(t)\right)^2+\sum_{i=1}^{m}\frac{1}{2D_{q,i}}\left(Q_i(t)-Q_i^*+q_i(t)\right)^2+\sum_{i=1}^{m}\sum_{j\in N_i}\frac{1}{4}\left(\frac{p_i(t)}{D_{p,i}}-\frac{p_j(t)}{D_{p,j}}\right)^2+\sum_{i=1}^{m}\sum_{j\in N_i}\frac{1}{4}\left(\frac{q_i(t)}{D_{q,i}}-\frac{q_j(t)}{D_{q,j}}\right)^2+\sum_{i=m+1}^{n}\frac{1}{2D_{p,Li}}(P_i(t)+P_{Li})^2+\sum_{i=m+1}^{n}\frac{1}{2D_{q,Li}}(Q_i(t)+Q_{Li})^2\leq\sum_{i=1}^{m}(\eta_{p,i}+\eta_{q,i})\}$. *This implies that the frequency and voltage magnitude of the microgrid starting from a neighborhood of the initial equilibrium converge into an arbitrate small neighborhood of the equilibrium since the parameters* $\eta_{p,i}$ *and* $\eta_{q,i}$ *for* $i\in\{1,...,m\}$ *can be selected small enough.*

Moreover, it can also be obtained that $|\frac{p_i(t)}{D_{p,i}}-\frac{p_j(t)}{D_{p,j}}|\leq 2\left(\sum_{i=1}^{m}(\eta_{p,i}+\eta_{q,i})\right)^{\frac{1}{2}}$ *for* $j\in N_i$ *in set* $\overline{S}$. *Defining the variable* $\xi_{p,i}(t)\triangleq P_i(t)-P_i^*+p_i(t)$, *we have* $|\xi_{p,i}(t)|\leqslant(2D_{p,i}\sum_{k=1}^{m}(\eta_{p,k}+\eta_{q,k}))^{\frac{1}{2}}$ *in set* $\overline{S}$. *Since*

$$
\begin{aligned}
&|\frac{p_i(t)}{D_{p,i}}-\frac{p_j(t)}{D_{p,j}}|\\
=&|\frac{P_i^*-P_i(t)-\xi_{p,i}(t)}{D_{p,i}}-\frac{P_j^*-P_j(t)-\xi_{p,j}(t)}{D_{p,j}}|\\
=&|\frac{P_i^*}{D_{p,i}}\left(1-\frac{P_i(t)}{P_i^*}-\frac{\xi_{p,i}(t)}{P_i^*}\right)-\frac{P_j^*}{D_{p,j}}\left(1-\frac{P_j(t)}{P_j^*}-\frac{\xi_{p,j}(t)}{P_j^*}\right)|\\
=&\frac{P_i^*}{D_{p,i}}|-\frac{P_i(t)}{P_i^*}-\frac{\xi_{p,i}(t)}{P_i^*}+\frac{P_j(t)}{P_j^*}+\frac{\xi_{p,j}(t)}{P_j^*}|\\
\geq&\frac{P_i^*}{D_{p,i}}\left(|\frac{P_i(t)}{P_i^*}-\frac{P_j(t)}{P_j^*}|-|\frac{\xi_{p,i}(t)}{P_i^*}-\frac{\xi_{p,j}(t)}{P_j^*}|\right)
\end{aligned}
$$

we can get the following inequation for $j\in N_i$ *in set* $\overline{S}$

$$
\begin{aligned}
&|\frac{P_i(t)}{P_i^*} - \frac{P_j(t)}{P_j^*}| \\
\le &\frac{D_{p,i}}{P_i^*}|\frac{p_i(t)}{D_{p,i}} - \frac{p_j(t)}{D_{p,j}}| + |\frac{\xi_{p,i}(t)}{P_i^*} - \frac{\xi_{p,j}(t)}{P_j^*}| \\
\le &\frac{2D_{p,i}}{P_i^*}\left(\sum_{i=1}^{m}(\eta_{p,i}+\eta_{q,i})\right)^{\frac{1}{2}} + \frac{(2D_{p,i}\sum_{k=1}^{m}(\eta_{p,k}+\eta_{q,k}))^{\frac{1}{2}}}{P_i^*} + \frac{(2D_{p,j}\sum_{k=1}^{m}(\eta_{p,k}+\eta_{q,k}))^{\frac{1}{2}}}{P_j^*} \\
= &\left(\frac{2D_{p,i}}{P_i^*} + \frac{(2D_{p,i})^{\frac{1}{2}}}{P_i^*} + \frac{(2D_{p,j})^{\frac{1}{2}}}{P_j^*}\right)\left(\sum_{i=1}^{m}(\eta_{p,i}+\eta_{q,i})\right)^{\frac{1}{2}}
\end{aligned} \tag{13.45}
$$

Since the parameters $\eta_{p,i}$ and $\eta_{q,i}$ for $i \in \{1, ..., m\}$ can be selected small enough, (13.45) indicates that the active power utilization ratios of all power sources achieve bounded consensus. The result about the reactive power utilization ratio can be obtained by a similar analysis and we omit it here. This concludes the proof.

Remark 13.3 In the practical microgrid, the frequency, voltage magnitude and utilization ratio are not necessary to strictly satisfy asymptotically stability as shown in Theorem 13.1. Instead, it is allowed that these variables fluctuate in a certain range around the equilibrium. Therefore, although the restoration mechanism under Event-triggered mechanism 2 can only obtain the uniform boundedness result as described in Theorem 13.3, it can still fulfill the operational requirements of isolated microgrid.

Theorem 13.4 *The lower bounds of the inter-event time intervals for the triggering conditions (13.40) and (13.41) are described as*

$$
\begin{aligned}
t_{g+1}^i - t_g^i > \frac{1}{\overline{M}_{p,i}}\Bigg(&\frac{D_{p,i}^2}{4|N_i|}\sum_{j\in N_i}\left(\frac{p_j(t^j_{g'(t^i_{g+1})})}{D_{p,j}} - \frac{p_i(t_g^i)}{D_{p,i}}\right)^2 \\
&+ \frac{D_{p,i}}{2|N_i|}(P_i^* - P_i(t_{g+1}^i) - p_i(t_{g+1}^i))^2 + \frac{D_{p,i}^2}{|N_i|}\eta_{p,i}\Bigg)^{\frac{1}{2}}
\end{aligned} \tag{13.46}
$$

and

$$
\begin{aligned}
\tau_{h+1}^i - \tau_h^i > \frac{1}{\overline{M}_{q,i}}\Bigg(&\frac{D_{q,i}^2}{4|N_i|}\sum_{j\in N_i}\left(\frac{q_j(\tau^j_{h'(\tau^i_{h+1})})}{D_{q,j}} - \frac{q_i(\tau_h^i)}{D_{q,i}}\right)^2 \\
&+ \frac{D_{q,i}}{2|N_i|}(Q_i^* - Q_i(\tau_{h+1}^i) - q_i(\tau_{h+1}^i))^2 + \frac{D_{q,i}^2}{|N_i|}\eta_{q,i}\Bigg)^{\frac{1}{2}}
\end{aligned} \tag{13.47}
$$

with the positive constants $\overline{M}_{p,i}$ and $\overline{M}_{q,i}$, and the Zeno behavior can be avoided.

Proof 13.4 *The triggering condition (13.40) is considered here. We have $|e_{p,i}(t)| \le \overline{M}_{p,i}(t - t_g^i)$ and $|e_{p,i}(t)| > \left(\frac{D_{p,i}^2}{4|N_i|}\sum_{j\in N_i}\left(\frac{p_j(t^j_{g'(t)})}{D_{p,j}} - \frac{p_i(t_g^i)}{D_{p,i}}\right)^2 + \frac{D_{p,i}}{2|N_i|}\right.$*

$\left(P_i^* - P_i(t) - p_i(t)\right)^2 + \frac{D_{p,i}^2}{|N_i|}\eta_{p,i}\Big)^{\frac{1}{2}}$ *for* $t \in [t_g^i, t_{g+1}^i)$ *by similar proof in Theorem 13.2, where* $\overline{M}_{p,i}$ *is a positive constant, and thus (13.46) holds. Furthermore, the uniform positive lower bound* $\frac{D_{p,i}\eta_{p,i}^{\frac{1}{2}}}{\overline{M}_{p,i}|N_i|^{\frac{1}{2}}}$ *of time intervals* $t_{g+1}^i - t_g^i$ *avoids the Zeno behavior. The lower bound (13.41) can be proved by a similar analysis.*

Remark 13.4 Theorem 13.4 shows that the inter-event time intervals of the triggering conditions (13.40) and (13.41) are longer than the constants $\frac{D_{p,i}\eta_{p,i}^{\frac{1}{2}}}{\overline{M}_{p,i}|N_i|^{\frac{1}{2}}}$ and $\frac{D_{q,i}\eta_{q,i}^{\frac{1}{2}}}{\overline{M}_{q,i}|N_i|^{\frac{1}{2}}}$ respectively. This implies that the potential high communication frequency can be excluded by Event-triggered mechanism 2 when the control purposes are almost completed.

Remark 13.5 Theorems 13.3 and 13.4 imply that the regulation of parameters $\eta_{p,i}$ and $\eta_{q,i}$ can influence the fluctuation range and the triggering frequency. Increasing these parameters can reduce the the trigging frequency and increase the fluctuation range of the frequency, voltage magnitude and utilization ratio, and vice versa. This suggests a trade-off between the control precision and the communication network bandwidth. The higher control precision requires the smaller parameters $\eta_{p,i}$ and $\eta_{q,i}$, which however, potentially makes the communication frequency higher. In practical application, these parameters should be chosen with weighting the factors of the fluctuation tolerant level and the communication network bandwidth limitation.

13.3 Simulation

This section simulates a test system of isolated inverter-based microgrid contained six power source inverters and two loads as shown in Fig. 13.1, where the solid and dash lines denote the electric and communication lines respectively.

The detailed parameters are presented as follows

- The system voltage is three-phase 220 V and 50 Hz.
- The rated power of each power source inverter is 3 KVA.
- The reactance of each electric line is given in Table 13.1.
- The droop coefficients of all the power source inverters are set as $D_{p,i} = 4 \times 10^4\,\text{W/rad}\cdot\text{s}^{-1}$ and $D_{q,i} = 160\,\text{Var/V}\cdot\text{s}^{-1}$ for $i \in \{1, ..., 6\}$.
- The expected disturbance is that the two loads increase the total power demand from $1.32\,\text{kW} + j0.94\,\text{kVar}$ to $2.55\,\text{kW} + j1.58\,\text{kVar}$ at 1 s.

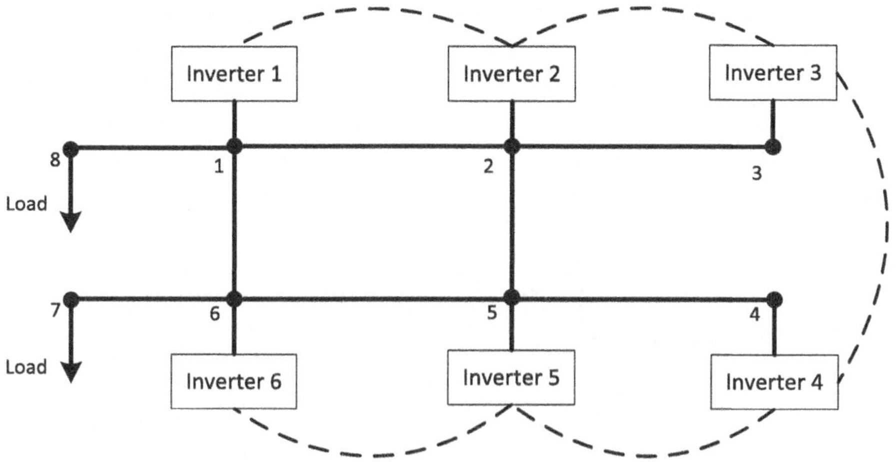

Fig. 13.1 The test microgrid

Table 13.1 Line reactance

From bus	To bus	Reactance ($m\Omega$)
1	2	650
1	6	572
1	8	342
2	3	487
2	5	391
4	5	370
5	6	509
6	7	414

13.3.1 Implementation with Event-Triggered Mechanism 1

The simulation result under the distributed restoration (13.11)–(13.12) with Event-triggered mechanism 1 are shown in Figs. 13.2, 13.3, 13.4, 13.5, 13.6, 13.7. Figures 13.2, 13.3 shows that the coincident inverter active and reactive power outputs can be achieved under the case of load change disturbance. It indicates that the fair utilization is realized since the rated powers of all power source inverter are set to be the same. Figures 13.4, 13.5 shows that the frequency and voltage magnitude of each bus are adjusted at their specific values respectively, which ensures the safe and stable operation of the isolated microgrid. Figures 13.6 and 13.7 depict the broadcast periods of inverter 1 under Event-triggered mechanism 1, where the abscissa and ordinate of crosse signify the triggering time and the spending time since the last triggering, respectively. The restoration mechanisms $p_i(t)$ and $q_i(t)$ are triggered 11 and 8 times respectively. Compared to the conventional restoration method based on continuous information transmission, the communication burdens is

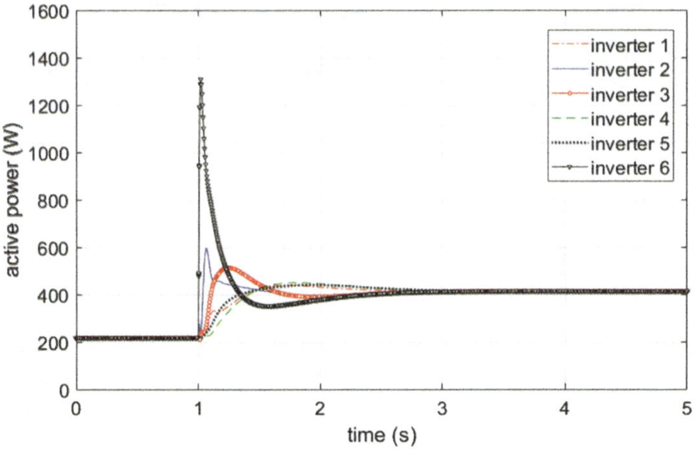

Fig. 13.2 Active power output under Event-triggered mechanism 1

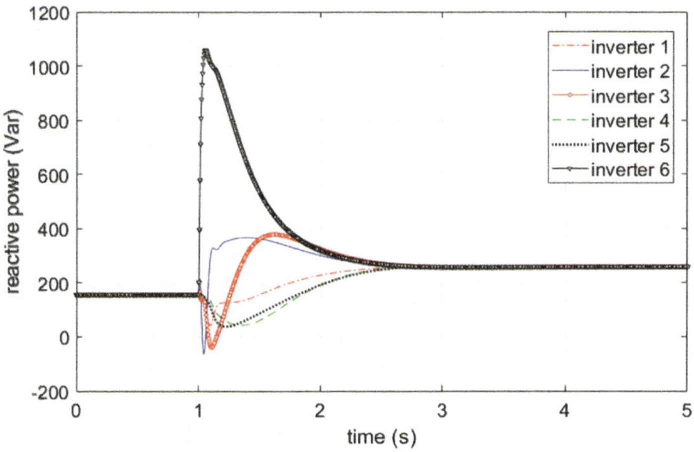

Fig. 13.3 Reactive power output under Event-triggered mechanism 1

reduced. Moreover, the unequal broadcast periods demonstrates that the information demand-sending character of event-triggered mechanism and the ability in adjusting communication frequency based on the system's state.

13.3.2 Implementation with Event-Triggered Mechanism 2

Figures 13.8, 13.9, 13.10, 13.11, 13.12, 13.13 show the result under the distributed restoration (13.19)–(13.20) with Event-triggered mechanism 2, where the parame-

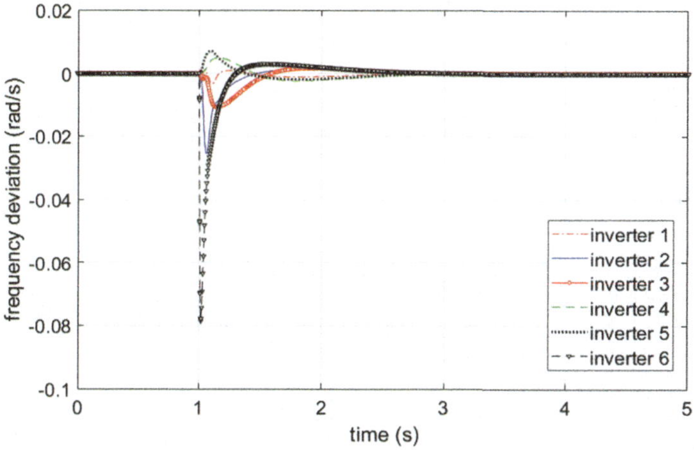

Fig. 13.4 Frequency deviation under Event-triggered mechanism 1

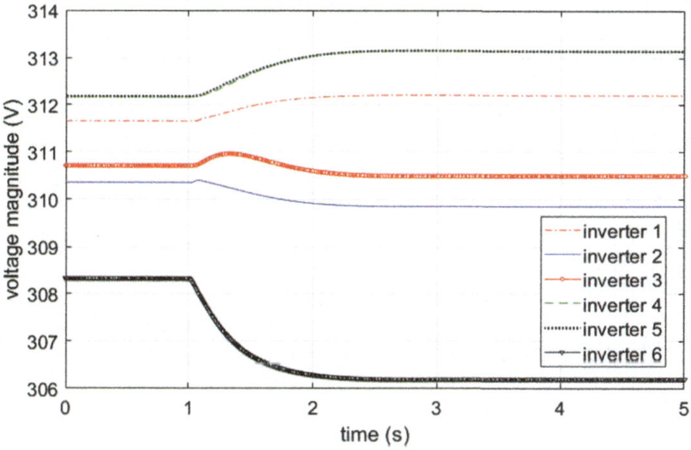

Fig. 13.5 Voltage magnitude under Event-triggered mechanism 1

ters are set as $\eta_{p,i} = 0.001$ and $\eta_{p,i} = 0.001$ for $i \in \{1, ..., 6\}$. Figures 13.8, 13.9, 13.10, 13.11 shows that the frequencies and voltage magnitudes can be restored and the utilization ratios of all power sources achieve consensus. The operational requirements of isolated microgrid can still be fulfilled under the modified event-triggered mechanism. It can be seen that the convergence rate here is lower than that under Event-triggered 1 as shown in Sect. 13.3.1, and there exists fluctuation in a certain range around the equilibrium. Either of the restoration mechanisms $p_i(t)$ and $q_i(t)$ is triggered 7 times, which indicates less information transmission than that in Event-triggered mechanism 1.

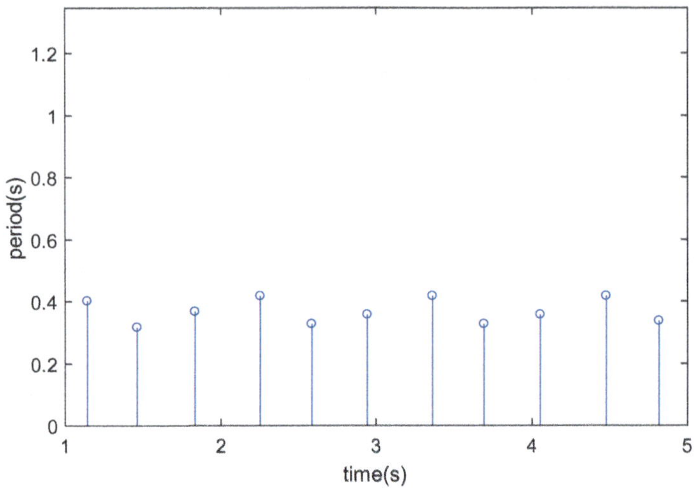

Fig. 13.6 Broadcast period of p_1 under Event-triggered mechanism 1

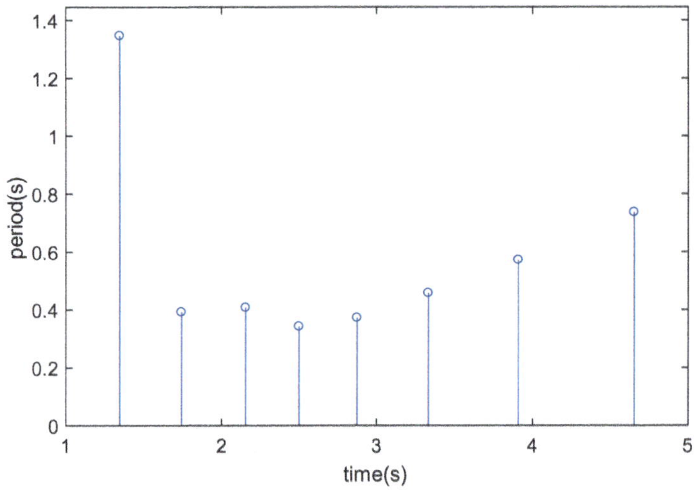

Fig. 13.7 Broadcast period of q_1 under Event-triggered mechanism 1

13.4 Conclusion

We have examined the problem of distributed frequency and voltage cooperative control in inverter-based isolated microgrid. The event-triggered mechanism is employed in the designed restoring laws to reduce the communication burdens. The proposed distributed event-triggered restoration mechanism can restore the frequency and voltage magnitude of microgrid, and realize the fair utilization of all power sources. Our future work concerns the lossy electric line case and the event-triggered coopera-

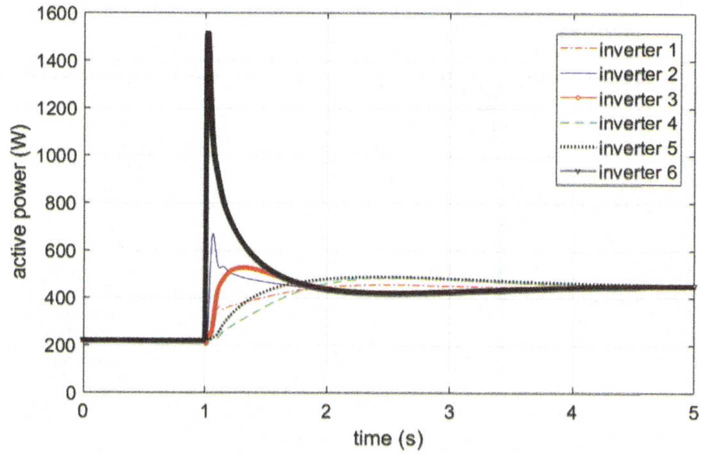

Fig. 13.8 Active power output under Event-triggered mechanism 2

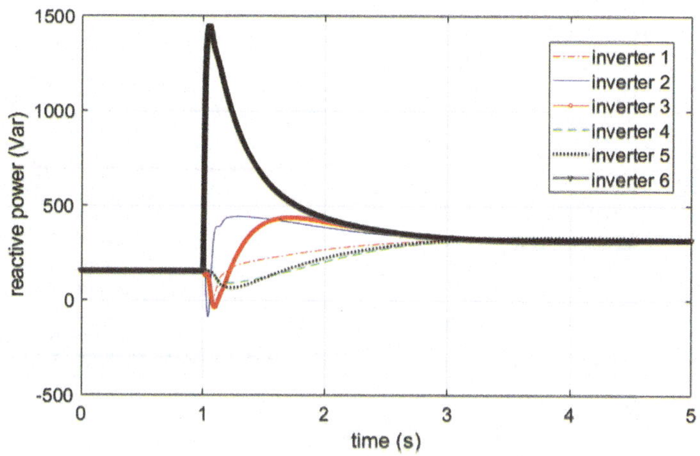

Fig. 13.9 Reactive power output under Event-triggered mechanism 2

tive control problem of virtual synchronous generators in isolated microgrid. The issue of power output limitation of generation will also be considered in the near future. Moreover, the nonideal signal transmission case in practical communication networks ia also planned.

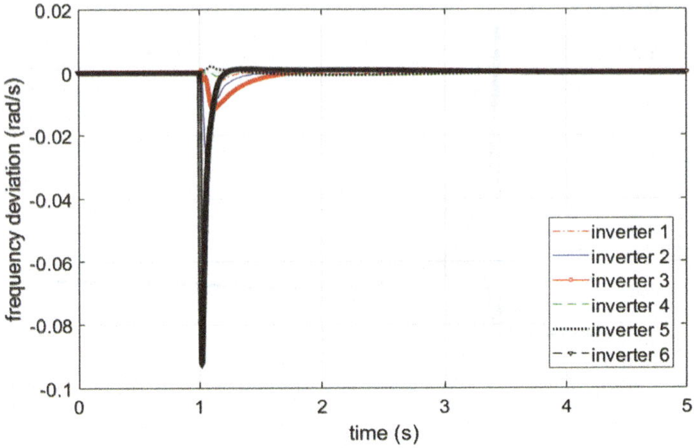

Fig. 13.10 Frequency deviation under Event-triggered mechanism 2

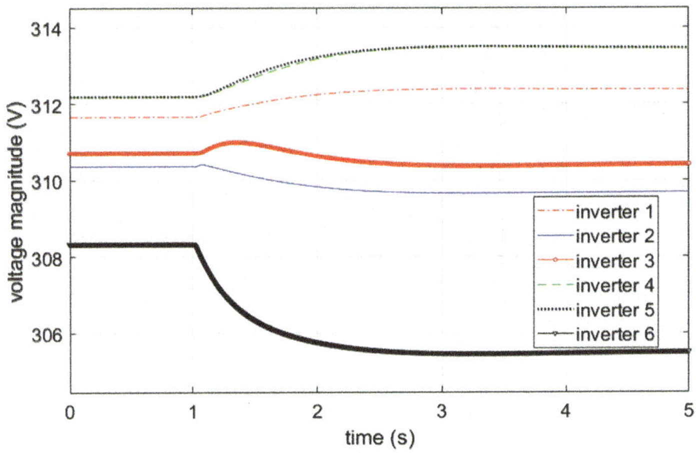

Fig. 13.11 Voltage magnitude under Event-triggered mechanism 2

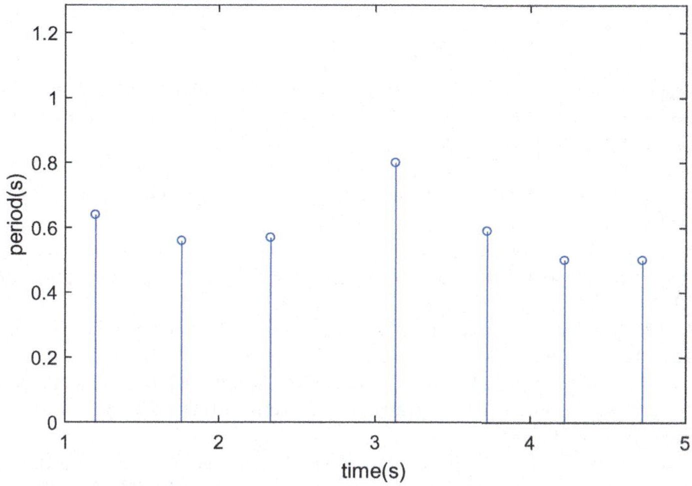

Fig. 13.12 Broadcast period of p_1 under Event-triggered mechanism 2

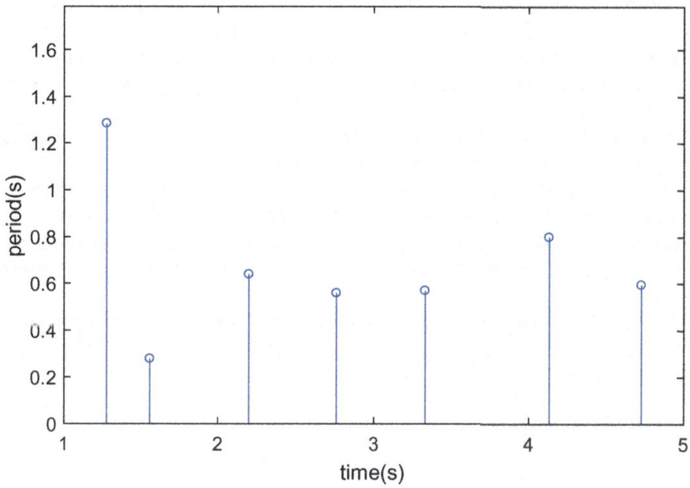

Fig. 13.13 Broadcast period of q_1 under Event-triggered mechanism 2

References

1. A.R. Bergen, *Power Systems Analysis*. Pearson Education India, India (2009)
2. C.K. Sao, P.W. Lehn, Autonomous load sharing of voltage source converters. IEEE Trans. Power Deliv. **20**(2), 1009–1016 (2005)
3. K. De Brabandere, B. Bolsens, J. Van den Keybus, A. Woyte, J. Driesen, R. Belmans, A voltage and frequency droop control method for parallel inverters. IEEE Trans. Power Electron. **22**(4), 1107–1115 (2007)
4. C. Lee, C. Chu, P. Cheng, A new droop control method for the autonomous operation of distributed energy resource interface converters. IEEE Trans. Power Electron. **28**(4), 1980–1993 (2013)
5. L. Lu, C. Chu, Consensus-based droop control synthesis for multiple dics in isolated micro-grids. IEEE Trans. Power Syst. **30**(5), 2243–2256 (2015)
6. G. Diaz, C. Gonzalez-Moran, J. Gomez-Aleixandre, A. Diez, Composite loads in stand-alone inverter-based microgrids—modeling procedure and effects on load margin. IEEE Trans. Power Syst. **25**(2), 894–905 (2010)
7. H.-D. Chang, C.-C. Chu, G. Cauley, Direct stability analysis of electric power systems using energy functions: theory, applications, and perspective. Proc. IEEE **83**(11), 1497–1529 (1995)
8. N. Tsolas, A. Arapostathis, P. Varaiya, A structure preserving energy function for power system transient stability analysis. IEEE Trans. Circuits Syst. **32**(10), 1041–1049 (1985)
9. J.W. Simpson-Porco, F. Dorfler, F. Bullo, Synchronization and power sharing for droop-controlled inverters in islanded microgrids. Automatica **49**(9), 2603–2611 (2013)
10. R. Majumder, G. Ledwich, A. Ghosh, S. Chakrabarti, F. Zare, Droop control of converter-interfaced microsources in rural distributed generation. IEEE Trans. Power Deliv. **25**(4), 2768–2778 (2010)
11. R. Majumder, G. Bag, K.H. Kim, Power sharing and control in distributed generation with wireless sensor networks. IEEE Trans. Smart Grid **3**(2), 618–634 (2012)
12. R. Olfati-Saber, R.M. Murray, Consensus problems in networks of agents with switching topology and time-delays. IEEE Trans. Autom. Control **49**(9), 1520–1533 (2004)
13. N. Ainsworth, S. Grijalva, A structure-preserving model and sufficient condition for frequency synchronization of lossless droop inverter-based ac networks. IEEE Trans. Power Syst. **28**(4), 4310–4319 (2013)
14. L. Lu, C. Chu, Consensus-based secondary frequency and voltage droop control of virtual synchronous generators for isolated ac micro-grids. IEEE J. Emerg. Select. Top. Circuits Syst. **5**(3), 443–455 (2015)
15. H. Li, X. Liao, T. Huang, W. Zhu, Event-triggering sampling based leader-following consensus in second-order multi-agent systems. IEEE Trans. Autom. Control **60**(7), 1998–2003 (2015)

Chapter 14
Event-Triggered Mechanism Based Distributed Optimal Frequency Regulation of Power Grid

Considering the limitation of communication network, the event-triggered mechanism based distributed optimal frequency regulation is proposed, which can restore the frequency and retain the economic efficiency of power grid simultaneously under certain load disturbances. Compared with the traditional distributed optimal frequency regulation based on continuous or periodic sampled-data information transmission mechanism, the communication burden among the areas in power grid is reduced since two novel types of event-triggered mechanisms, which possess the information demand-transmission property, are constructed and introduced in the regulation method in this chapter. Moreover, Zeno behavior is proved to be avoided in order to guarantee the availability of event-triggered mechanisms in practical power grid. The effectiveness of theoretical result of this chapter is verified by the simulation study.

14.1 Problem Formulation

14.1.1 Power Grid Model

The considered power grid is partitioned into n smaller areas according to the coherency and aggregation techniques [1, 2], and then the power network can be represented by an undirected and connected graph $\mathcal{G} \triangleq (\mathcal{V}, \mathcal{E})$. The node set $\mathcal{V} \triangleq \{1, 2, ..., n\}$ and edge set $\mathcal{E} \triangleq \{(j, i)\} \subseteq \mathcal{V} \times \mathcal{V}$ describe the areas and transmission lines of power grid respectively, where the edge (j, i) means that there exists the transmission line connecting area j and i directly. The set of all the areas connected to area i by transmission lines is expressed as N_i. Assuming that there are a total of m transmission lines between the n areas in power grid, the incidence

D. Yue et al., *Cooperative Optimal Control of Hybrid Energy Systems*,
https://doi.org/10.1007/978-981-33-6722-7_14

matrix of graph $\mathcal{G}$ can be defined as $D \triangleq \{d_{ik}\} \in R^{n\times m}$, where $d_{ik} = 1$ if node i is the positive end of edge k, and $d_{ik} = -1$ if node i is the negative end of edge k, otherwise $d_{ik} = 0$. Based on the detail derivation shown in [3], the dynamical behavior of each area i can be regarded as an equivalent single generator and is represented as follows

$$\frac{d}{dt}\delta_i = \omega_i^b \tag{14.1}$$

$$\begin{aligned} M_i \frac{d}{dt}\omega_i^b =& u_i - \sum_{j\in N_i} V_i V_j B_{ij} sin(\delta_i - \delta_j) \\ &- A_i(\omega_i^b - \omega^n) - P_i^l \end{aligned} \tag{14.2}$$

$$\begin{aligned} \frac{T_{odi}}{X_{di} - X'_{di}}\frac{d}{dt}V_i =& \frac{E_{fi}}{X_{di} - X'_{di}} - \frac{1 - B_{ii}(X_{di} - X'_{di})}{X_{di} - X'_{di}} V_i \\ &+ \sum_{j\in N_i} V_j B_{ij} cos(\delta_i - \delta_j) \end{aligned} \tag{14.3}$$

where the state variables δ_i, ω_i^b and V_i represent the voltage angle, frequency and voltage magnitude, the parameters ω^n, M_i, A_i, T_{doi}, X_{di} and X'_{di} denote the nominal frequency, moment of inertia, damping constant, direct axis transient open-circuit constant, direct synchronous reactance and direct synchronous transient reactance, B_{ij} denotes the susceptance of the transmission line connecting area i and j, the uncontrollable input P_i^l is the power demand, the controllable inputs E_{fi} and u_i describe the exciter voltage and power generation respectively. Mentioned that the parameters ω^n, M_i, A_i, T_{doi}, X_{di}, X'_{di} and B_{ij} are all constants in the power grid model.

Remark 14.1 The issue of time-varying voltage is considered in the nonlinear power grid model by introducing the differential equation (14.3), which makes the model more accurately on describing the dynamic behavior of power grid. Mentioned that the conductance is neglected since this chapter considers the high voltage transmission network. Moreover, the exciter voltage E_{fi} is assumed to be constant and do not explicitly include exciter dynamics in order to keep the analysis concise.

The power grid model (14.1)–(14.3) can be equivalently re-written in the following compact form

$$\frac{d}{dt}\eta = D^T\omega \tag{14.4}$$

$$M\frac{d}{dt}\omega = u - D\Gamma(V)\mathbf{sin}(\eta) - A\omega - P^l \tag{14.5}$$

$$T\frac{d}{dt}V = -E(\eta)V + \overline{E}_{fd} \tag{14.6}$$

where the vectors $\omega = (\omega_1^b - \omega^n, ..., \omega_n^b - \omega^n)^T$, $u = (u_1, ..., u_n)^T$, $V = (V_1, ..., V_n)^T$, $P^l = (P_1^l, ..., P_n^l)^T$,

$$\overline{E}_{fd} = \left(\frac{E_{f1}}{X_{d1} - X'_{d1}}, \ldots, \frac{E_{fn}}{X_{dn} - X'_{dn}} \right)^T, \tag{14.7}$$

D is the incidence matrix as mentioned before and $\eta = D^T \delta$ with $\delta = (\delta_1, \ldots, \delta_n)^T$, $\mathbf{sin}(\eta) = (sin(\eta_1), \ldots, sin(\eta_m))^T$, the matrix $M = diag\{M_1, \ldots, M_n\}$, $A = diag\{A_1, \ldots, A_n\}$ and $T = diag\{\frac{T_{od1}}{X_{d1}-X'_{d1}}, \ldots, \frac{T_{odn}}{X_{dn}-X'_{dn}}\}$, $\Gamma(V) = diag\{\gamma_1, \ldots, \gamma_m\}$ with $\gamma_k = V_i V_j B_{ij} = V_j V_i B_{ji}$ and the matrix $E(\eta)$ contains

$$E_{ii} = \frac{1 - B_{ii}(X_{di} - X'_{di})}{X_{di} - X'_{di}}$$

and $E_{ij} = -B_{ij} cos(\eta_k)$ for $i \neq j$ where the index k denotes the edge (i, j). By the analysis in [4], the matrix $E(\eta)$ is positive definite in a realistic power grid.

14.1.2 Communication Network

The communication network, whose topology can be described by an undirected and connected graph $\mathcal{G}^c \triangleq (\mathcal{V}, \mathcal{E}^c)$ similar to the power network described in Sect. 14.2.1, is introduced to implement the optimal frequency regulation in a distributed architecture. The edge (j, i) in set $\mathcal{E}^c$ implies that area j can transmit its own information to area i through the communication network. $C^c \triangleq \{c^c_{ij}\} \in R^{n \times n}$ is the adjacency matrix of graph $\mathcal{G}^c$, where $c^c_{ij} = 1$ if and only if $(i, j) \in \mathcal{E}^c$, otherwise $c^c_{ij} = 0$, and it is set that $c^c_{ii} = 0$. The set $N^c_i \triangleq \{j \in \mathcal{V} | (j, i) \in \mathcal{E}^c\}$ contains all the neighbours of area i in the communication network, and its cardinal number is denoted as $|N^c_i|$. The Laplacian matrix of graph $\mathcal{G}^c$ is $L^c = diag\{|N^c_1|, \ldots, |N^c_n|\} - C^c$, which is semi-positive definite since the graph $\mathcal{G}^c$ is undirected and connected [5]. Note that the communication network $\mathcal{G}^c$ does not necessarily have to coincide with the power network $\mathcal{G}$.

14.1.3 Control Purpose

When the load patterns are changed caused by certain disturbances, the non-zero frequency deviation of areas in power grid will occur. The main purpose of frequency regulation is restoring the non-zero frequency deviation ω in (14.4)–(14.6) to the steady state 0 by adjusting the controllable power generation u. This means that the stability of steady state of system (14.4)–(14.6) should be concerned.

For a given constant power generation $u = \overline{u}$, the constant steady state $(\overline{\eta}, \overline{\omega}, \overline{V})$ of system (14.4)–(14.6) satisfies

$$0 = D^T \overline{\omega} \tag{14.8}$$

$$0 = \overline{u} - D\Gamma(\overline{V})\mathbf{sin}(\overline{\eta}) - A\overline{\omega} - P^l \tag{14.9}$$

$$0 = -E(\overline{\eta})\overline{V} + \overline{E}_{fd} \tag{14.10}$$

By simple calculation of (14.8)–(14.10), it holds that

$$\overline{\omega} = \frac{(\overline{u} - P^l)^T \mathbf{1}}{\mathbf{1}^T A \mathbf{1}} \mathbf{1} = \frac{\sum_{i \in \mathcal{V}} (\overline{u}_i - P_i^l)}{\sum_{i \in \mathcal{V}} A_i} \mathbf{1},$$

where the n dimensional vector $\mathbf{1} = (1, ..., 1)^T$. This means that the zero frequency deviation steady state, i.e. $\overline{\omega} = 0$, can be achieved when the balance between the total power demand and supply is obtained, i.e. $(\overline{u} - P^l)^T \mathbf{1} = 0$. It implies that the frequency restoration purpose is regulating the controllable power generation u such that $(u - P^l)^T \mathbf{1} = 0$ at the steady state. Moreover, there usually exists the requirement to reduce the generation cost in order to coordinate the generation in an economically efficient way. As a consequence, considering the frequency restoration and economic efficiency purposes simultaneously, the optimal frequency regulation problem is characterized as follows

$$min_{\overline{u}} C(\overline{u}) = min_{\overline{u}} \sum_{i \in \mathcal{V}} C_i(\overline{u}_i) \tag{14.11}$$

$$s.t. \quad (\overline{u} - P^l)^T \mathbf{1} = 0 \tag{14.12}$$

where the cost function $C_i(\overline{u}_i) = \frac{1}{2} q_i \overline{u}_i^2$ with $q_i > 0$. Obviously, $C(\overline{u}) = \frac{1}{2} \overline{u}^T Q \overline{u}$ with the positive definite matrix $Q = diag\{q_1, ..., q_n\}$. The distributed optimal frequency regulation (14.13)–(14.14) was constructed in [4] to addressed the problem (14.11)–(14.12)

$$u_i(t) = \frac{\theta_i(t)}{q_i} \tag{14.13}$$

$$\frac{d}{dt}\theta_i(t) = \sum_{j \in N_i^c} (\theta_j(t) - \theta_i(t)) - \frac{\omega_i(t)}{q_i} \tag{14.14}$$

The variable θ_i is an internal dynamic variable of regulation (14.13)–(14.14), which exists physically in the form of the data stored and computed in operation device in practical application.

However, it should be mentioned that the implementation of regulation method (14.13)–(14.14) requires each area i to transmit its own information θ_i to the neighboring areas $j (j \in N_i^c)$ continuously, which is not realistic since the transmission is physically realized discretely by using node-to-node wired/wireless communication. Although the continuous information transmission can make the discretization implementation by the periodic sampled-data communication scheme, it generates the high data flows in communication network usually due to the fixed worst-case sampling rate. The event-triggered mechanism, whose information transmission obeys the demand-transmission character, will be introduced to reduce the communication burden in distributed optimal frequency regulation in this chapter. The event-triggered mechanism is configured at each area to determine the information broadcasting time of the local area, and the corresponding decision is made by the designed event-triggering condition.

14.2 Main Results

14.2.1 Distributed Optimal Frequency Regulation Based on Event-Triggered Sampling Data

Inspired by the idea of event-triggered mechanism, denote the triggering time sequence corresponding to area i as $\{t_0^i, t_1^i, ..., t_k^i, ...\}$ where $t_0^i < t_1^i < ... < t_k^i < ...$, and then the distributed optimal frequency regulation based on event-triggered information is constructed as follows for $t \in [t_k^i, t_{k+1}^i)$.

$$u_i(t) = \frac{\theta_i(t)}{q_i} \tag{14.15}$$

$$\frac{d}{dt}\theta_i(t) = \sum_{j \in N_i^c} (\theta_j(t^j_{k'(t)}) - \theta_i(t^i_k)) - \frac{\omega_i(t)}{q_i} \tag{14.16}$$

where $t^j_{k'(t)}$ is the latest triggering time corresponding to area j at time t. Note that (14.15)–(14.16) can be equivalently re-written as

$$u(t) = Q^{-1}\theta(t) \tag{14.17}$$

$$\frac{d}{dt}\theta(t) = -L^c\widehat{\theta}(t) - Q^{-1}\omega(t) \tag{14.18}$$

where $u(t) = (u_1(t), ..., u_n(t))^T, \theta(t) = (\theta_1(t), ..., \theta_n(t))^T$ and $\widehat{\theta}(t) = (\theta_1(t^1_{k'(t)}), ..., \theta_n(t^n_{k'(t)}))^T$ for $t \in [t_k^i, t_{k+1}^i)$.

Remark 14.2 The implementation of optimal frequency regulation (14.15)–(14.16) only requires the local information $\theta_j(t^j_{k'(t)})$ of the neighboring areas in N_i^c instead of the global information, which implies that it is compliant with the distributed control architecture. Moreover, area i in the power grid only broadcasts its discrete triggered information at the triggering times when the triggering events occur, and area i obtains the triggered information $\theta_j(t^j_{k'(t)})$ and remains it unchanged until the next triggered information is transmitted from area j. In other word, area i only uses its neighbors' latest triggered information $\theta_j(t^j_{k'(t)})$ instead of their current exact information $\theta_j(t)$ to implement the regulation method, which demonstrates that less information transmission is required under (14.15)–(14.16) compared with (14.13)–(14.14). It is expected that the communication burden can be reduced.

According to the method of Lagrange multipliers on convex optimization, the constant $\overline{u} = Q^{-1}\frac{\mathbf{1}\mathbf{1}^T P^l}{\mathbf{1}^T Q^{-1}\mathbf{1}}$ is the solution for the optimal frequency regulation problem (14.11)–(14.12). It is assumed that the solution $(\overline{\eta}, \overline{\omega}, \overline{V})$ of Eqs. (14.8)–(14.10) with the given $\overline{u}$ exists, and this guarantees the existence of steady state $(\overline{\eta}, \overline{\omega}, \overline{V}, \overline{\theta})$ of the closed-loop system given by (14.4)–(14.6) and (14.17)–(14.18), where $\overline{\omega} = 0$, $\overline{\theta} = \frac{\mathbf{1}\mathbf{1}^T P^l}{\mathbf{1}^T Q^{-1}\mathbf{1}}$, $\overline{\eta}$ and $\overline{V}$ are some constants. In order to analyze the stability of steady state $(\overline{\eta}, \overline{\omega}, \overline{V}, \overline{\theta})$ of the closed-loop system, the triggering time sequence should be chosen properly since the discrete triggered information is used to implement the regulation method (14.15)–(14.16) instead of the exact information. As a consequence, the next step is to design the event-triggered mechanisms for each area $i \in \{1, 2, ...n\}$ to generate the proper event triggering time sequence $\{t_k^i\}_{k=0}^{\infty}$.

14.2.2 Static Event-Triggered Mechanism

Define the measurement error between the last transmitted value $\theta_i(t_k^i)$ and the current value $\theta_i(t)$ for area i as

$$e_i(t) = \theta_i(t_k^i) - \theta_i(t), \qquad t \in [t_k^i, t_{k+1}^i) \tag{14.19}$$

The static event-triggered mechanism for area i is designed as follows.

Static event-triggered mechanism: Construct the triggering condition for area i as

$$e_i^2(t) > \frac{1}{|N_i^c|}\left(\frac{\sigma_i^s}{4}\sum_{j\in N_i^c}\left(\theta_j\left(t_{k'(t)}^j\right) - \theta_i\left(t_k^i\right)\right)^2 + \xi_i\right) \tag{14.20}$$

where the constants $\sigma_i^s \in (0, 1)$ and $\xi_i > 0$. Denoting the last triggering time for θ_i as t_k^i, the next triggering occurs if the triggering condition (14.20) is satisfied at time t, then denote the new triggering time $t_{k+1}^i = t$ and transmit the value of the sampled information $\theta_i(t_{k+1}^i)$ to the areas in N_i^c.

Remark 14.3 The event-triggered condition (14.20) is named 'static' since it does not contain any other extra dynamic variables except the closed-loop system variables $\theta_i(t)$, $\theta_i(t_k^i)$ and $\theta_j(t_{k'(t)}^j)$ for $j \in N_i^c$. It also shows that the information which is used to check condition (14.20) can be obtained locally for area i through the communication network, and this illustrates that the constructed event-triggered mechanism is compatible with distributed control architecture. Mentioned that theoretical foundation for obtaining the triggering condition (14.20) is given in the proof of Theorem 14.1 below.

Based on the condition (14.20), area i determines the triggering time sequence $\{t_k^i\}_{k=0}^{\infty}$ by

$$t_{k+1}^i = \inf_{t>t_k^i}\left\{t|e_i^2(t) > \frac{1}{|N_i^c|}\left(\frac{\sigma_i^s}{4}\sum_{j\in N_i^c}\left(\theta_j\left(t_{k'(t)}^j\right)-\theta_i\left(t_k^i\right)\right)^2+\xi_i\right)\right\}$$

The information of area i will be broadcast to its neighbors only at the triggering time sequence $\{t_k^i\}_{k=0}^{\infty}$ under the static event-triggered mechanism. In this sense, the number of interactive communications of the areas in power grid can be significantly reduced, and less communication burden can be expected.

The following assumption quoted from [4] is introduced to guarantee the existence of a suitable Lyapunov function. This assumption is mild for practical power system and the similar ones can be found in [6, 7]. It will be used to prove the stability of closed-loop system.

Assumption 14.1 The steady states $\overline{\eta} \in (-\frac{\pi}{2}, \frac{\pi}{2})^m$ and $\overline{V} \in R_{>0}^n$ satisfy $E(\overline{\eta}) - diag(\overline{V})^{-1}D^+\Gamma(\overline{V})diag(\mathbf{sin}(\overline{\eta}))diag(\mathbf{cos}(\overline{\eta}))^{-1}$
$diag(\mathbf{sin}(\overline{\eta}))(D^+)^T diag(\overline{V})^{-1} > 0$, where the elements in matrix D^+ are the absolute values of those in incidence matrix D.

Theorem 14.1 *The distributed optimal frequency regulation (14.15)–(14.16) [or (14.17)–(14.18) equivalently] with the static event-triggered mechanism (14.20) guarantees the solution of closed-loop system (14.4)–(14.6) and (14.17)–(14.18), which starts in a neighborhood of the steady state* $(\overline{\eta}, \overline{\omega}, \overline{V}, \overline{\theta})$, *to converge into an arbitrary small neighborhood of the new steady state* $(\widetilde{\eta}, \overline{\omega}, \widetilde{V}, \overline{\theta})$ *where* $E(\widetilde{\eta})\widetilde{V} - \overline{E}_{fd} = 0$ *for constant* $\widetilde{\eta}$ *and* $\widetilde{V}$.

Proof 14.1 *Construct the Lyapunov function for the closed-loop system as*

$$\begin{aligned} U(\eta, \omega, V, \theta) =& W(\eta, V) + \frac{1}{2}(\omega - \overline{\omega})^T M(\omega - \overline{\omega}) \\ &+ \frac{1}{2}(\theta - \overline{\theta})^T(\theta - \overline{\theta}) \end{aligned} \tag{14.21}$$

where the function

$$\begin{aligned} W(\eta, V) \triangleq& -\mathbf{1}^T\Gamma(V)\mathbf{cos}(\eta) + \mathbf{1}^T\Gamma(\overline{V})\mathbf{cos}(\overline{\eta}) - \left(\Gamma(\overline{V})\mathbf{sin}(\overline{\eta})\right)^T \\ &(\eta - \overline{\eta}) - \overline{E}_{fd}(V - \overline{V}) + \frac{1}{2}V^T F V - \frac{1}{2}\overline{V}^T F\overline{V}, \end{aligned} \tag{14.22}$$

and F is a diagonal matrix with

$$F_{ii} = \frac{1 - B_{ii}(X_{di} - X'_{di})}{X_{di} - X'_{di}}$$

Noted that the gradient of $W(\eta, V)$ is 0 at the steady state $(\overline{\eta}, \overline{V})$, and the Hessian matrix of $W(\eta, V)$ at $(\overline{\eta}, \overline{V})$ is positive definite under Assumption 14.1 as shown in [4]. This implies that $W(\eta, V)$ has a strict local minimum at $(\overline{\eta}, \overline{V})$, which further means that the Lyapunov function $U(\eta, \omega, V, \theta)$ has the local minimum point $(\overline{\eta}, \overline{\omega}, \overline{V}, \overline{\theta})$.

By the help of (14.4)–(14.6), (14.8), (14.9), and the fact that $u = Q^{-1}\theta$, $\overline{\omega} = 0$ and $\frac{d}{dt}(\theta - \overline{\theta}) = -L^c(\widehat{\theta} - \overline{\theta}) - Q^{-1}\omega$, the derivation of Lyapunov function U with respect to time t along the solution of the closed-loop system is

$$\begin{aligned}\frac{d}{dt}U =& -(\omega - \overline{\omega})^T A(\omega - \overline{\omega}) - \|\nabla_V W_2\|^2_{T^{-1}} \\ & -(\theta - \overline{\theta})^T L^c(\widehat{\theta} - \overline{\theta})\end{aligned} \tag{14.23}$$

where the notation

$$\|\nabla_V W_2\|_{T^{-1}} \triangleq \big(E(\eta)V - \overline{E}_{fd}\big)^T T^{-1}\big(E(\eta)V - \overline{E}_{fd}\big). \tag{14.24}$$

For $t \in [t^i_k, t^i_{k+1})$, we have

$$\begin{aligned}&\frac{1}{2}\sum_{i=1}^{n}\sum_{j=1}^{n} l^c_{ij}\big(\theta_i(t^i_k) - \theta_j(t^j_{k'(t)})\big)^2 \\ =&\frac{1}{2}\sum_{i=1}^{n}\sum_{j=1}^{n} l^c_{ij}\theta_j^2(t^j_{k'(t)}) + \frac{1}{2}\sum_{i=1}^{n}\sum_{j=1}^{n} l^c_{ij}\theta_i^2(t^i_k) \\ &-\sum_{i=1}^{n}\sum_{j=1}^{n}\theta_i(t^i_k) l^c_{ij}\theta_j(t^j_{k'(t)}) \\ =&-\sum_{i=1}^{n}\theta_i(t^i_k)\sum_{j=1}^{n}\big(l^c_{ij}\theta_j(t^j_{k'(t)})\big)\end{aligned} \tag{14.25}$$

since $\sum_{i=1}^n\sum_{j=1}^n l^c_{ij}\theta_j^2(t^j_{k'(t)}) = \mathbf{1}^T L^c\widehat{\theta^2} = 0$ and $\sum_{i=1}^n\sum_{j=1}^n l^c_{ij}\theta_i^2(t^i_k) = \sum_{i=1}^n\theta_i^2(t^i_k)\sum_{j=1}^n l^c_{ij} = 0$ where $\widehat{\theta^2} \triangleq (\theta_1^2(t^1_{k'(t)}), \ldots, \theta_n^2(t^n_{k'(t)}))^T$.

According to (14.25), it can be obtained that

$$\begin{aligned}&-(\theta - \overline{\theta})^T L^c(\widehat{\theta} - \overline{\theta}) \\ =&-\theta^T L^c\widehat{\theta} \\ =&-\sum_{i=1}^{n}\big(\theta_i(t^i_k) - e_i(t)\big)\sum_{j=1}^{n}\big(l^c_{ij}\theta_j(t^j_{k'(t)})\big) \\ =&\frac{1}{2}\sum_{i=1}^{n}\sum_{j=1}^{n} l^c_{ij}\big(\theta_i(t^i_k) - \theta_j(t^j_{k'(t)})\big)^2\end{aligned}$$

$$-\sum_{i=1}^{n}\sum_{j=1}^{n} e_i(t)l_{ij}^c\big(\theta_i(t_k^i)-\theta_j(t_{k'(t)}^j)\big) \tag{14.26}$$

By the help of Young's inequality, we have

$$\begin{aligned}
&-\sum_{i=1}^{n}\sum_{j=1}^{n} e_i(t)l_{ij}^c\big(\theta_i(t_k^i)-\theta_j(t_{k'(t)}^j)\big)\\
\le&\sum_{i=1}^{n}\sum_{j=1,j\neq i}^{n}\Big(-l_{ij}^c e_i^2(t)-\frac{l_{ij}^c}{4}\big(\theta_i(t_k^i)-\theta_j(t_{k'(t)}^j)\big)^2\Big)\\
=&\sum_{i=1}^{n}|N_i^c|e_i^2(t)-\sum_{i=1}^{n}\sum_{j=1}^{n}\frac{l_{ij}^c}{4}\big(\theta_i(t_k^i)-\theta_j(t_{k'(t)}^j)\big)^2
\end{aligned} \tag{14.27}$$

It is mentioned that the triggering condition (14.20) implies that for $t\in[t_k^i,t_{k+1}^i)$,

$$e_i^2(t)\le\frac{\sigma_i^s}{4|N_i^c|}\sum_{j\in N_i^c}\Big(\theta_j(t_{k'(t)}^j)-\theta_i(t_k^i)\Big)^2+\frac{\xi_i}{|N_i^c|} \tag{14.28}$$

Combining (14.23), (14.26)–(14.28) leads to

$$\begin{aligned}
\frac{d}{dt}U\le&-(\omega-\overline{\omega})^TA(\omega-\overline{\omega})-\|\nabla_V W_2\|_{T^{-1}}^2+\sum_{i=1}^{n}|N_i^c|e_i^2(t)\\
&-\frac{1}{4}\sum_{i=1}^{n}\sum_{j\in N_i^c}\big(\theta_i(t_k^i)-\theta_j(t_{k'(t)}^j)\big)^2\\
\le&-(\omega-\overline{\omega})^TA(\omega-\overline{\omega})-\|\nabla_V W_2\|_{T^{-1}}^2+\sum_{i=1}^{n}\xi_i\\
&-\sum_{i=1}^{n}\frac{1-\sigma_i^s}{4}\sum_{j\in N_i^c}\big(\theta_i(t_k^i)-\theta_j(t_{k'(t)}^j)\big)^2\\
\le&-(\omega-\overline{\omega})^TA(\omega-\overline{\omega})-\|\nabla_V W_2\|_{T^{-1}}^2+\sum_{i=1}^{n}\xi_i\\
&-\frac{1-\sigma_{max}^s}{2}\widehat{\theta}^TL^c\widehat{\theta}
\end{aligned} \tag{14.29}$$

where (14.25) is used here and $\sigma_{max}^s\triangleq max\{\sigma_1^s,\dots,\sigma_n^s\}$.

Denoting $e(t)=(e_1(t),\dots,e_n(t))^T$, *since*

$$
\begin{aligned}
\|e\|^2 &= \sum_{i=1}^{n} e_i^2 \\
&\le \sum_{i=1}^{n} \left(\frac{\sigma_i^s}{4|N_i^c|} \sum_{j\in N_i^c} \left(\theta_j(t^j_{k'(t)}) - \theta_i(t^i_k)\right)^2 + \frac{\xi_i}{|N_i^c|} \right) \\
&\le \frac{\sigma^s_{max}}{2N^c_{min}} \sum_{i=1}^{n} \sum_{j\in N_i^c} \frac{1}{2} \left(\theta_j(t^j_{k'(t)}) - \theta_i(t^i_k)\right)^2 + \sum_{i=1}^{n} \frac{\xi_i}{|N_i^c|} \\
&= \frac{\sigma^s_{max}}{2N^c_{min}} \widehat{\theta}^T L^c \widehat{\theta} + \sum_{i=1}^{n} \frac{\xi_i}{|N_i^c|}
\end{aligned}
$$

where $N^c_{min} \triangleq min\{|N_1^c|, ..., |N_n^c|\}$, *we have*

$$
\begin{aligned}
&(\theta - \overline{\theta})^T L^c (\theta - \overline{\theta}) \\
=&(\widehat{\theta} - e)^T L^c (\widehat{\theta} - e) \\
\le& 2\widehat{\theta}^T L^c \widehat{\theta} + 2\|L^c\| \|e\|^2 \\
\le& \left(2 + \frac{\sigma^s_{max}\|L^c\|}{N^c_{min}}\right) \widehat{\theta}^T L^c \widehat{\theta} + 2\|L^c\| \sum_{i=1}^{n} \frac{\xi_i}{|N_i^c|}
\end{aligned}
\tag{14.30}
$$

Based on (14.30), it can be obtained from (14.29) that

$$
\begin{aligned}
\frac{d}{dt}U \le& -(\omega - \overline{\omega})^T A(\omega - \overline{\omega}) - \|\nabla_V W_2\|^2_{T^{-1}} \\
&-c_1(\theta - \overline{\theta})^T L^c (\theta - \overline{\theta}) + c_2 \sum_{i=1}^{n} \xi_i
\end{aligned}
\tag{14.31}
$$

where the positive constant

$$
c_1 = \frac{N^c_{min}(1 - \sigma^s_{max})}{2(2N^c_{min} + \|L^c\|\sigma^s_{max})} and c_2 = \frac{2\|L^c\|c_1}{N^c_{min}} + 1
$$

Equation (14.31) means that the solution of the closed-loop system converges into the set

$$
S \triangleq \left\{(\eta, \omega, V, \theta) | (\omega - \overline{\omega})^T A(\omega - \overline{\omega}) + \|\nabla_V W_2\|^2_{T^{-1}} + c_1(\theta - \overline{\theta})^T L^c (\theta - \overline{\theta}) \le c_2 \sum_{i=1}^{n} \xi_i \right\}
\tag{14.32}
$$

This implies that the solution of closed-loop system, which starts in a neighborhood of the steady state $(\overline{\eta}, \overline{\omega}, \overline{V}, \overline{\theta})$, *converges into an arbitrary small neighborhood of the new steady state* $(\widetilde{\eta}, \overline{\omega}, \widetilde{V}, \overline{\theta})$ *where* $E(\widetilde{\eta})\widetilde{V} - \overline{E}_{fd} = 0$ *for constant* $\widetilde{\eta}$ *and* $\widetilde{V}$ *by choosing the arbitrary small constants* ξ_i *for* $i = 1, ..., n$. *This concludes the proof.*

Remark 14.4 The dynamical scenario of optimal frequency regulation for power grid can be depicted as follows. When the balance between the total power demand and supply is broken due to certain disturbances in power grid, the power flow which is operated at the initial steady state $(\overline{\eta}, \overline{\omega}, \overline{V}, \overline{\theta})$ will be mismatched. Suddenly, the optimal frequency regulation (14.15)–(14.16) will guarantee the power grid to operate in an arbitrary small neighborhood $\{(\eta, \omega, V, \theta)|(\omega - \overline{\omega})^T A(\omega - \overline{\omega}) + (E(\eta)V - \overline{E}_{fd})^T T^{-1}(E(\eta)V - \overline{E}_{fd}) + c_1(\theta - \overline{\theta})^T L^c(\theta - \overline{\theta}) \leq c_2 \sum_{i=1}^n \xi_i\}$ around another new steady state $(\widetilde{\eta}, \overline{\omega}, \widetilde{V}, \overline{\theta})$, where c_1 and c_2 are some constants, and $E(\widetilde{\eta})\widetilde{V} - \overline{E}_{fd} = 0$ for constant $\widetilde{\eta}$ and $\widetilde{V}$. As shown in [4, 8], the new steady state $(\widetilde{\eta}, \overline{\omega}, \widetilde{V}, \overline{\theta})$ is an accepted power grid stable operation condition, and the requirement of optimal frequency regulation problem (14.11)–(14.12) is satisfied at the new steady state.

Remark 14.5 The positive constant ξ_i makes the solution of closed-loop system to converge into set S, which is a neighborhood of steady state $(\widetilde{\eta}, \overline{\omega}, \widetilde{V}, \overline{\theta})$, as shown in the proof of Theorem 14.1. If $\xi_i = 0$ for all i, we have that $S = \{(\eta, \omega, V, \theta)|\eta = \widetilde{\eta}, \omega = \overline{\omega}, V = \widetilde{V}, \theta = \overline{\theta}\}$ and the asymptotic convergence result can be obtained. However, taking ξ_i to 0 may cause Zeno behavior, which will be explained in Theorem 14.2 and Remark 14.6. Although the asymptotic convergence result for the new steady state $(\widetilde{\eta}, \overline{\omega}, \widetilde{V}, \overline{\theta})$ can not be achieved under the distributed optimal frequency regulation (14.15)–(14.16) based on static event-triggered mechanism (14.20), the state variables (i.e. frequency deviation, voltage magnitude, power generation and so on) are allowed to fluctuate in a certain range around the steady state in practical power grid. Mentioned that the neighborhood $\{(\eta, \omega, V, \theta)|(\omega - \overline{\omega})^T A(\omega - \overline{\omega}) + (E(\eta)V - \overline{E}_{fd})^T T^{-1}(E(\eta)V - \overline{E}_{fd}) + c_1(\theta - \overline{\theta})^T L^c(\theta - \overline{\theta}) \leq c_2 \sum_{i=1}^n \xi_i\}$, which the system state $(\eta, \omega, V, \theta)$ converges into, can be arbitrary small by selecting the parameters $\xi_i (i = 1, ..., n)$ suitably. This implies that the distributed optimal frequency regulation (14.15)–(14.16) based on the static event-triggered mechanism (14.20) can drive the system state $(\eta, \omega, V, \theta)$ into an allowable range around the new steady state $(\widetilde{\eta}, \overline{\omega}, \widetilde{V}, \overline{\theta})$ by adjusting the parameters $\xi_i (i = 1, ..., n)$ in the static event-triggered mechanism (14.20).

One main role of event-triggered mechanism is to rule out the possibility of Zeno behavior, which indicates that an infinite number of events will be triggered at an accumulation time T. Zeno behavior is problematic and impractical for the physical power grid implementation since it leads the area to update and transmit the information an infinite number of times in a finite time period. The following theorem shows that Zeno behavior can be excluded by analyzing the inter-event time interval $t_{k+1}^i - t_k^i$.

Theorem 14.2 *The inter-event time interval for the static event-triggered mechanism (14.20) is lower bounded as*

$$t_{k+1}^i - t_k^i > \frac{\xi_i^{\frac{1}{2}}}{M_i |N_i^c|^{\frac{1}{2}}} \tag{14.33}$$

with the positive constant M_i, and the Zeno behavior can be avoided.

Proof 14.2 *For $t \in [t_k^i, t_{k+1}^i)$, according to (14.16) and (14.19), we have*

$$\begin{aligned}
\frac{d}{dt}|e_i(t)| &\le |\frac{d}{dt}e_i(t)| = |\frac{d}{dt}\theta_i(t)| \\
&= |\sum_{j\in N_i^c}\left(\theta_j(t_{k'(t)}^j) - \theta_i(t_k^i)\right) - \frac{\omega_i(t)}{q_i}| \le M_i
\end{aligned} \tag{14.34}$$

where the last inequation holds since the continuously differentiable function $\sum_{j\in N_i^c}(\theta_j(t) - \theta_i(t)) - \frac{\omega_i(t)}{q_i}$ converges into a bounded set as shown in Theorem 14.1. Integrating both sides of (14.34) yields $|e_i(t)| - |e_i(t_k^i)| \le M_i(t - t_k^i)$, and then $|e_i(t)| \le M_i(t - t_k^i)$ since $e_i(t_k^i) = 0$ for $t \in [t_k^i, t_{k+1}^i)$. It can be seen from (14.20) that the next triggering time $t = t_{k+1}^i$ occurs when $|e_i(t)| > \left(\frac{\sigma_i^s}{4|N_i^c|}\sum_{j\in N_i^c}(\theta_j(t_{k'(t)}^j) - \theta_i(t_k^i))^2 + \frac{\xi_i}{|N_i^c|}\right)^{\frac{1}{2}}$. This implies that $t_{k+1}^i - t_k^i > \frac{1}{M_i}\left(\frac{\sigma_i^s}{4|N_i^c|}\sum_{j\in N_i^c}(\theta_j(t_{k'(t)}^j) - \theta_i(t_k^i))^2 + \frac{\xi_i}{|N_i^c|}\right)^{\frac{1}{2}} > \frac{\xi_i^{\frac{1}{2}}}{M_i|N_i^c|^{\frac{1}{2}}}$. Since the uniform positive lower bound of inter-event time interval $t_{k+1}^i - t_k^i$ is obtained as $\frac{\xi_i^{\frac{1}{2}}}{M_i|N_i^c|^{\frac{1}{2}}}$, it ensures that it is impossible to trigger infinite times at any accumulation time, which indicates that Zeno behavior can be excluded.

Remark 14.6 In order to ensure the existence of uniform positive lower bound of inter-event time interval, the positive constant ξ_i is introduced in (14.20). Moreover, formula (14.33) implies that the triggering frequency is influenced by the parameter ξ_i. Increasing this parameter can enlarge the lower bound of inter-event time interval and thus lower the triggering frequency, which reduce the communication burden further. However, as illustrated in Remark 14.5, increasing the parameter ξ_i may enlarge the fluctuation range of the system state $(\eta, \omega, V, \theta)$ around the steady state $(\widetilde{\eta}, \overline{\omega}, \widetilde{V}, \overline{\theta})$, which may violate the optimal frequency regulation requirement and the stable operation of power grid. On the contrary, decreasing ξ_i could reduce the fluctuation range, which however, potentially makes the triggering frequency higher. In practical application, the parameter ξ_i should be chosen with weighting the factors of the fluctuation tolerant level and the communication resource limitation.

14.2.3 *Dynamic Event-Triggered Mechanism*

The parameter ξ_i in static event-triggered mechanism (14.20) avoids the Zeno behavior while hampering the asymptotically convergence of steady state. Moreover, as shown in Remark 14.5, in order to drive the state variables converge into an allowable range around the steady state, the parameter ξ_i should be selected by obtaining some global information, such as the matrix A, T, L^c and so on, which may be difficult for the practical power grid with plenty of heterogeneous devices and complex structure. In this section, the dynamic event-triggered mechanism is constructed aiming to obtain the asymptotically stability result without the difficulty on turning the parameters.

Dynamic event-triggered mechanism: Construct the triggering condition for area i as

$$e_i^2(t) > \frac{1}{|N_i^c|}\left(\frac{\sigma_i^d}{4}\sum_{j\in N_i^c}\left(\theta_j(t^j_{k'(t)}) - \theta_i(t^i_k)\right)^2 + \frac{\varphi_i(t)}{\alpha_i}\right) \tag{14.35}$$

with the dynamic behavior of internal dynamic variable $\varphi_i(t)$ as

$$\frac{d}{dt}\varphi_i(t) = -\beta_i\varphi_i(t) + \gamma_i\left(\frac{\sigma_i^d}{4}\sum_{j\in N_i^c}\left(\theta_j(t^j_{k'(t)}) - \theta_i(t^i_k)\right)^2 - |N_i^c|e_i^2(t)\right) \tag{14.36}$$

where the measurement error $e_i(t)$ is defined in (14.19), the constants $\alpha_i > 0$, $\gamma_i \in (0, 1)$, $\sigma_i^d \in (0, 1)$ and $\beta_i > \frac{1-\gamma_i}{\alpha_i} + \frac{1-\sigma^d_{max}}{2}$ with $\sigma^d_{max} \triangleq max\{\sigma_1^d, ..., \sigma_n^d\}$, and the initial value $\eta_i(0) > 0$. Denoting the last triggering time for θ_i as t^i_k, the next triggering occurs if the triggering condition (14.35) is satisfied at time t, then denote the new triggering time $t^i_{k+1} = t$ and transmit the value of the sampled information $\theta_i(t^i_{k+1})$ to the areas in N_i^c.

Remark 14.7 The implementation of dynamic event-triggered mechanism (14.35)–(14.36) only requires the discrete triggered information $\theta_j(t^j_{k'(t)})$ of the neighboring areas j for ($j \in N_i^c$) instead of the global information, which demonstrates that it is compatible with the distributed control architecture.

Lemma 14.1 *The internal dynamic variable $\varphi_i(t)$ in dynamic event-triggered mechanism (14.35)–(14.36) satisfies*

$$\varphi_i(t) \geq \varphi_i(0)e^{-((\beta_i + \frac{\gamma_i}{\alpha_i})t}} \tag{14.37}$$

Proof 14.3 *The triggering condition (14.35) implies that for $t \in [t^i_k, t^i_{k+1})$*

$$e_i^2(t) \leq \frac{1}{|N_i^c|}\left(\frac{\varphi_i(t)}{\alpha_i} + \frac{\sigma_i^d}{4}\sum_{j\in N_i^c}\left(\theta_j(t^j_{k'(t)}) - \theta_i(t^i_k)\right)^2\right) \tag{14.38}$$

Combining (14.36) and (14.38) yields $\frac{d}{dt}\varphi_i(t) \geq -(\beta_i + \frac{\gamma_i}{\alpha_i})\varphi_i(t)$, then the proof is concluded according to the comparison lemma.

Based on the triggering condition (14.35), area i determines the triggering time sequence $\{t_k^i\}_{k=0}^{\infty}$ by

$$t_{k+1}^i = \inf_{t>t_k^i}\left\{t|e_i^2(t) > \frac{1}{|N_i^c|}\left(\frac{\sigma_i^d}{4}\sum_{j\in N_i^c}\left(\theta_j(t_{k'(t)}^j) - \theta_i(t_k^i)\right)^2 + \frac{\varphi_i(t)}{\alpha_i}\right)\right\}$$

Remark 14.8 The event-triggered mechanism (14.35)–(14.36) is named 'dynamic' since an additional internal dynamic variable $\varphi_i(t)$ is introduced to determine the triggering time sequence. Based on Lemma 14.1, it gets that $\varphi_i(t) > 0$ for $t \geq 0$ since we choose the positive initial value $\varphi_i(0)$. Mentioned that, the positive variable $\frac{\varphi_i(t)}{\alpha_i}$ is used in the dynamic triggering condition (14.35) to replace the positive constant ξ_i, which hampers the asymptotically convergence of steady state, in the static triggering condition (14.20). Compared with the static event-triggered mechanism, a better result can be obtained under the dynamic one, since the variable $\varphi_i(t)$ is adjusted according to the system state as shown in (14.36).

Theorem 14.3 *The distributed optimal frequency regulation (14.15)–(14.16) [or (14.17)–(14.18) equivalently] with the dynamic event-triggered mechanism (14.35)–(14.36) guarantees the solution of closed-loop system (14.4)–(14.6) and (14.17)–(14.18), which starts in a neighborhood of steady state* $(\overline{\eta}, \overline{\omega}, \overline{V}, \overline{\theta})$, *to converge asymptotically to the new steady state* $(\widetilde{\eta}, \overline{\omega}, \widetilde{V}, \overline{\theta})$ *where* $E(\widetilde{\eta})\widetilde{V} - \overline{E}_{fd} = 0$ *for constant* $\widetilde{\eta}$ *and* $\widetilde{V}$.

Proof 14.4 *Construct the Lyapunov function as*

$$\widetilde{U}(\eta, \omega, V, \theta, \varphi) = U(\eta, \omega, V, \theta) + \sum_{i=1}^{n}\varphi_i(t) \tag{14.39}$$

where the function $U(\eta, \omega, V, \theta)$ *is defined as (14.21) and* $\varphi = (\varphi_1, ..., \varphi_n)^T$. *Lemma 14.1 and the fact that the function* $U(\eta, \omega, V, \theta)$ *has the local minimum point* $(\overline{\eta}, \overline{\omega}, \overline{V}, \overline{\theta})$ *as mentioned in the proof of Theorem 14.1 yield that* $(\overline{\eta}, \overline{\omega}, \overline{V}, \overline{\theta}, 0)$ *is one of the local minimum point of the Lyapunov function (14.39) for the closed-loop system.*

Since the distributed optimal frequency regulation (14.15)–(14.16) keeps the same under both of the static and dynamic event-triggered mechanism, the derivative of U *in (14.39) satisfies the first inequation of (14.29), and then the derivative of* $\widetilde{U}$ *with respect to time* t *satisfies*

$$\begin{aligned}\frac{d}{dt}U \leq& -(\omega-\overline{\omega})^T A(\omega-\overline{\omega}) - \|\nabla_V W_2\|^2_{T^{-1}} + \sum_{i=1}^n \frac{d}{dt}\varphi_i(t)\\
&-\frac{1}{4}\sum_{i=1}^n \sum_{j\in N_i^c} \left(\theta_i(t_k^i)-\theta_j(t^j_{k'(t)})\right)^2 + \sum_{i=1}^n |N_i^c| e_i^2(t)\\
=& -(\omega-\overline{\omega})^T A(\omega-\overline{\omega}) - \|\nabla_V W_2\|^2_{T^{-1}} - \sum_{i=1}^n \beta_i\varphi_i(t)\\
&-\frac{1}{4}\sum_{i=1}^n (1-\sigma_i^d\gamma_i)\sum_{j\in N_i^c} \left(\theta_i(t_k^i)-\theta_j(t^j_{k'(t)})\right)^2\\
&+\sum_{i=1}^n (1-\gamma_i)|N_i^c| e_i^2(t)\end{aligned} \tag{14.40}$$

Combining (14.38) and (14.40) obtains

$$\begin{aligned}\frac{d}{dt}U \leq& -(\omega-\overline{\omega})^T A(\omega-\overline{\omega}) - \|\nabla_V W_2\|^2_{T^{-1}}\\
&-\frac{1}{4}\sum_{i=1}^n (1-\sigma_i^d)\sum_{j\in N_i^c} \left(\theta_i(t_k^i)-\theta_j(t^j_{k'(t)})\right)^2\\
&-\sum_{i=1}^n \left(\beta_i-\frac{1-\gamma_i}{\alpha_i}\right)\varphi_i(t)\\
\leq& -(\omega-\overline{\omega})^T A(\omega-\overline{\omega}) - \|\nabla_V W_2\|^2_{T^{-1}}\\
&-\frac{1-\sigma_{max}^d}{4}\sum_{i=1}^n \sum_{j\in N_i^c} \left(\theta_i(t_k^i)-\theta_j(t^j_{k'(t)})\right)^2\\
&-\sum_{i=1}^n \left(\beta_i-\frac{1-\gamma_i}{\alpha_i}\right)\varphi_i(t)\\
=& -(\omega-\overline{\omega})^T A(\omega-\overline{\omega}) - \|\nabla_V W_2\|^2_{T^{-1}}\\
&-\frac{1-\sigma_{max}^d}{2}\widehat{\theta}^T L^c \widehat{\theta} - \sum_{i=1}^n \left(\beta_i-\frac{1-\gamma_i}{\alpha_i}\right)\varphi_i(t)\end{aligned} \tag{14.41}$$

where (14.25) is used to obtain the last equation.

Since

$$
\begin{aligned}
\|e\|^2 =& \sum_{i=1}^{n} e_i^2 \\
\leq& \sum_{i=1}^{n} \frac{1}{|N_i^c|} \left(\frac{\varphi_i(t)}{\alpha_i} + \frac{\sigma_i^d}{4} \sum_{j \in N_i^c} \left(\theta_j(t_{k'(t)}^j) - \theta_i(t_k^i)\right)^2 \right) \\
\leq& \frac{\sigma_{max}^d}{4N_{min}^c} \sum_{i=1}^{n} \sum_{j \in N_i^c} \left(\theta_j(t_{k'(t)}^j) - \theta_i(t_k^i)\right)^2 \\
&+ \frac{1}{\alpha_{min} N_{min}^c} \sum_{i=1}^{n} \varphi_i(t)
\end{aligned}
$$

where $\alpha_{min} \triangleq min\{\alpha_1, ..., \alpha_n\}$, *we can get that*

$$
\begin{aligned}
&(\theta - \overline{\theta})^T L^c (\theta - \overline{\theta}) \\
\leq& 2\widehat{\theta}^T L^c \widehat{\theta} + 2\|L^c\| \|e\|^2 \\
\leq& 2\widehat{\theta}^T L^c \widehat{\theta} + \frac{\sigma_{max}^d \|L^c\|}{2N_{min}^c} \sum_{i=1}^{n} \sum_{j \in N_i^c} \left(\theta_j(t_{k'(t)}^j) - \theta_i(t_k^i)\right)^2 \\
&+ \frac{2\|L^c\|}{\alpha_{min} N_{min}^c} \sum_{i=1}^{n} \varphi_i(t) \\
=& \left(2 + \frac{\sigma_{max}^d \|L^c\|}{N_{min}^c}\right) \widehat{\theta}^T L^c \widehat{\theta} + \frac{2\|L^c\|}{\alpha_{min} N_{min}^c} \sum_{i=1}^{n} \varphi_i(t) r \\
\leq& c \left(\widehat{\theta}^T L^c \widehat{\theta} + \sum_{i=1}^{n} \varphi_i(t) \right)
\end{aligned}
$$

where $c \triangleq max\{2 + \frac{\sigma_{max}^d \|L^c\|}{N_{min}^c}, \frac{2\|L^c\|}{\alpha_{min} N_{min}^c}\}$, *and thus*

$$
\widehat{\theta}^T L^c \widehat{\theta} \geq \frac{1}{c} (\theta - \overline{\theta})^T L^c (\theta - \overline{\theta}) - \sum_{i=1}^{n} \varphi_i(t) \tag{14.42}
$$

Combining (14.41) and (14.42) yields

$$
\begin{aligned}
\frac{d}{dt} U \leq& -(\omega - \overline{\omega})^T A (\omega - \overline{\omega}) - \|\nabla_V W_2\|_{T^{-1}}^2 \\
&- \frac{1 - \sigma_{max}^d}{2c} (\theta - \overline{\theta})^T L^c (\theta - \overline{\theta}) \\
&- \sum_{i=1}^{n} \left(\beta_i - \frac{1 - \gamma_i}{\alpha_i} - \frac{1 - \sigma_{max}^d}{2} \right) \varphi_i(t)
\end{aligned} \tag{14.43}
$$

It follows that $\frac{d}{dt}U \leq 0$ *since* $\beta_i > \frac{1-\gamma_i}{\alpha_i} + \frac{1-\sigma^d_{max}}{2}$ *and* $\varphi_i(t) \geq 0$. *According to Lasalle's invariance principle, Eq. (14.43) indicates that the system state* $(\eta, \omega, V, \theta)$ *converges into the set* $\widetilde{S} \triangleq \{(\eta, \omega, V, \theta)|\omega = \overline{\omega} = 0, \|\nabla_V W_2\|_{T^{-1}} = 0, \theta = \overline{\theta}\}$, *which concludes the proof.*

Remark 14.9 Theorem 14.3 shows that the system state $(\eta, \omega, V, \theta)$ asymptotically converges to the new steady state $(\widetilde{\eta}, \overline{\omega}, \widetilde{V}, \overline{\theta})$ under the distributed optimal frequency regulation (14.15)–(14.16) with the dynamic event-triggered mechanism (14.35)–(14.36), which means that the optimal frequency regulation problem (14.11)–(14.12) is addressed. Compared with the static event-triggered mechanism, the computation complexity of the dynamic one is increased. However, the asymptotically stability of steady state can be achieved, and the difficulty of tuning the parameter ξ_i for some complex power grid, which is illustrated before, can be eliminated.

Theorem 14.4 *Zeno behavior is excluded for the dynamic event-triggered mechanism (14.35)–(14.36).*

Proof 14.5 *According to the asymptotically stability result shown in Theorem 14.3, it can be obtain that* $|\theta_i(t) - \overline{\theta}_i| \leq M_1$ *and* $|\frac{d}{dt}\theta_i(t)| \leq M_2$ *with some positive constants* M_1 *and* M_2 *for* $t \geq 0$. *The theorem is then proved by contradiction.*

Assume that there exists Zeno behavior, then we have $\lim_{k\to\infty} t^i_k = T_a$ *for some* $i \in \{1, ..., n\}$ *with an accumulation time* T_a. *Based on the definition of limitation, it means that for the given constant* $\epsilon_a = \frac{(\varphi_i(0))^{\frac{1}{2}}}{2M_2(\alpha_i|N^c_i|)^{\frac{1}{2}}} e^{-\frac{1}{2}(\beta_i + \frac{\gamma_i}{\alpha_i})T_a} > 0$, *there exists a positive integer* $N(\epsilon_a)$, *which is related to* ϵ_a, *such that*

$$t^i_k \in [T_a - \epsilon_a, T_a), \qquad \forall k \geq N(\epsilon_a) \tag{14.44}$$

Consider the following equation for $t > t^i_k$

$$M_2(t - t^i_k) \leq \frac{(\varphi_i(0))^{\frac{1}{2}}}{(\alpha_i|N^c_i|)^{\frac{1}{2}}} e^{-\frac{1}{2}(\beta_i + \frac{\gamma_i}{\alpha_i})T_a} \tag{14.45}$$

then we have $\int_{t^i_k}^t |\frac{d}{dt}\theta_i| \leq \frac{(\varphi_i(0))^{\frac{1}{2}}}{(\alpha_i|N^c_i|)^{\frac{1}{2}}} e^{-\frac{1}{2}(\beta_i + \frac{\gamma_i}{\alpha_i})T_a}$ *since* $\int_{t^i_k}^t |\frac{d}{dt}\theta_i|$ $\leq M_2(t - t^i_k)$ *based on the fact that* $|\frac{d}{dt}\theta_i| \leq M_2$ *mentioned before. This further implies that* $|\int_{t^i_k}^t \frac{d}{dt}\theta_i| \leq \frac{(\varphi_i(0))^{\frac{1}{2}}}{(\alpha_i|N^c_i|)^{\frac{1}{2}}} e^{-\frac{1}{2}(\beta_i + \frac{\gamma_i}{\alpha_i})T_a}$, *and then we can get*

$$|e_i(t)| = |\theta_i(t) - \theta_i(t^i_k)| \leq \frac{(\varphi_i(0))^{\frac{1}{2}}}{(\alpha_i|N^c_i|)^{\frac{1}{2}}} e^{-\frac{1}{2}(\beta_i + \frac{\gamma_i}{\alpha_i})T_a} \tag{14.46}$$

Equations (14.37) and (14.46) demonstrate that $e^2_i(t) \leq \frac{\varphi_i(t)}{\alpha_i|N^c_i|}$ *which implies*

$$e_i^2(t) \le \frac{1}{|N_i^c|}\left(\frac{\varphi_i(t)}{\alpha_i} + \frac{\sigma_i^d}{4}\sum_{j\in N_i^c}\left(\theta_j(t_{k'(t)}^j) - \theta_i(t_k^i)\right)^2\right) \tag{14.47}$$

Therefore, Eq. (14.45) is one sufficient condition of Eq. (14.47), and thus we have

$$e_i^2(t) > \frac{1}{|N_i^c|}\left(\frac{\varphi_i(t)}{\alpha_i} + \frac{\sigma_i^d}{4}\sum_{j\in N_i^c}\left(\theta_j(t_{k'(t)}^j) - \theta_i(t_k^i)\right)^2\right) \tag{14.48}$$

is one sufficient condition of

$$M_2(t - t_k^i) > \frac{(\varphi_i(0))^{\frac{1}{2}}}{(\alpha_i|N_i^c|)^{\frac{1}{2}}} e^{-\frac{1}{2}(\beta_i + \frac{\gamma_i}{\alpha_i})T_a} \tag{14.49}$$

Letting $k = N(\epsilon_a)$ and $t = t_{N(\epsilon_a)+1}^i$, based on (14.35), we have

$$e_i^2(t_{N(\epsilon_a)+1}^i) > \frac{1}{|N_i^c|}\left(\frac{\varphi_i(t_{N(\epsilon_a)+1}^i)}{\alpha_i} + \frac{\sigma_i^d}{4}\sum_{j\in N_i^c}\left(\theta_j(t_{k'(t_{N(\epsilon_a)+1}^i)}^j) - \theta_i(t_{N(\epsilon_a)}^i)\right)^2\right)$$

since $t_{N(\epsilon_a)}^i$ and $t_{N(\epsilon_a)+1}^i$ are two neighboring triggering time instants for area i. As a consequence, according to the relationship between (14.48) and (14.49) mentioned before, it get that

$$M_2(t_{N(\epsilon_a)+1}^i - t_{N(\epsilon_a)}^i) > \frac{(\varphi_i(0))^{\frac{1}{2}}}{(\alpha_i|N_i^c|)^{\frac{1}{2}}} e^{-\frac{1}{2}(\beta_i + \frac{\gamma_i}{\alpha_i})T_a}$$

which means

$$t_{N(\epsilon_a)+1}^i - t_{N(\epsilon_a)}^i > \frac{(\varphi_i(0))^{\frac{1}{2}}}{M_2(\alpha_i|N_i^c|)^{\frac{1}{2}}} e^{-\frac{1}{2}(\beta_i + \frac{\gamma_i}{\alpha_i})T_a} = 2\epsilon_a \tag{14.50}$$

It is noted that (14.50) is in contradiction with (14.44), and this implies that Zeno behavior existence assumption is invalid, which concludes the proof.

Remark 14.10 Mentioned that the static event-triggered mechanism (14.20) with $\xi_i = 0$, which may cause Zeno behavior based on the analysis of Theorem 14.2, can be seen as a limit case of the dynamic event-triggered mechanism (14.35) when α_i approaches to infinite. This illustrates that Zeno behavior may occur in the absence of internal dynamical variable φ_i. The introducing of positive variable φ_i excludes Zeno behavior by dynamically tuning the threshold value of the triggering condition

(14.35), which can guarantees the asymptotically stability of steady state simultaneously.

Remark 14.11 Although the static and dynamic event-triggered mechanisms are also considered in [9, 10], there does exist some differences. Firstly, the purpose of event-triggered mechanism in [9, 10] is saving the energy of actuator or estimator, while the purpose of (14.20) and (14.35) is reducing burdens in communication network. Secondly, The analysis of Zeno behavior is not necessary in [9, 10] since the discrete system is considered and event-triggered mechanism is constructed based on discrete sampling data. However, Zeno behavior must be analyzed for (14.20) and (14.35) since the continuous system described power system dynamics is considered and the event-triggered mechanism of each area is designed based on the continuous information comparison of the local area. Moreover, the construction of event-triggered mechanism (14.20) and (14.35) must meet the requirement of neighbour-to-neighbour communication and be compatible with distributed control architecture, which is not necessary for that in [9, 10].

14.3 Simulation

This section simulates a test system of power grid contained four areas, which is equivalent to IEEE New England 39-bus system as shown in [11], to verify the effectiveness of event-triggered based distributed optimal frequency regulation.

Fig. 14.1 Test power grid

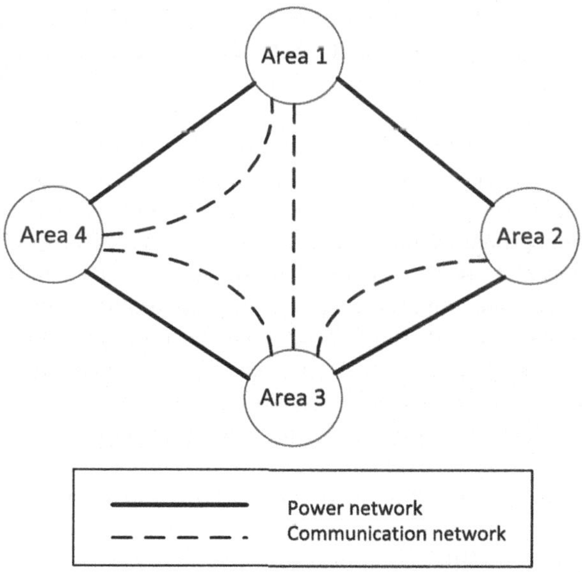

Table 14.1 The value of parameters in simulation

	Area 1	Area 2	Area 3	Area 4
M_i (p.u.)	5.22	3.98	4.49	4.22
A_i (p.u.)	1.60	1.22	1.38	1.42
T_{doi} (s)	5.54	7.41	6.11	6.22
X_{di} (p.u.)	1.84	1.62	1.80	1.94
X'_{di} (p.u.)	0.25	0.17	0.36	0.44
E_{fdi} (p.u.)	4.41	4.20	4.37	4.45
B_{ii} (p.u.)	−49.61	−61.66	−52.17	−40.18
q_i (p.u.)	1.00	0.75	1.50	0.50

Figure 14.1 gives the topology of power network and communication network for power grid. The susceptances of the transmission lines are given as $B_{14} = 21.0$, $B_{12} = 25.6$, $B_{23} = 33.1$ and $B_{34} = 16.6$, and other parameters are shown in Table 14.1. The expected disturbance is that the loads increase the power demand in the four areas from $P^l = (2.00, 1.00, 1.50, 1.00)^T$ to $P^l = (2.20, 1.05, 1.55, 1.10)^T$ at 20 s.

14.3.1 Implementation with Static Event-Triggered Mechanism

This section shows the simulation result of the distributed optimal frequency regulation (14.15)–(14.16) with static event-triggered mechanism (14.20). The parameters in static event-triggered mechanism are set as $\sigma_i^s = 0.9$ and $\xi_i = 0.03$ for all areas. From the state response of system shown in Fig. 14.2a, it can be seen that the frequency deviation is driven in a neighborhood of 0 and the voltage magnitudes do not deviate much from their nominal value of 1 p.u. under the static event-triggered mechanism based regulation method when the power demand changes. Moreover, the power generation is allocated to minimize the cost function, which implies that the optimal frequency regulation problem (14.11)–(14.12) is addressed. Mentioned that although the states of power grid fluctuate in a small range around the new steady state illustrated in Fig. 14.2a the requirement of optimal frequency regulation problem is still fulfilled in practical power grid. Figure 14.3a gives the triggering time sequences of the four areas. The unequal triggering periods demonstrate the information demand-transmission character of event-triggered mechanism and the ability in adjusting communication frequency based on the system's state.

In order to explain the advantage of event-triggered mechanism, the control performance and communication burden comparisons between the static event-triggered mechanism and periodic sampling one are shown. The discretization implementation of continuous information transmission in (14.13)–(14.14) is carried out by the

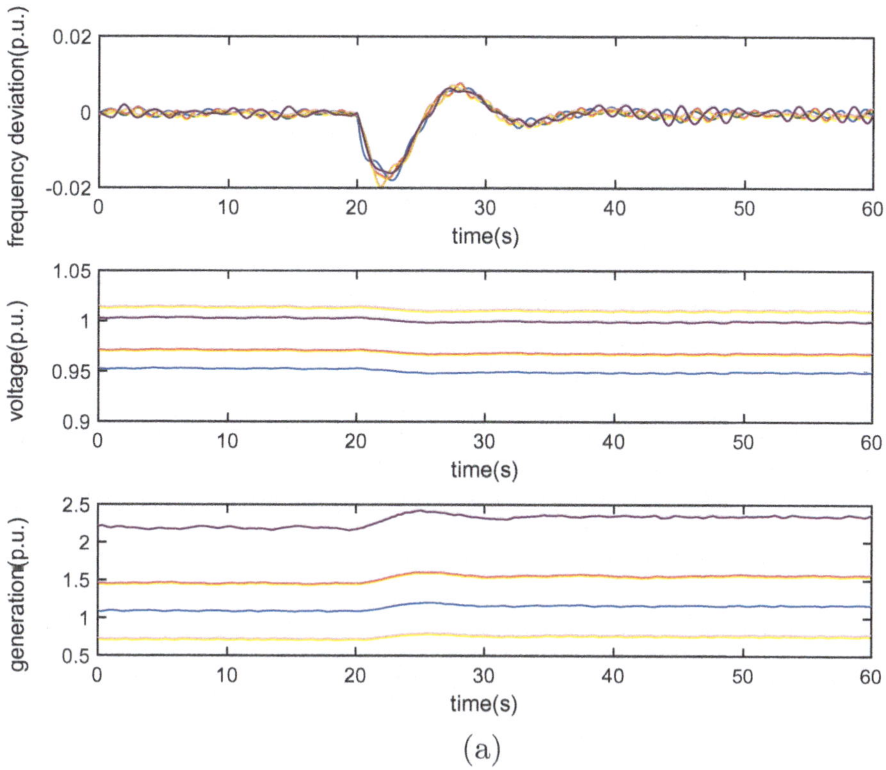

Fig. 14.2 Control performance comparison. System response under: **a** static event-triggered mechanism; **b**: periodic sampling mechanism

periodic sampling scheme, and distributed optimal frequency regulation based on periodic sampling mechanism is constructed as follows for $t \in [kh, (k+1)h)$

$$u_i(t) = \frac{\theta_i(t)}{q_i} \tag{14.51}$$

$$\frac{d}{dt}\theta_i(t) = \sum_{j \in N_i^c} (\theta_j(kh) - \theta_i(kh)) - \frac{\omega_i(t)}{q_i} \tag{14.52}$$

where the non-negative integer k is the sampling number and positive constant h is the fixed sampling period. Figure 14.2b illustrates the state response of system under periodic sampling mechanism with the sampling period h=0.5 s. Compared with Fig. 14.2a, it can seen that nearly identical control performance can be ensured under these two types of mechanisms. However, according to the number of communication comparison shown in Table 14.2 and Fig. 14.3b, the static event-triggered mechanism can lead to less number of sampling and communication than the periodic sampling one, which implies that the communication burden can be reduced.

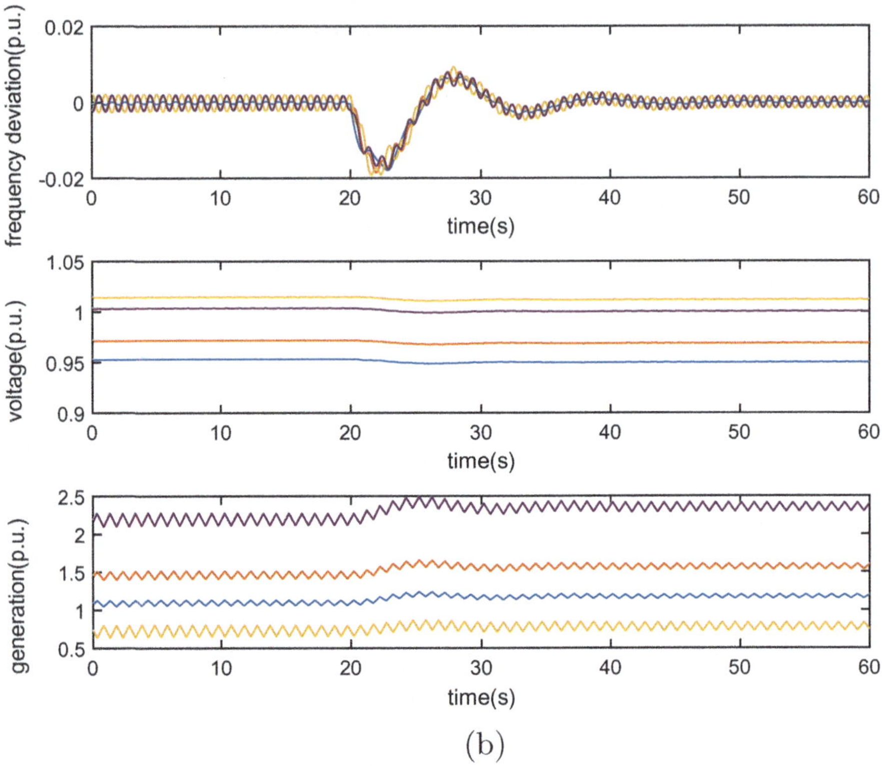

Fig. 14.2 (continued)

Table 14.2 Number of communication in Fig. 14.3b (static event-triggered mechanism (S-ETM) versus periodic sampling mechanism (PSM))

	Area 1	Area 2	Area 3	Area 4
S-ETM	43	19	64	35
PSM	120	120	120	120
Rate	35.83%	15.83%	53.33%	29.17%

14.3.2 Implementation with Dynamic Event-Triggered Mechanism

This section gives the simulation result for the distributed optimal frequency regulation based on dynamic event-triggered mechanism. The parameters in event-triggered mechanism (14.35)–(14.36) are set as $\alpha_i = 0.5$, $\beta_i = 2.1$, $\gamma_i = 0.05$ and $\sigma_i^d = 0.9$ for all areas. The response of system state under dynamic event-triggered mechanism is illustrated in Fig. 14.4a, where the frequency deviation ω, voltage magnitude

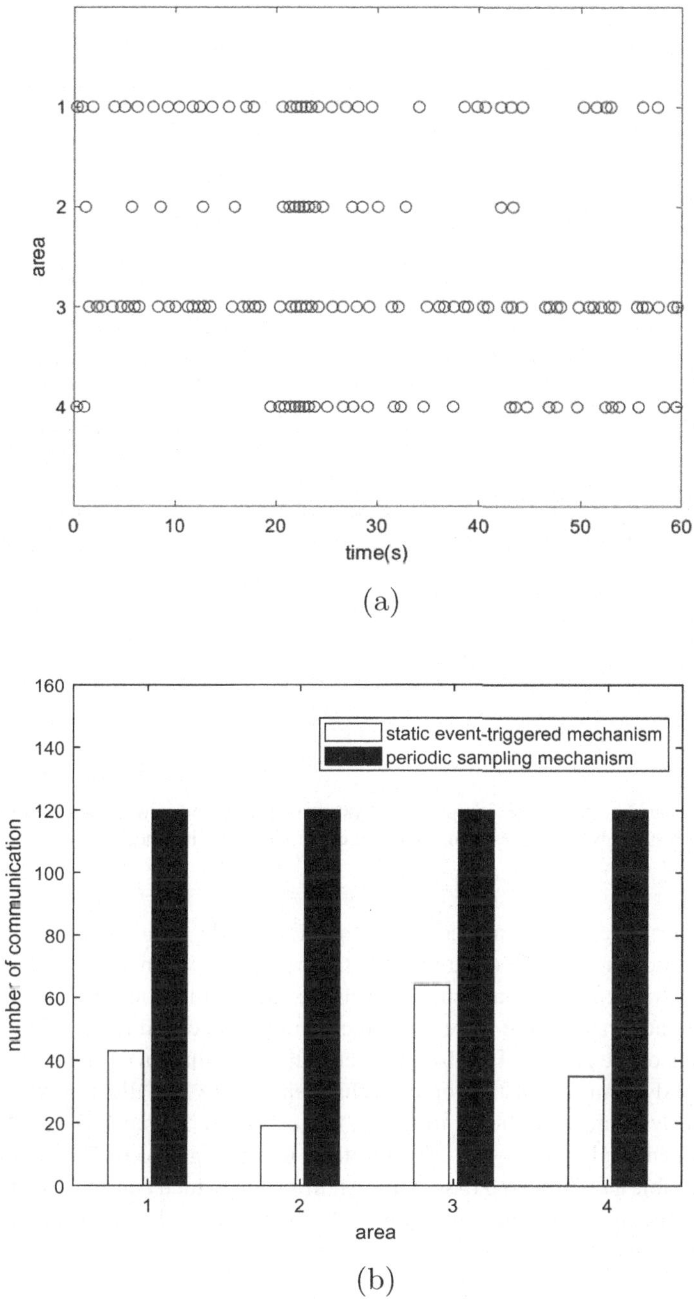

Fig. 14.3 Communication burden comparison. **a** Triggering instants under static event-triggered mechanism; **b** Number of communication comparison between static event-triggered and periodic sampling mechanism

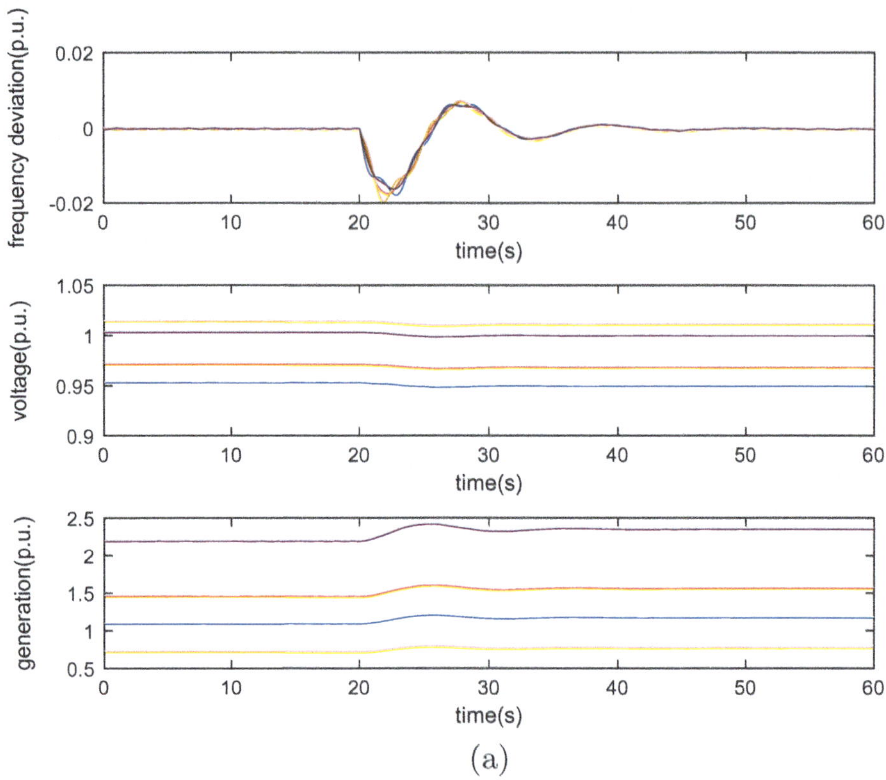

Fig. 14.4 Control performance comparison. System response under: **a** dynamic event-triggered mechanism; **b** static event-triggered mechanism; **c** periodic sampling mechanism

V and power generation u are driven to the new steady state when the disturbance happens in power grid. Compared with the result shown in Fig. 14.2a under the static event-triggered mechanism, the system state asymptotically converges to the steady state accurately here instead of fluctuating in a certain range. The triggering time sequences are given in Fig. 14.5a to verify the information demand-transmission character of dynamic event-triggered mechanism. Figure 14.6 illustrates the response of internal dynamic variable φ_i in (14.35). It can be seen that this variable abrupt changes when loads increase at 20 s and then it converges to 0. This demonstrates that the variable φ_i is adjusted based on system's state, which can be expected a better effect in reducing communication burden compared with the static event-triggered mechanism.

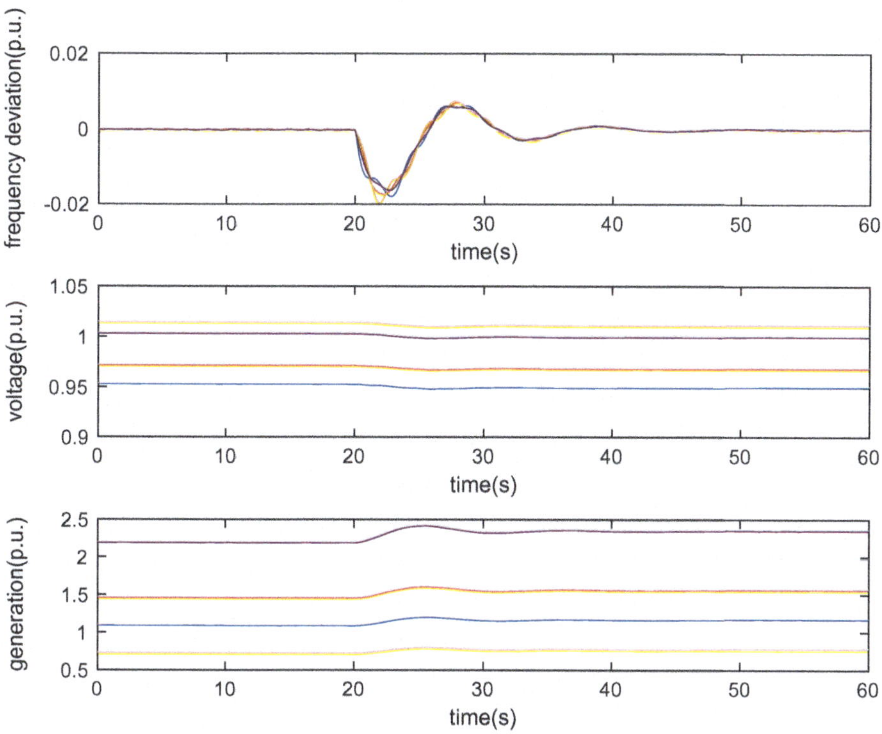

(b)

Fig. 14.4 (continued)

The comparison is given to present the advantages of dynamic event-triggered mechanism over the static event-triggered mechanism and period sampling mechanism. In order to get the similar system performance as shown in Fig. 14.4, the parameters should be chosen as $\sigma_i^s = 0.9$ and $\xi_i = 0.005$ in static event-triggered mechanism (14.20), and the sampling period is selected h=0.4 s in regulation method (14.51)–(14.52). Although the control performance under the three types of mechanism is nearly the same, the comparison of communication number illustrated in Tables 14.3, 14.4, Figs. 14.5c and 14.5b implies that the communication burden can be reduced and less communication resource is utilized under the dynamic event-triggered mechanism.

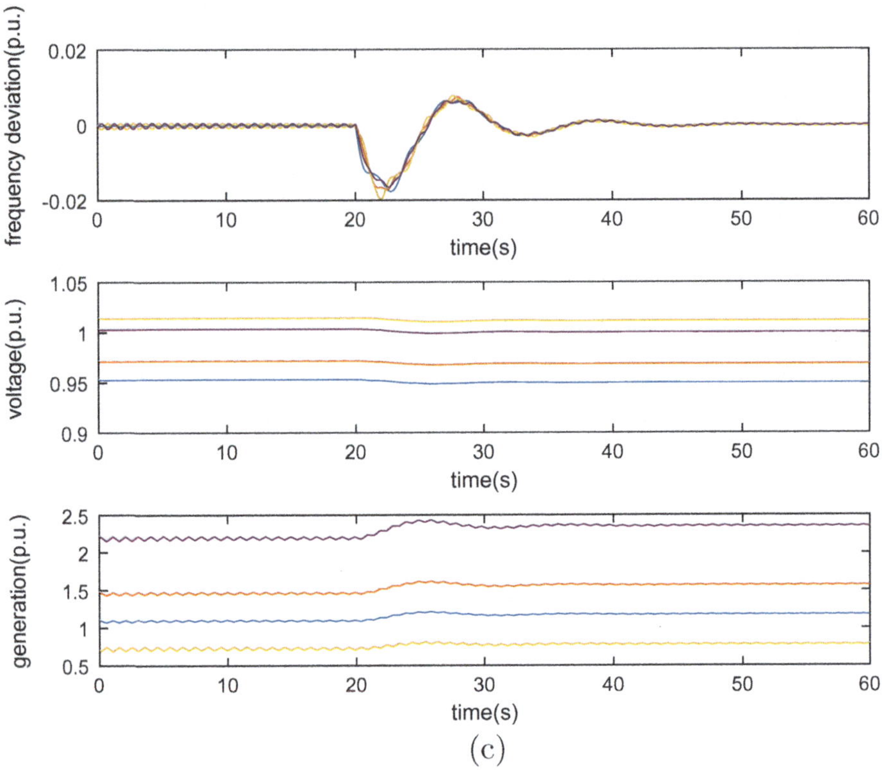

Fig. 14.4 (continued)

14.4 Conclusion

Considering the limitation of communication network, the event-triggered mechanism based distributed optimal frequency regulation of power grid is investigated to restore the frequency and retain the economic efficiency simultaneously. Two types of event-triggered mechanisms are constructed in this chapter. The static event-triggered mechanism causes the system state to fluctuate in a allowable range, while the dynamic one guarantees the asymptotically stability of steady state by an increasing computation complexity. Since the introduction of event-triggered mechanism, the communication burden is reduced and less communication resource is utilized compared with the regulation method based on continuous information transmission or periodic sampling mechanism. Moreover, Zeno behavior is avoided, which implies that the constructed event-triggered mechanisms are reasonable and practicable for physical power grid implementation. Our future work concerns the time-varying power demand and the lossy electric line case, and considers some other event-triggered mechanism to reduce the communication burden further. Furthermore, the event-triggered frequency regulation under cyber-attack is also planned.

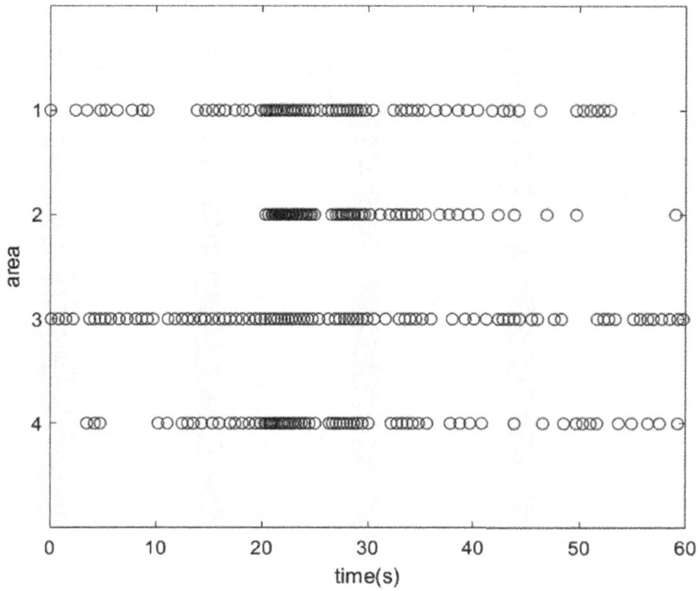

(a)

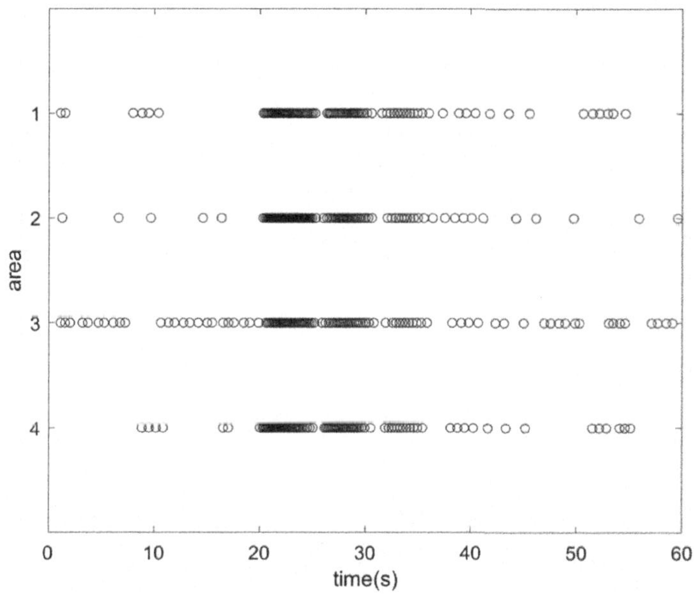

(b)

Fig. 14.5 Communication burden comparison. **a** Triggering instants under dynamic event-triggered mechanism; **b** Triggering instants under static event-triggered mechanism; **c** Number of communication comparison between three types of mechanism

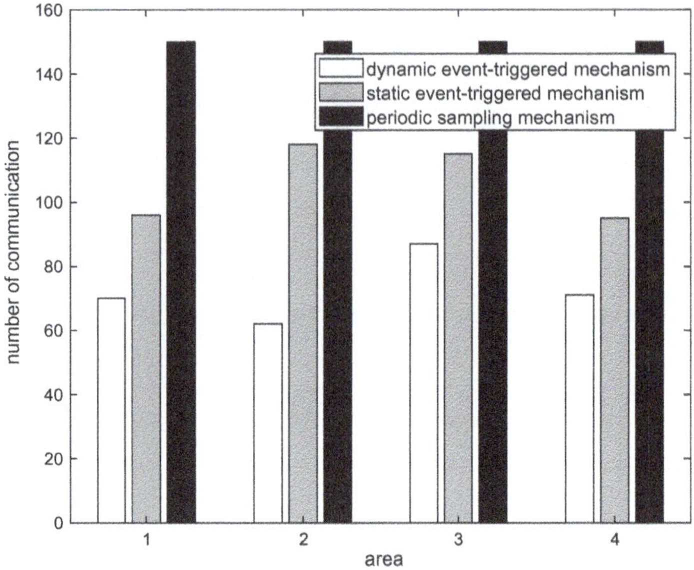

(c)

Fig. 14.5 (continued)

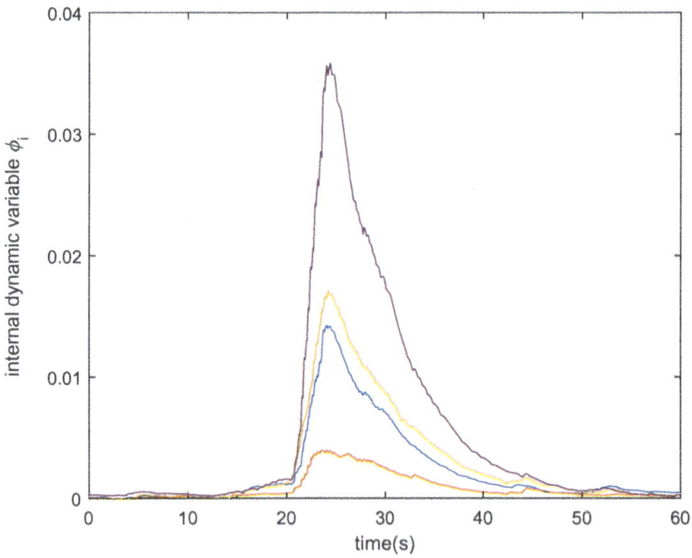

Fig. 14.6 Response of internal dynamic variable φ_i in dynamic event-triggered mechanism

Table 14.3 Number of communication in Fig. 14.5c (dynamic event-triggered mechanism (D-ETM) versus periodic sampling mechanism (PSM))

	Area 1	Area 2	Area 3	Area 4
D-ETM	70	62	87	71
PSM	150	150	150	150
Rate	46.67%	41.33%	58.00%	47.33%

Table 14.4 Number of communication Fig. 14.5c (dynamic event-triggered mechanism (D-ETM) versus static event-triggered mechanism (S-ETM))

	Area 1	Area 2	Area 3	Area 4
D-ETM	70	62	87	71
S-ETM	96	118	115	95
Rate	72.92%	52.54%	75.65%	74.74%

References

1. A. Chakrabortty, J.H. Chow, A. Salazar, A measurement-based framework for dynamic equivalencing of large power systems using wide-area phasor measurements. IEEE Trans. Smart Grid **2**(1), 68–81 (2011)
2. M.L. Ourari, L.A. Dessaint, V.-Q. Do, Dynamic equivalent modeling of large power systems using structure preservation technique. IEEE Trans. Power Syst. **21**(3), 1284–1295 (2006)
3. J. Machowski, J. Bialek, J. Bumby, *Power system dynamics: stability and control* (Wiley, Berlin, 2011)
4. S. Trip, M. Burger, C.D. Persis, An internal model approach to (optimal) frequency regulation in power grids with time-varying voltages. Automatica **64**, 240–253 (2016)
5. R. Olfati-Saber, R.M. Murray, Consensus problems in networks of agents with switching topology and time-delays. IEEE Trans. Autom. Control **49**(9), 1520–1533 (2004)
6. J. Schiffer, D. Goldin, J. Raisch, T. Sezi, Synchronization of droop-controlled microgrids with distributed rotational and electronic generation, in *52nd IEEE Conference on Decision and Control* (IEEE, Piscataway, 2013), pp. 2334–2339
7. F. Dörfler, J.W. Simpson-Porco, F. Bullo, Breaking the hierarchy: distributed control and economic optimality in microgrids. IEEE Trans. Control Netw. Syst. **3**(3), 241–253 (2015)
8. L. Lu, C. Chu, Consensus-based droop control synthesis for multiple dics in isolated microgrids. IEEE Trans. Power Syst. **30**(5), 2243–2256 (2015)
9. B. Shen, Z. Wang, H. Qiao, Event-triggered state estimation for discrete-time multidelayed neural networks with stochastic parameters and incomplete measurements. IEEE Trans. Neural Netw. Learn. Syst. **28**(5), 1152–1163 (2016)
10. Q. Li, B. Shen, Z. Wang, T. Huang, J. Luo, Synchronization control for a class of discrete time-delay complex dynamical networks: a dynamic event-triggered approach. IEEE Trans. Cybern. **49**(5), 1979–1986 (2019)
11. S. Nabavi, A. Chakrabortty, Topology identification for dynamic equivalent models of large power system networks, in *2013 American Control Conference*, pp. 1138–1143 (2013)

Chapter 15
A Virtual Complex Impedance Based $P - \dot{V}$ Droop Method for Parallel-Connected Inverters in Low-Voltage AC Microgrids

Due to the high R/X ratio and mismatched feeder impedance of low-voltage microgrids (LVMGs), conventional droop method is no longer able to decouple the active and reactive power of distributed generators (DGs) and the power sharing accuracy is degraded. In this chapter, a virtual complex impedance based $P - \dot{V}$ droop method is proposed to decouple the powers and improve the power sharing accuracy among DGs. With the virtual impedance method, the equivalent impedance between virtual power source (VPS) and point of common coupling (PCC) is shaped to be purely resistive. Then, a $P - \dot{V}$ strategy is adopted to alleviate the effect of mismatched line impedance, where the virtual powers rather than the ordinary P/Q are used in the droop equation. In case the output voltage violates the operation code, a restoration mechanism is proposed to reset $\dot{V}$ to zero. Compared with existing virtual impedance and $Q - \dot{V}$ droop methods, the proposed method combines the advantages of both. Besides, a modified $P - \dot{V}$ strategy is also presented to accelerate the restoration process and improve the active power sharing accuracy at the same time. Simulation results validate the effectiveness of the proposed methods.

15.1 Islanded Microgrid Structure, Modeling and Control

In this study, two parallel connected DGs with LCL filters are analyzed, where the passive damping method is taken for robustness, as shown in Fig. 15.1. Besides, the DC voltage is considered as constant for simplicity. The local balanced load consists of resistors and inductors is shared by the two DGs. To analyze the influence of mismatched feeder impedance, the line impedance is set to be different. The detailed analysis of the control loop is presented as follows.

D. Yue et al., *Cooperative Optimal Control of Hybrid Energy Systems*,
https://doi.org/10.1007/978-981-33-6722-7_15

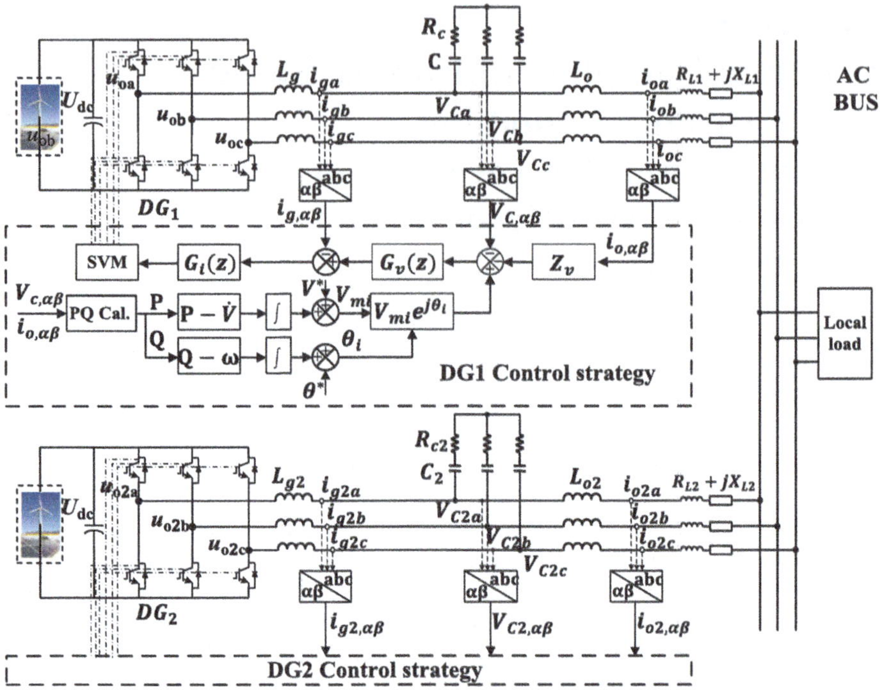

Fig. 15.1 Configuration of the investigated MG

15.1.1 Voltage and Current Control Loop in Stationary Frame

The detailed block diagram of the inner loop is shown as Fig. 15.2, where PR controller is adopted in the outer voltage control loop while P controller is preferred in the inner current control loop. z^{-1} is incorporated to model the computation delay. Besides, the resistor parasitics of DG and grid side inductor are ignored for the most unstable cases.

The discrete formulation of PR controller is shown as [1, 2].

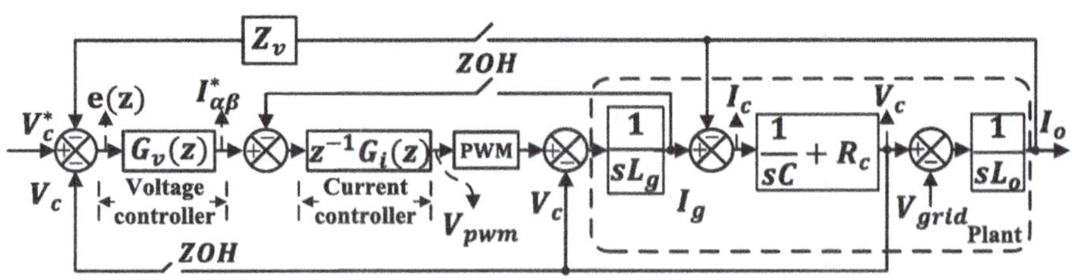

Fig. 15.2 Inner control loop

$$G_v(z) = G_{PR}(z) = K_p + K_i \cdot R_{tp}(z) \tag{15.1}$$

where K_p is the proportional coefficient and K_i is the resonant one, and

$$\begin{aligned} R_{tp}(z) =& \frac{0.5(1 - z^{-2})cos(\varphi)sin(\omega_e T_s)}{\omega_e(1 - 2z^{-1}cos(\omega_e T_s) + z^{-2})} \\ & - \frac{(1 + 2z^{-1} + z^{-2})sin(\varphi)\{sin\frac{\omega_e T_s}{2}\}^2}{\omega_e(1 - 2z^{-1}cos(\omega_e T_s) + z^{-2})} \end{aligned} \tag{15.2}$$

where T_s is the sampling rate, φ is the leading angle which should be properly designed to compensate for the delay of controlled plant, ω_e is the controlled frequency, which is equal to the angular frequency derived by droop equation.

The proportional term $G_i(z) = K_I$ is suitable for the inner current controller to provide extra virtual damping for the LCL filter and the PCC voltage is seen as disturbance. Then, according to Fig. 15.2, the following equation is derived.

$$\begin{cases} \left(V_{pwm}(s) - V_c(s)\right)\frac{1}{sL_g} = I_g \\ (I_g(s) - I_o(s))\left(\frac{1}{sC} + R_c\right) = V_c(s) \end{cases} \tag{15.3}$$

where V_{pwm} is the modulated voltage reference while V_c is the capacitor voltage. L_g is the inverter-side inductor. I_g and I_o are the inverter-side and grid-side current. C is the filter capacitor and R_c is the resistor used for passive damping.

From Eq. (15.3), the following formulation is derived.

$$\begin{cases} V_c(s) = G_1(s)V_p(s) - G_2(s)I_o(s) \\ I_g(s) = G_3(s)V_p(s) + G_1(s)I_o(s) \end{cases} \tag{15.4}$$

where V_{pwm} is simply denoted as V_p and $G_1(s) = \frac{sCR_c+1}{s^2L_gC+sCR_c+1}$, $G_2(s) = \frac{s^2L_gCR_c+sL_g}{s^2L_gC+sCR_c+1}$, $G_3(s) = \frac{sC}{s^2L_gC+sCR_c+1}$.

Considering the characteristic of zero-hold (ZOH), $G_{1,2,3}(s)$ are discretized as [1]

$$\begin{aligned} G_1(z) =& \frac{e^{-2aT_s} + z - (z+1)e^{-aT_s}cos(\omega_r T_s)}{z^2 - 2ze^{-aT_s}cos(\omega_r T_s) + e^{-2aT_s}} \\ & + \frac{\frac{R_c(z-1)}{2L_g\omega_r}(z-1)e^{-aT_s}sin(w_r T_s)}{z^2 - 2ze^{-aT_s}cos(\omega_r T_s) + e^{-2aT_s}} \\ G_2(z) =& \frac{R_c(z-1)(z - e^{-aT_s}cos(\omega_r T_s))}{z^2 - 2ze^{-aT_s}cos(\omega_r T_s) + e^{-2aT_s}} \\ & + \frac{\frac{2L_g - R_c^2C}{2L_gC\omega_r}(z-1)e^{-aT_s}sin(\omega_r T_s)}{z^2 - 2ze^{-aT_s}cos(\omega_r T_s) + e^{-2aT_s}} \\ G_3(z) =& \frac{(z-1)e^{-aT_s}sin(\omega_r T_s)}{z^2 - 2ze^{-aT_s}cos(\omega_r T_s) + e^{-2aT_s}} \cdot \frac{1}{\omega_r L_g} \end{aligned} \tag{15.5}$$

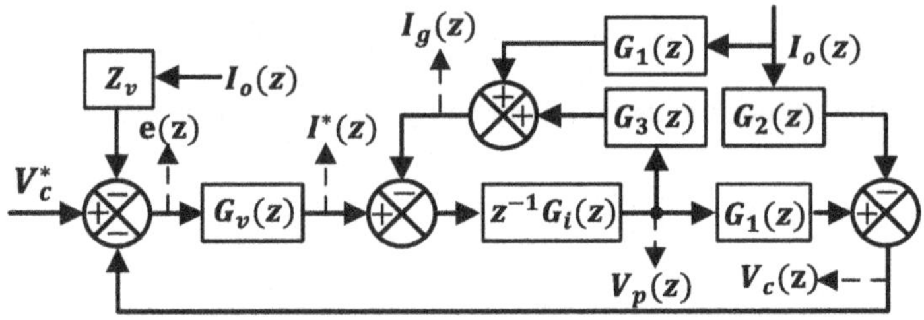

Fig. 15.3 Inner control loop in z domain

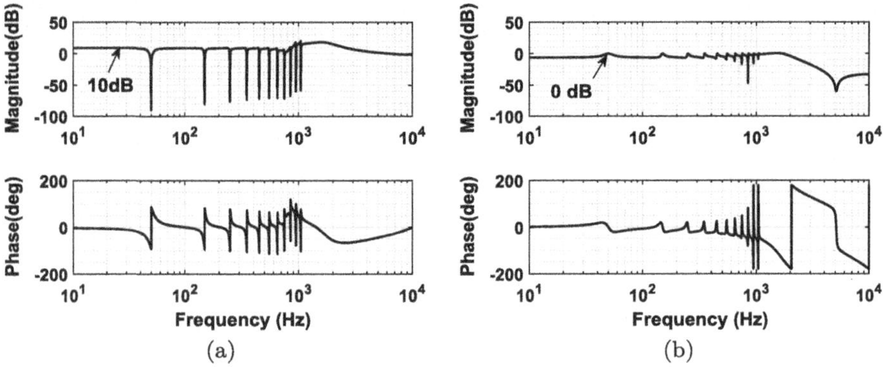

Fig. 15.4 Bode plot of (**a**) $Z_o(z)$, **b** $G_4(z)$

where $a = \frac{R_c}{2L_g}$, $\omega_r = \sqrt{\frac{1}{L_gC} - \frac{R_c^2}{4L_g^2}}$. Then the block diagram of the inner loop is redrawn as Fig. 15.3.

According to Fig. 15.3, the following equation is derived

$$\begin{cases} [V_c^*(z) - V_c(z)]G_v(z) = I^*(z) \\ [I^*(z) - I_g(z)]G_i(z)z^{-1} = V_p(z) \end{cases} \tag{15.6}$$

Combining Eqs. (15.4), (15.5), (15.6), one has

$$V_c(z) = G_4(z)V_c^*(z) - Z_o(z)I_o(z) \tag{15.7}$$

where $G_4(z) = \frac{G_1(z)G_i(z)G_v(z)}{z+G_1(z)G_v(z)G_i(z)+G_3(z)G_i(z)}$, $Z_o(z) = \frac{G_2(z)G_3(z)G_i(z)+G_1^2(z)G_i(z)+zG_2(z)}{z+G_1(z)G_v(z)G_i(z)+G_3(z)G_i(z)}$

To further investigate the characteristic of the DG, the bode plot of $G_4(z)$ and $Z_o(z)$ is provided in Figs. 15.4a, b respectively.

The output impedance of the DGs under the PR controller differs from that under PI controller [3] where the output impedance is nearly inductive at the fundamental frequency. According to Fig. 15.4a, the output impedance is nearly resistive (10 dB) around the controlled frequency. The magnitude gain of $Z_o(z)$ at resonant frequencies is negative infinity dB, which means that the output impedance at resonant frequencies is zero. Thus, the output voltage will not be influenced by the output current at resonant frequencies.

Since both the magnitude gain (in dB) and phase gain of $G_4(z)$ are zero at resonant frequencies, as shown in Fig. 15.4b, the transfer function $G_4(z)$ has no influence on the voltage reference $V_c^*(z)$ at resonant frequencies. Combining the influence of $Z_o(z)$ and $G_4(z)$, the output voltage $V_c(z)$ is able to track the reference $V_c^*(z)$ without steady errors at resonant frequencies.

15.1.2 Virtual Complex Impedance Strategy

Although zero impedance characteristic is an advantage from the aspect of tracking performance, it will exaggerate the circulate current between DGs [4, 5]. Besides, the stability of the system is endangered since the disturbance of output voltage $V_c(z)$ is undamped.

Considering the near resistive characteristic within the controlled frequencies and the effect of grid-side inductors and/or the transformers, the following virtual complex impedance is adopted in this study.

$$Z_{vi} = R_{vi} + \frac{1}{sC_{vi}} = R_{vi} - jX_{vi} \tag{15.8}$$

where R_{vi} is the virtual resistor and C_{vi} is the virtual capacitor. X_{vi} is the equivalent reactance and Z_{vi} is the virtual complex impedance. Thus, output voltage of DGs is modified as

$$\begin{aligned} V_{ci}(z) &= G_4(z)\left[V_{ci}^*(z) - I_{oi}(z)Z_{vi})\right] - Z_o(z)I_{oi}(z) \\ &= G_4(z)V_{ci}^*(z) - Z_{oi}^{'}(z)I_{oi}(z) \end{aligned} \tag{15.9}$$

where $Z_{oi}^{'}(z) = G_4(R_{vi} - jX_{vi}) + Z_o(z)$ is equivalent output impedance.

According to the analysis of the Fig. 15.4a, b, equivalent output impedance is simplified as $Z_{oi}^{'}(z) = Z_{vi} = R_{vi} - jX_{vi}$. The virtual capacitor will alleviate the impact of reactance, especially when the line reactance X_{Li} between DGs and PCC is known or can be estimated. The final impedance $Z_i^{'}$ is reshaped as purely resistive at fundamental frequency as shown in Fig. 15.5, if X_{vi} is set to be equal to X_{Li}.

$$\begin{aligned} Z_i^{'} &= Z_{vi} + Z_{Li} = R_{vi} - jX_{vi} + R_{Li} + jX_{Li} \\ &= R_{vi} + R_{Li} = R_{vi}^{'} \end{aligned} \tag{15.10}$$

where R_{vi} of different DGs is set to be equal, R_{Li} is typically unequal representing the mismatch of feeder impedance, $R_{vi}^{'}$ is the equivalent impedance between VPS and PCC.

In Fig. 15.5, $V_i\angle\varphi_i = G_4(z)V_{ci}^*(z)$ is defined as the virtual power source (VPS). $P_i^{'}$ and $Q_i^{'}$ are the virtual active and reactive power injected by $V_i\angle\varphi_i.Z_{oi} + Z_{vi}$ is

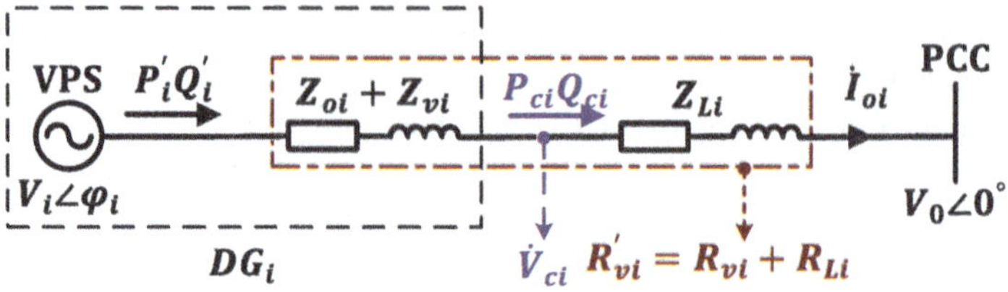

Fig. 15.5 Equivalent circuit of DG under virtual complex impedance

the equivalent output impedance. Z_{Li} is the line impedance. P_{ci} and Q_{ci} is the active and reactive power injected by capacitor voltage $\dot{V}_{ci}$. $V_0\angle 0°$ is the PCC voltage. $\dot{I}_i$ is the output current of DG_i.

15.1.3 Modified Droop Control Equation

Since the equivalent impedance between VPS and PCC is shaped as purely resistive, the virtual active power P' and reactive power Q' can be expressed as

$$\begin{aligned} P_i' &= \frac{V_i(V_i - V_o)}{R_{vi}'}cos(\varphi_i) = \frac{V_i(V_i - V_o)}{R_{vi}'} \\ Q_i' &= -\frac{V_i V_o}{R_{vi}'}sin(\varphi_i) = -\frac{V_i V_o}{R_{vi}'}\varphi_i \end{aligned} \tag{15.11}$$

where φ_i is assumed to be small enough.

Thus, the $P - V$ and $Q - \omega$ droop scheme can be adopted to regulate the frequency and amplitude of the VPS output-voltage reference.

$$\begin{aligned} V_i &= V^* - m_i P_i' \\ \omega_i &= \omega^* + n_i Q_i' \end{aligned} \tag{15.12}$$

where m_i and n_i are the droop coefficient. V_i and ω_i are the output voltage magnitude and angular frequency command.

It should be noted that P_{ci} and Q_{ci} are not decoupled since the line impedance between DG_i and PCC is complex. Different from most existing papers [3, 6–9], where P_{ci} and Q_{ci} calculated with the sampled voltage of the filter capacitor and grid-side current are used in the droop Eq. (15.12), the virtual active power P_i' and reactive power Q_i' are preferred in this chapter. The difference between $P_{ci}Q_{ci}$ and $P_i'Q_i'$ is the power consumed by Z_{oi} and Z_{vi} which may not be negligible.

15.2 Proposed $P - \dot{V}$ Droop Control Method

Since the virtual active power $P_i^{'}$ sharing accuracy is often degraded under the traditional methods owing to the mismatched feeder impedance, the following original and modified $P - \dot{V}$ droop control method are proposed to improve the active power sharing accuracy.

15.2.1 *Original $P - \dot{V}$ Droop Control Method*

In this section, a $P - \dot{V}$ droop control method is proposed, where $\dot{V}$ represents the time rate of change of the output voltage reference.

$$\begin{aligned} \dot{V}_{oi} &= m_i(P_{oi}^{'} - P_i^{'}) \\ V_i &= V^* + \int_t \dot{V}_{oi} d\tau \end{aligned} \tag{15.13}$$

where $P_{oi}^{'}$ is the virtual active power set point at the nominal V_i, $\dot{V}_{oi}$ is the changing rate of V_i and $V^* = 311\,\text{V}$ is the nominal voltage reference.

The operation of proposed $P - \dot{V}$ method is illustrated in Fig. 15.6. Supposing that the rated power of DG_2 is half of that of DG_1, then $P_{N1} = 2P_{N2}, m_2 = 2m_1$. At the beginning of operation time t_0, both the DGs are operated in steady state and

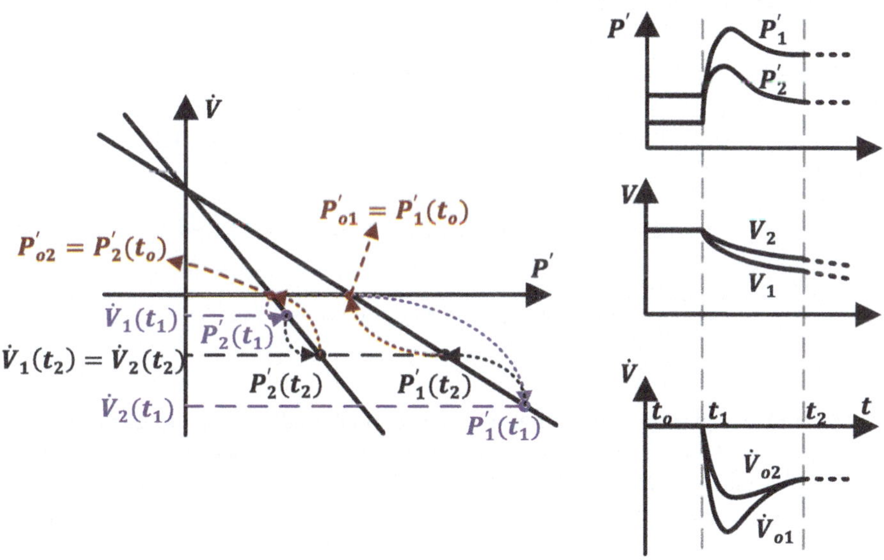

Fig. 15.6 Operation of proposed $P' - \dot{V}$ droop controller

the initial virtual active power is shared proportionally, $m_1 P_1'(t_0) = m_2 P_2'(t_0)$. As the loads increase at time t_1, the output power of DG_1 increases sharply to $P_1'(t_1)$ to pick up the loads while that of DG_2 rises marginally to $P_2'(t_1)$. Since $P_1'(t_1)$ is larger than $P_2'(t_1)$, $\dot{V}_{o1}(t_1)$ is more negative than $\dot{V}_{o2}(t_1)$ according to proposed droop control Eq. (15.13). During the interval t_1 and t_2, both $\dot{V}_{o1}$ and $\dot{V}_{o2}$ decreases at first and then increases until they reach the balanced point $\dot{V}_{o1} = \dot{V}_{o2}$. The changing rate of P_i' is opposite to the case of $\dot{V}_{oi}$. The voltage magnitude reference decreases rapidly at the beginning and then slows down since $\dot{V}_{oi}$ increases.

At time t_2, the virtual active power P' has been shared proportionally and $\dot{V}_{o1} = \dot{V}_{o2}$. The resulting V_1 and V_2 decrease at the same pace without changing their relative magnitude. According to (15.13), the following equation holds: $m_1(P_1' - P_{o1}') = m_2(P_2' - P_{o2}') = \cdots m_n(P_n' - P_{on}')$. Considering the initial steady state at t_0, $m_1 P_{o1}' = m_2 P_{o2}' = \cdots m_n P_{on}'$, then the virtual active power of each DG satisfies $m_1 P_1' = m_2 P_2' = \cdots m_n P_n'$, which means the virtual active power has been shared proportionally.

It should be mentioned that V_i and P_i' still decreases after time t_2 as long as $P_i' \neq P_{oi}'$. However, this tendency is neglected here since the following introduced restoration mechanism will force the $\dot{V}_{oi}$ to be zero.

15.2.2 Restoration Mechanism

After $\dot{V}_{oi}$ reaches the balanced point, there must exist a restoration mechanism bringing $\dot{V}_{oi}$ back to zero to prevent the output magnitude varying. The $\dot{V}_i$ restoration mechanism is designed as follows:

$$\dot{P}_{oi}' = -k_{res} P_{Ni}' \dot{V}_{oi} \tag{15.14}$$

where k_{res} is the restoration gain and P_{Ni}' is the rated virtual active power capacity and $m_1 P_{N1}' = m_2 P_{N2}' = \cdots m_n P_{Nn}'$.

Assuming that the time constant of the restoration control is much longer than that of $P - \dot{V}$ droop control. The following relationship is derived.

$$\frac{d(\dot{V}_{oi})}{dt} = m_i \dot{P}_{oi}' = -m_i k_{res} P_{Ni}' \dot{V}_{oi} = -k_{res}' \dot{V}_{oi} \tag{15.15}$$

where $k_{res}' = m_i k_{res} P_{Ni}'$.

The above equation shows that $\dot{V}_{oi}$ will exponentially decay to zero after the restoration mechanism is executed. With the same k_{res} adopted for different DGs, $\dot{V}_{oi}$ of different DGs will decay at the same pace.

The proposed $P - \dot{V}$ together with the restoration mechanism is illustrated as Fig. 15.7.

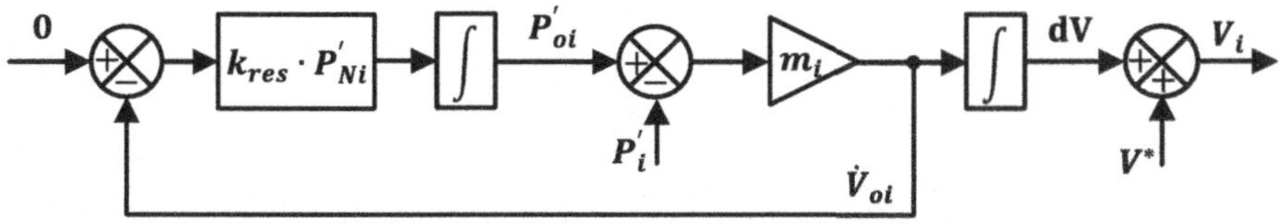

Fig. 15.7 Block diagram of the restoration mechanism

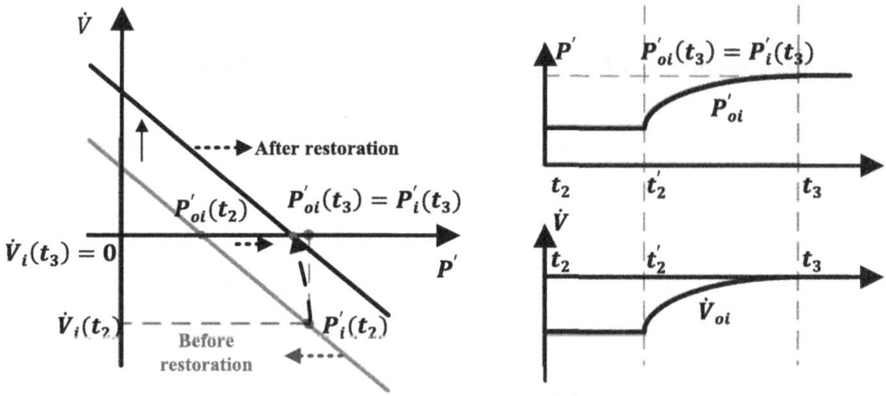

Fig. 15.8 Operation of $\dot{V}$ restoration process

Figure 15.8 demonstrates the operation of $\dot{V}$ restoration. Assuming that the restoration starts at time $t = t_2'$, $\dot{V}_i$ and P_{oi}' hold the same value while P_i' decreases at the same pace due to the continuing reduction of voltage magnitude V_i during t_2 and t_2'. After t_2', the proposed restoration mechanism continuously drives $\dot{V}_{oi}$ to the 0 as can be observed in Fig. 15.7. At $t = t_3$, $\dot{V}_{oi}$ is reset to be zero and then V_i keeps constant. During t_2' and t_3, the proposed restoration mechanism raises up P_{oi}' until it is equal to P_i'. Comparing with $t = t_2'$, $P_i'(t_3)$ is slightly smaller than $P_i'(t_2)$ owing to the continuing decrease of V_i and the difference between them is affected by the convergence rate of the restoration mechanism. After $t = t_3$, the system has entered into the steady state.

It should be noted that both Figs. 15.6, 15.8 only give the case of abrupt increase of loads. The opposite case can be analyzed accordingly and is omitted here.

15.2.3 Modified $P - \dot{V}$ Droop Control Method

Although the original $P - \dot{V}$ droop method is able to improve the active power sharing accuracy, its transient and steady performance can be further optimized. A modified $P - \dot{V}$ droop control method is proposed as follows, where the restoration mechanism remains the same.

$$\begin{aligned} \dot{V}_{oi} &= m_i(P'_{oi} - P'_i) \\ V_i &= V^* + S_p \cdot \int_t \dot{V}_{oi} d\tau \end{aligned} \tag{15.16}$$

where $S_p > 1$ is a proportional coefficient and the difference between the original and modified $P - \dot{V}$ strategy will be compared in the following section.

15.3 Discussion on the Effects of Relative Coefficients

In this section, the effect of relative coefficients is discussed. According to Eq. (15.16), one can have

$$\frac{d\dot{V}_{oi}}{dt} = m_i(\dot{P}'_{oi} - \dot{P}'_i) \tag{15.17}$$

Considering the restoration mechanism (15.14), it can be transformed into

$$\frac{d\dot{V}_{oi}}{dt} = -m_i k_{res} P'_{Ni} \dot{V}_{oi} - m_i \dot{P}'_i \tag{15.18}$$

The differential form of the P'_i can be derived from Eq. (15.11)

$$\begin{aligned} \dot{P}'_i &= \frac{\dot{V}_i(V_i - V_o) + V_i(\dot{V}_i - \dot{V}_o)}{R'_{vi}} \\ &\approx \frac{2(1 - k_i)V_i \dot{V}_i}{R'_{vi}} \end{aligned} \tag{15.19}$$

where V_o is assumed to be proportional to V_i,$V_o \approx k_i V_i$.

With Eqs. (15.16) and (15.19), one can have

$$\dot{P}'_i \approx \frac{2(1 - k_i)V^*}{R'_{vi}} \cdot S_p \dot{V}_{oi} \tag{15.20}$$

where the actual voltage deviation $S_p \cdot \int \dot{V}_{oi} d\tau$ is smaller than allowed value $\delta V_{max} = 0.05 V^*$ and is thus ignored here. Combine the Eqs. (15.18) and (15.20)

$$\frac{d\dot{V}_{oi}}{dt} \approx -m_i \left(k_{res} P'_{Ni} + \frac{2(1 - k_i)V^*}{R'_{vi}} \cdot S_p \right) \cdot \dot{V}_{oi} \tag{15.21}$$

Thus, the convergency rate is

$$C_{rmi} = C_{roi} + \frac{2(1-k_i)V^*}{R'_{vi}} \cdot m_i S_p \tag{15.22}$$

where $C_{roi} = m_i k_{res} P'_{Ni}$ is the convergency rate of the original droop method (15.13).

Since the second term of (15.22) is positive, C_{rmi} is larger than C_{roi}. Thus, the convergency rate of the modified droop method is faster than the original one and the difference between them can be adjusted by k_{res}, S_p, m_i and R'_{vi}. Since m_i and R'_{vi} are predefined in this chapter, only k_{res} and S_p are discussed in this section.

The ultimate expression of $\dot{V}_{oi}$ is listed as follows.

$$\dot{V}_{oi}(t) = \dot{V}_{oi}(0) \cdot e^{-C_{rmi} \cdot t} \tag{15.23}$$

The total voltage deviation whenever load changes can be calculated as

$$\begin{aligned} \delta V_{roi} &= 1 \cdot \int \dot{V}_{oi}(0) \cdot e^{-C_{roi} \cdot t} dt \\ &= \left(P'_{oi}(0) - P'_i(0)\right)\left(k_{res} P'_{Ni}\right)^{-1} \\ \delta V_{rmi} &= S_p \cdot \int \dot{V}_{oi}(0) \cdot e^{-C_{rmi} \cdot t} dt \\ &= \left(P'_{oi}(0) - P'_i(0)\right)\left(\frac{k_{res} P'_{Ni}}{S_p} + \frac{2(1-k_i)V^*}{R'_{vi}}\right)^{-1} \end{aligned} \tag{15.24}$$

where $P'_i(0)$ and $P'_{oi}(0)$ are the initial value and initial set point of virtual active power, respectively, when the load changes.

With $S_p = 5$ which are defined later, $k_{res} P'_{Ni} > \frac{k_{res} P'_{Ni}}{S_p} + \frac{(1-k_i)V^*}{R'_{vi}}$, thus $\delta V_{roi} < \delta V_{rmi}$, which means the total voltage deviation of the modified method is larger than the original one. However, as long as δV_{rmi} is constrained in an allowable level, the modified droop method is preferred for its transient performance.

Apart from the convergency rate and voltage difference, the difference of P'_{oi} between DGs is another indicator of the performance of the proposed droop method. According to Eq. (15.14), the steady state active power sharing difference $\delta P'_{oio}$ of the original $P - \dot{V}$ method and $\delta P'_{oim}$ of the modified one are derived as follows.

$$\begin{aligned} \delta P'_{oio} &= \frac{P'_1(0) - P'_2(0) + \frac{2(1-k_i)V^*}{P'_N R'_v k_{res}}\left[P'_{01}(0) - P'_{02}(0)\right]}{1 + \frac{2(1-k_i)V^*}{P'_N R'_v k_{res}}} \\ \delta P'_{oim} &= \frac{P'_1(0) - P'_2(0) + \frac{2(1-k_i)V^* \cdot S_p}{P'_N R'_v k_{res}}\left[P'_{01}(0) - P'_{02}(0)\right]}{1 + \frac{2(1-k_i)V^* \cdot S_p}{P'_N R'_v k_{res}}} \end{aligned} \tag{15.25}$$

With simple manipulation, it can be derived that if

$$P_{01}^{'}(0) - P_{02}^{'}(0) < P_1^{'}(0) - P_2^{'}(0) \tag{15.26}$$

holds, Then, $\delta P_{oim}^{'} < \delta P_{oio}^{'}$. Since condition (15.26) is easier to be fulfilled when the load changes, the sharing accuracy of the modified droop method (15.26) is better than that of original method (15.13). In addition, if $P_{01}^{'}(0) = P_{02}^{'}(0)$, Eq. (15.25) can be simplified as

$$\begin{aligned} \delta P_{oio}^{'} &= \frac{P_1^{'}(0) - P_2^{'}(0)}{1 + \frac{2(1-k_i)V^*}{P_N^{'} R_v^{'}} \cdot \frac{1}{k_{res}}} \\ \delta P_{oim}^{'} &= \frac{P_1^{'}(0) - P_2^{'}(0)}{1 + \frac{2(1-k_i)V^*}{P_N^{'} R_v^{'}} \cdot \frac{S_p}{k_{res}}} \end{aligned} \tag{15.27}$$

It is more clear that $\delta P_{oim}^{'} < \delta P_{oio}^{'}$ as long as $S_p > 1$. In fact, the original droop method (15.13) is a special case of the modified one (15.16).

Remark 15.1 Since droop coefficient m_i is not included in Eqs. (15.24), (15.25), only the convergency rate of the restoration process will be affected by m_i. In other words, the voltage drop δV and virtual active power sharing accuracy $\delta P_{oi}^{'}$ are immune to the variation of m_i.

Remark 15.2 k_{res} and S_p should be carefully designed to reach a compromise between δV and $\delta P_{oi}^{'}$. Considering all the three Eqs. (15.22), (15.24), (15.27), if S_p keeps constant, the bigger k_{res} is, the faster the convergency rate C_{rmi} will be. Besides, the total voltage difference δV will be smaller at the expense of a larger $\delta P_{oi}^{'}$, which means the active power sharing accuracy is deteriorated. Let k_{res} be constant, with the increasing of S_p, it is interesting that the speed of restoration process is faster and the active power sharing accuracy is improved at the same time at the cost of larger δV.

To ensure the δV_{rmi} is constrained in an acceptable level, the common code of 5% at most is adopted.

$$\begin{aligned} &\delta V_{max} = \Delta P_{max} \left(\frac{k_{res} P_{Ni}^{'}}{S_p} + \frac{(1-k_i)V^*}{R_{vi}^{'}} \right)^{-1} < 0.05 V^* \\ &\Rightarrow \\ &\frac{k_{res}}{S_p} > \frac{20}{V^*} - \frac{2(1-k_i)V^*}{R_{vi}^{'} P_N^{'}} \approx \frac{20}{V^*} \approx 0.065 \end{aligned} \tag{15.28}$$

In this study, the main consideration is the tradeoff between the voltage drop δV and active power sharing accuracy $\delta P_{oi}^{'}$. With the instruction of Remark 15.2 and several experiments, $S_p = 5$ and $k_{res} = 0.325$ are adopted.

15.4 Simulation Results

To verify the effectiveness of the proposed $P - \dot{V}$ method and the modified version, simulations based on MG of Fig. 15.1 is presented in Sects. 15.4.1 and 15.4.2. A complex MG case as shown in Fig. 15.11 is also analyzed in Sect. 15.4.3. The line impedance and controller parameters of Fig. 15.1 are listed as follows.

(1) The system voltage, frequency, switching frequency are $V^* = 311\text{V}$, $f_0 = 50\,\text{Hz}$ and $f_{sw} = 10\,\text{kHz}$;
(2) The converter-side and grid-side inductor, filter capacitor and damping resistor are $L_{gi} = 1\,\text{mH}$, $L_{oi} = 40\,\text{uH}$, $C_i = 30\,\text{uF}$ and $R_{ci} = 1.6\Omega$. The mismatched line impedance are $R_{L1} = 0.1\Omega$, $L_{L1} = 40\,\text{uH}$, $R_{L2} = 0.3\Omega$, $L_{L2} = 1\,\text{mH}$;
(3) The controller parameters are $K_p = 0.15$, $K_i = 200$, $K_I = 5.55$ and virtual resistor is $R_{vi} = 1\Omega$.

The different parameters of Fig. 15.11 compared with Fig. 15.1 are listed as follows.

(1) The line impedance are $R_{12} = 0.1\Omega$, $L_{12} = 0.4\,\text{mH}$, $R_{23} = 0.2\Omega$, $L_{23} = 0.6\,\text{mH}$, $R_{34} = 0.4\Omega$ and $L_{34} = 0.8\,\text{mH}$.

15.4.1 Performance Comparison of P-V and $P - \dot{V}$ Droop Method

DG_1 starts at $t = 0\,\text{s}$ while DG_2 starts at $t = 0.2\,\text{s}$ and synchronizes with DG_1 through PLL. At $t = 0.5\,\text{s}$, DG_2 is connected to PCC and both DGs are controlled under traditional $P - V$ droop method. The initial regulation process is deliberately presented in Figs. 15.9 and 15.10 for completeness.

Since the line impedance of DGs are different, it is obvious that the generated active power of DGs is unequal although their rated power is set to be same, which is the intrinsic flaw of traditional $P - V$ droop strategy [10].

Figure 15.9a, b are the calculated virtual power while Fig. 15.9c, d the power injected into PCC. It shows that the virtual power is larger than the injected power. For example, at $t = 2\,\text{s}$, the injected active powers are $P_1 = 3069W$, $P_2 = 1397W$ while the virtual active powers are $P_1^{'} = 3281W$, $P_2^{'} = 1580W$. The injected reactive powers are $Q_1 = Q_2 = 4140$ Var while the virtual reactive powers are $Q_1^{'} = Q_2^{'} = 4132$ Var. This example verifies the opinion in Sect. 15.2.3 that the virtual power is different from the injected power and thus the decoupling of virtual power is not equal to the decoupling of injected power.

At $t = 3.5\,\text{s}$, the proposed $P - \dot{V}$ droop method is activated. Compared with $t = 2\,\text{s}$, the virtual active powers of DGs at $t = 8\,\text{s}$ are $P_1^{'} = 2945W$, $P_2^{'} = 1973W$. Thus, the active power sharing has been improved from 48% at $t = 2\,\text{s}$ to 67% at $t = 8\,\text{s}$. Besides, the reactive power is accurately shared among the regulation process.

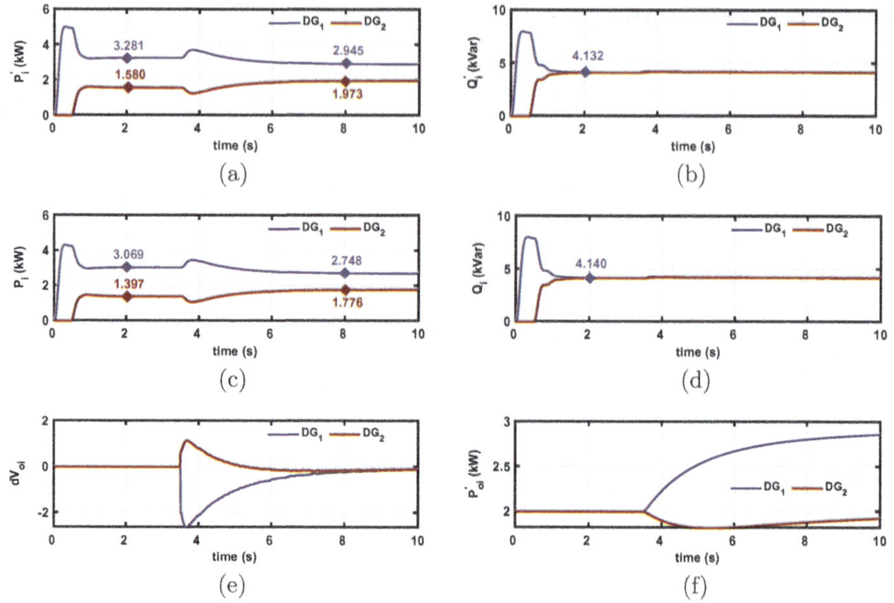

Fig. 15.9 Performance comparison of $P - V$ and $P - \dot{V}$ droop method. **a** Virtual active power $P^{'}$. **b** Virtual reactive power $Q^{'}$. **c** Injected active power P. **d** Injected reactive power Q. **e** Changing rate of V_{oi}. **f** Virtual active power Reference $P_{oi}^{'}$

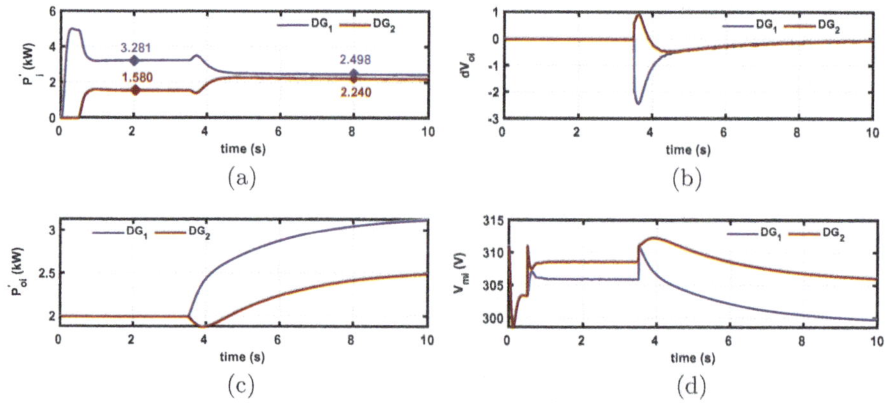

Fig. 15.10 Performance comparison of $P - V$ and modified $P - \dot{V}$ droop method. **a** Virtual active power $P^{'}$. **b** Changing rate of V_{oi}. **c** Virtual active power Reference $P_{oi}^{'}$. **d** Voltage magnitude V_{mi}

According to Fig. 15.9e, $\dot{V}_{oi}$ is zero before $t = 3.5$ s. After the $P - \dot{V}$ is activated, $\dot{V}_{oi}$ regulate the magnitude of voltage according to (15.13). After the initial dynamic regulation, $\dot{V}_{o1}$ and $\dot{V}_{o2}$ reach the same value and turn back to zero at the same pace. During the restoration process, the virtual active power reference $P_{oi}^{'}$ increases until $\dot{V}_{oi}$ reaches zero, as shown in Fig. 15.9f. The difference between $P_{o1}^{'}$ and $P_{o2}^{'}$ is the root cause that the active power cannot be shared equally at the same power rating.

Remark 15.3 Before $t = 3.5$ s, the traditional droop Eq. (15.12) is adopted and the voltage reference V_i deviates from its nominal value V^*. After $t = 3.5$ s, the droop method (15.12) is substituted by (15.13) and thus the initial voltage reference V_i is equal to V^* since the integral term is set to be zero initially. Therefore,there exists a sudden change of output power at $t = 3.5$ s as shown in Figs. 15.9 and 15.10.

15.4.2 Improved Performance Brought by Modified Droop Method

As analyzed in Sect. 15.4, the transient performance of the modified $P - \dot{V}$ droop method can be improved with the proportional term S_p. With $S_p = 5$ and $k_{res} = 0.325$, the simulation result is presented in Fig. 15.10.

With the modified $P - \dot{V}$ droop method, the virtual active power sharing accuracy at $t = 8$ s is 89%, which witnesses a 32% improvement compared with that in Fig. 15.9a. The total time consumed before $\dot{V}_{oi}$ reaches the same value since $t = 3.5$ s in Fig. 15.10b is smaller than that of Fig. 15.9e, which means that the transient performance of the modified method is better than that of the original one.

According to Fig. 15.10c, the difference between $P_{o1}^{'}$ and $P_{o2}^{'}$ at steady state decreases compared with that of Fig. 15.9f, which means the steady performance of the modified method is also improved compared to the original one. After the regulation process, the voltage level still satisfy the operation code which requires that $295V < V_{mi} < 326V$, as shown in Fig. 15.10d.

The above observation is consistent with Remark 15.2.

15.4.3 Performance Comparison Based on a Complex Microgrid

Thanks to the anonymous reviewer's advice, a complex microgrid as shown in Fig. 15.11 has been established in Matlab/Simulink. The starting process is the same as that in Sect. 15.4.1 and 15.4.2. In Fig. 15.12, the traditional $P - V$ droop method is utilized for the decentralized control of four DGs. The regulation process of the proposed modified $P - \dot{V}$ droop method is shown as Fig. 15.13.

To test the plug-and-play capability, DG_4 is disconnect from microgrid at $t = 8$ s and is reconnected at $t = 13$ s. In Fig. 15.13, the microgrid is controlled under $P - V$

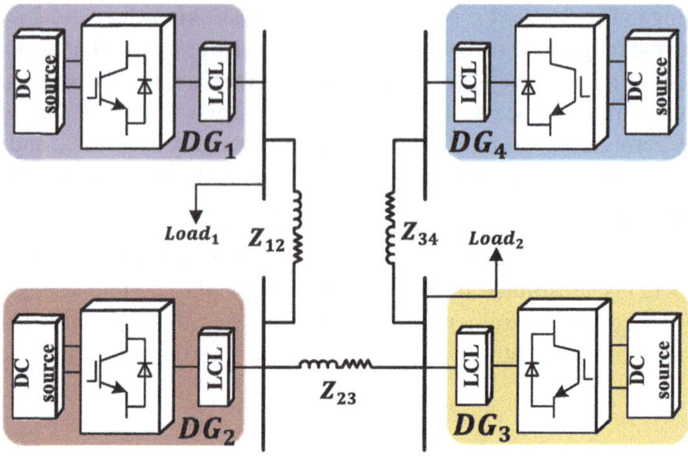

Fig. 15.11 A complex MG consists of four DGs and local loads

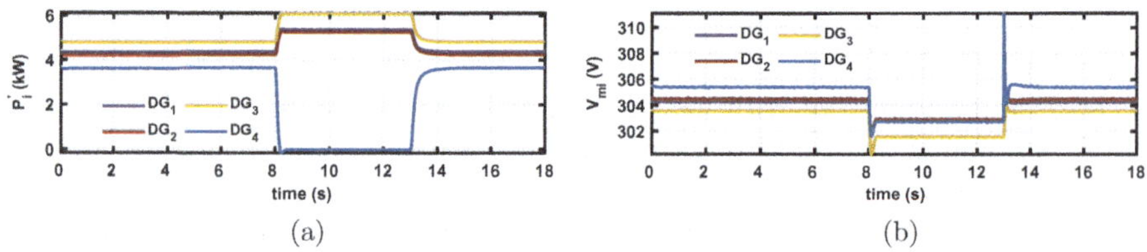

Fig. 15.12 Performance of $P - V$ droop method under a complex microgrid. **a** Virtual active power P'. **b** Voltage magnitude V_{mi}

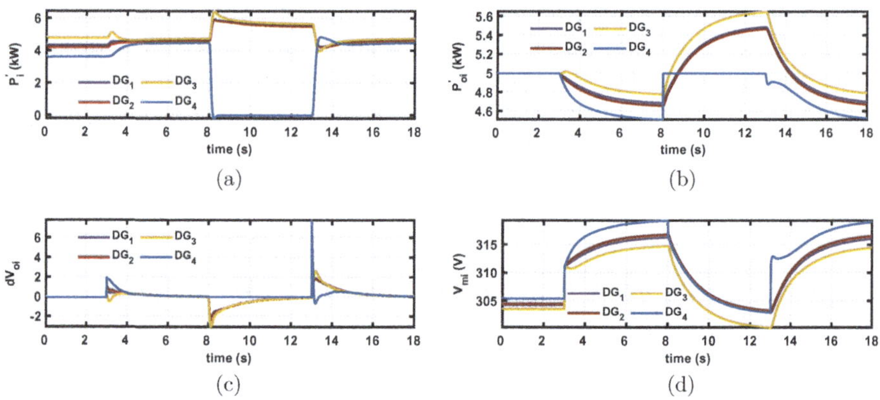

Fig. 15.13 Performance of modified $P - \dot{V}$ droop method under a complex microgrid. **a** Virtual active power P'. **b** Virtual active power Reference P_{oi}'. **c** Changing rate of V_{oi}. **d** Voltage magnitude V_{mi}

droop method before $t = 3$ s, then it exchanges to the modified $P - \dot{V}$ droop method. Comparing Fig. 15.12a with Fig. 15.13a, the active power sharing accuracy of DGs has been apparently improved. The initial $P_{oi}^{'}$ is set to be 5 kW. Thus, when DG_4 is disconnected from MG, its value is reset to be 5 kW as shown in Fig. 15.13b during $8\,\text{s} < t < 13\,\text{s}$. Figure 15.13c illustrates that whenever the load changes, $\dot{V}_{oi}$ will converge to a same value and then regulate the voltage at a same pace. During the whole regulation process, the voltage magnitudes satisfy that $295 < V_{mi} < 326$ as shown in Fig. 15.13d. However, the voltage difference of Fig. 15.13d is slightly larger than that of Fig. 15.12b, which is the cost of the improved active power sharing accuracy.

In conclusion, the proposed modified $P - \dot{V}$ droop method still promise improved power sharing accuracy under complex microgrid and is capable of dealing with plug-and-play incident.

15.5 Conclusion

In this chapter, a virtual positive resistor and negative inductor impedance strategy is adopted based on the analysis of the output impedance of DG in LVMG. With the equivalent impedance between VPS and PCC is reshaped to be purely resistive, an original and a modified $P - \dot{V}$ droop methods are proposed to improve the active power sharing accuracy, where the decoupled virtual power injected by VPS is adopted in droop equation rather than ordinary P/Q. A parameter S_p, which can accelerate the restoration process and reduce the sharing error at the same time, is incorporated in the modified $P - \dot{V}$ droop methods. Besides, the effect of m_i and k_{res} is also discussed. A simple MG consisting of two parallel DG with common load and a complex MG consisting of four DGs with local loads are constructed in Matlab/Simulink. The simulation results demonstrate that the strategies proposed here are effective.

References

1. A.G. Yepes, F.D. Freijedo, J. Doval-Gandoy, ó. López, J. Malvar, P. Fernandez-Comesaña, Effects of discretization methods on the performance of resonant controllers. IEEE Trans. Power Electron. **25**(7), 1692–1712 (2010)
2. Y. Yang, K. Zhou, M. Cheng, Phase compensation resonant controller for PWM converters. IEEE Trans. Ind. Inf. **9**(2), 957–964 (2013)
3. J.M. Guerrero, J. Matas, L. Garcia De Vicunagarcia De Vicuna, M. Castilla, J. Miret, Wireless-control strategy for parallel operation of distributed-generation inverters. IEEE Trans. Ind. Electron. **53**(5), 1461–1470 (2006)
4. Y. Han, P. Shen, X. Zhao, J.M. Guerrero, Control strategies for islanded microgrid using enhanced hierarchical control structure with multiple current-loop damping schemes. IEEE Trans. Smart Grid **8**(3), 1139–1153 (2017)

5. A. Eisapour-Moarref, M. Kalantar, M. Esmaili, Power sharing in hybrid microgrids using a harmonic-based multi-dimensional droop, in *IEEE Transactions on Industrial Informatics*, pp. 1–1 (2019)
6. J. He, Y.W. Li, J.M. Guerrero, F. Blaabjerg, J.C. Vasquez, An islanding microgrid power sharing approach using enhanced virtual impedance control scheme. IEEE Trans. Power Electron. **28**(11), 5272–5282 (2013)
7. Q. Zhong, Robust droop controller for accurate proportional load sharing among inverters operated in parallel. IEEE Trans. Ind. Electron. **60**(4), 1281–1290 (2013)
8. J. He, Y.W. Li, An enhanced microgrid load demand sharing strategy. IEEE Trans. Power Electron. **27**(9), 3984–3995 (2012)
9. C. Lee, C. Chu, P. Cheng, A new droop control method for the autonomous operation of distributed energy resource interface converters, in *2010 IEEE Energy Conversion Congress and Exposition*, pp. 702–709 (2010)
10. J.M. Guerrero, J. Matas, L. Garcia de Vicuna, M. Castilla, J. Miret, Decentralized control for parallel operation of distributed generation inverters using resistive output impedance. IEEE Trans. Ind. Electron. **54**(2), 994–1004 (2007)

The manufacturer's authorised representative in the EU is Springer Nature Customer Service Centre GmbH, Europaplatz 3, 69115 Heidelberg, Germany. If you have any concerns regarding our products, please contact ProductSafety@springernature.com

Printed and bound by CPI Group (UK) Ltd, Croydon, CR0 4YY
30/09/2026
02225293-0012